Behaving

What's Genetic, What's Not, and Why Should We Care?

KENNETH F. SCHAFFNER

OXFORD
UNIVERSITY PRESS

Oxford University Press is a department of the University of Oxford. It furthers the University's objective of excellence in research, scholarship, and education by publishing worldwide.Oxford is a registered trade mark of Oxford University Press in the UK and certain other countries.

Published in the United States of America by Oxford University Press
198 Madison Avenue, New York, NY 10016, United States of America.

First Edition published in 2016

Library of Congress Cataloging-in-Publication Data
Names: Schaffner, Kenneth F.
Title: Behaving : what's genetic, what's not, and why should we care? / Kenneth F. Schaffner.
Description: New York, NY : Oxford University Press, 2016. | Includes index.
Identifiers: LCCN 2015027801 | ISBN 978–0–19–517140–2 (hardcover : alk. paper)
Subjects: LCSH: Behavior genetics. | Nature and nurture.
Classification: LCC QH457 .S33 2016 | DDC 591.5—dc23 LC record available at
http://lccn.loc.gov/2015027801

9 8 7 6 5 4 3 2 1
Printed by Sheridan, USA

For some special people with whom I share genes and environment: Jeanette, Gabrielle, Laura, Michela, and Nancy

CONTENTS

FIGURES AND TABLES

FIGURES

TABLES

PREFACE

I conceived of this book in 2001. In the years 2000–2003 I was assisting with a project entitled "Tools for a Public Conversation about Behavioral Genetics" directed by the Hastings Center (HC), the major institution examining bioethics in the United States, working jointly with the American Association for the Advancement of Science (AAAS). (More information on the project can be found in chapter 8 of this book.) Several members of the project, including one of the principal investigators, asked me if there was any history or general philosophical introduction to human behavioral genetics. A search at that time indicated none existed, and I began to plan to write one. I received major support for the intended book from the National Science Foundation (via grants 9618229, 0324367, 0628825, and 0958270), as well as support from the NIH via the HC-AAAS Ethical, Legal, and Social Implications (ELSI) grant (R01 HG001873-03). Important earlier support for my research into schizophrenia was provided by the Greenwall Foundation. I am also grateful to both Donald Lehman, former Vice President for Academic Affairs at The George Washington University, and Dean N. John Cooper at the University of Pittsburgh for providing me with sabbaticals and research leaves during which the research for this book was pursued.

This book, and a successor book that plans to look more closely at the behavioral genetic issues in IQ and aggression/criminality studies, required that I talk to numerous contributors and critics of behavioral genetics. Those interviews—virtually all of them audio recorded—took place from 2003 through 2011. I am indebted to those many investigators who took the time to review and explain their work. They included Dean Hamer, David Reiss, David Goldman, Charles Murray, Kenneth Kendler, Michael Neale, Lyndon Eaves, Eric Turkheimer, Lee Ehrman, L. (Niki) Erlenmeyer-Kimling, Ming Tsuang, Richard Lewontin, Jon Beckwith, Diane Paul, Bernie Devlin, Michael Pogue-Geile, Karoly Mirnics, Irwin Waldman, Robert Cloninger, John Rice, Gar Allan, Elving Anderson, Tom Bouchard, Irving Gottesman, David Lykken, Matt McGue, Nick Martin, Greg Carey, John DeVries, John Hewitt, John Loehlin, Marc Feldman, Arthur Jensen, Neil Risch, Luigi Luca Cavalli-Sforza, Gardner Lindsay, Robert Plomin, Michael Rutter, Peter McGuffin, Terrie Moffitt, Steven Rose; Dorret Boomsma, Nancy Pedersen, Leon Kamin, and Daniel Weinberger.

I am also grateful to Erik Parens for encouraging me to write earlier versions of chapters 1 and 2. Philip Kitcher was instrumental in encouraging me to submit a target article that appeared in *Philosophy of Science* in 1998, and proving an opportunity for a reply to stimulating criticisms by Paul Griffiths, Bill Wimsatt, Rob Knight, Scott Gilbert, and Eric Jorgenson; both the article and reply comprise parts of chapters 3 and 4. Ken Kendler kindly read and commented on earlier versions of chapters 6 and 7, and Bob Krueger offered very valuable comments on several earlier versions of chapter 6. Beginning in the mid-1990s, I began a two-decades-long set of discussions and meetings with Irv Gottesman, Eric Turkheimer, Don Linszen, and Ken Kendler. Over the years, each of them has expertly and patiently guided me through the thickets of behavioral and psychiatric genetics. I also want to thank Juan Mezzich for inviting me to several major meetings to discuss the nature of psychiatric disorders and validity, including a foundational meeting of the committee on the International Classification of Diseases (ICD) mental disorders section in Toulouse, France, in 2005. I am also grateful for my colleagues and graduate students who were willing to read earlier draft chapters of this manuscript and offer comments on it. Among these, in addition to those mentioned in the list of interviewees above, were my Pitt colleagues Jim Bogen, Peter Machamer, Edouard Machery, Sandra Mitchell, and Jim Woodward; also Lindley Darden and Ilya Farber from my GWU and DC days, and more recently in Copenhagen Josef Parnas. I have had many discussions about these topics in these chapters with graduate students including Carl Craver, Thomas Cunningham, Lauren Ross, Kathryn Tabb, James Tabery and with Philosophy of Science Center postdocs and fellows Maria Kronfeldner, Marta Bertolaso, Marie Darrason, Serife Tekin, Carrie Figdor, and Bill Bechtel, as well as at conferences with Alex Rosenberg, Sahotra Sarkar, Jon Tsou, and Evert van Leeuwen. Finally, I am most grateful to my family members, to whom this book is dedicated.

By the end of 2005, when I returned to the University of Pittsburgh, I had a reasonably complete draft of these book chapters. But then the period 2006 until now coincided with a virtual revolution in behavioral and psychiatric genetics. As new and unsettling results from genome-wide association studies (GWAS) began to appear, those methods were applied to behavioral genetics, and heightened standards of replication were developed. The story of the effects of GWAS and increased replication standards on this field, and on its close relation, "psychiatric genetics," is told in a number of the chapters below, especially in 2, 6, and 7. GWAS (and now its related method of genome-wide complex trait analysis [GCTA]) thus required much of the book's manuscript to be recast and rewritten. More recently, the evolving DSM-5 process required changes in this text during 2012–2013, especially in chapters 6 and 7, as that major project wound down. Advances in simpler systems, such as the worm *C. elegans*, which is provided as a touchstone for how difficult it is to gain a complete hold on the relation of genes and behaviors, required somewhat less rewriting. But advances in the worm's neuroscience and its epigenetics did require significant updating over the course of the past five years. Though advances in both human and simpler model organisms will continue to occur, I think we are now on a methodological plateau, and that the analyses and conclusions about behavioral and psychiatric genetics presented here will be most useful for philosophers, historians, and scientists.

Last, I gratefully acknowledge that in this book I draw on parts of articles that I have published previously, though all those parts have been modified and updated to reflect more recent developments in the fields of behavioral and psychiatric genetics, as well as in the philosophy of science. As noted above, in their original form chapters 1 and 2 drew on pre-GWAS developments, which required significant changes in those chapters for this book. The earlier versions of chapters 1 and 2 appeared in *Wrestling with Behavioral Genetics: Ethics, Science, and Public Conversation* (Baltimore: Johns Hopkins University Press, 2005), 3–39, and 40–73, a book that was edited by Erik Parens, Audrey Chapman, and Nancy Press, and that grew out of an ELSI grant (R01 HG001873-03) to the Hastings Center. Chapters 3 and 4 depend in parts on two articles that appeared as "Genes, Behavior, and Developmental Emergentism: One Process, Indivisible?" and "Model Organisms and Behavioral Genetics: A Rejoinder," *Philosophy of Science* 65 (June 1998), 209–52 and 276–88. Chapter 5 was published in an earlier version in "Reduction: The Cheshire Cat Problem and a Return to Roots," *Synthese* 151 (3) (August 2006), 377–402. Finally, chapter 7 utilizes modified portions of "Reductionism and Psychiatry," which appeared in K. W. M. Fulford et al., eds., *Oxford Handbook of Philosophy and Psychiatry* (Oxford: Oxford University Press, 2013), 1003–22.

Behaving

Introduction

The general purpose of this book is to provide an overview of aspects of the recent history and methodology of behavioral genetics, as well as of psychiatric genetics, with which it shares both its history and methodology. The perspective is mainly philosophical and addresses a wide range of issues, including genetic reductionism and determinism, as well as the behavioral genetic implications for "free will." Some chapters can be easily read by individuals without a preexisting philosophical background, such as chapters 1, 2, 6, and 8. Other chapters dig deeper into both the genetics and the philosophy. Chapters 3, 4, and 5 examine criticisms of some "traditional" standard views of behavioral genetics, and do so by making use of what we have learned about the simplest biological system with a neural network, *Caenorhabditis elegans*, or "the worm," as it is often affectionately called.

As readers will soon discover, however, though the worm is a "simple system," the molecular and neurogenetics of it are far from simple. Many of the arguments of these three chapters (3, 4, and 5) are related to issues of reduction and the nature of explanations in genetics. Chapters 6 and 7 look closely at human behavioral genetics, both at normal behaviors and mental dispositions via personality theories and at psychopathology, especially at depression, psychosis, and schizophrenia. Chapters 6 and 7 also discuss recent critiques of candidate gene studies in the light of genome-wide association studies (GWAS), initially mentioned in chapter 2. Chapter 8 considers more general philosophical issues as to why there is so much concern about, and general interest in, behavioral and psychiatric genetics. These more general issues include introductions to issues of free will, the fundamentality of genetic explanations, and implications for ethical and legal dimensions of behavioral genetics.

The philosophy material in this book is mainly analytical and pragmatic in orientation, but is open to alternative philosophical perspectives, including the phenomenological approaches represented by Parnas's writings (in chapter 7). In chapters 1, 2, 6, and 7 the philosophy is largely implicit, akin to the approach that Ronald Dworkin took that he called doing "philosophy from the inside out" (Dworkin 1994, 28–29). Chapters 3, 4, 5, and 8 are written more in the explicit analytical tradition of philosophy of science.

Chapters 1 and 2 provide a common basis and a vocabulary for appreciating the contours of behavioral and psychiatric genetics. The material is largely introduced in the form of three dialogues between a behavioral geneticist and an appeals court judge who wishes to find out more about behavioral genetics. The standard but often unappreciated distinction between classical quantitative genetics and modern molecular genetics is introduced. In the former, *no* specific genes are identified, though heritability estimates are provided; in the latter, modern molecular genetics, both gene finding and biological pathways connecting genes and traits are developed. The material covered in chapter 2 includes results from a more recent GWAS approach, genome-wide complex trait analysis (GCTA), and analyses of the effects of what are termed copy number variants (CNVs) that seem to be having a significant effect on some behavioral disorders.

Chapters 3 through 5 are data-driven, bottom-up analyses of how easy, though actually quite hard, it might be to relate genes and behavioral traits. Here the aforementioned worm, *C. elegans*, is the focus and tool—one that has been investigated since the late 1960s for its genetic, neural, and behavioral properties by thousands of researchers. A specific theme that is followed in some detail in these three chapters is what accounts for different foraging and feeding behaviors in worm populations. I begin by framing the discussion about genes and behaviors and the nature-nurture controversy in terms of the "developmentalist challenge" to any simple genetic determinism of behaviors, and develop that challenge in terms, first, of 11 theses separating the developmentalists from traditional views, discerning eight "rules" connecting genes and behaviors, and then extracting five core concepts of developmentalism. Those core concepts are parity, nonpreformationism, contextualism, indivisibility, and unpredictability, and are defined in chapter 3 and also examined for their validity in connection with results from worm studies. Chapter 4 asks how and to what extent, and why, the worm can be used as a model organism for other organisms, including humans. The answer lies partly in the notion of deep homology and high connectivity among appropriate organisms. This chapter also introduces another theme that will be encountered in chapter 5 on types of reduction in biology, and chapter 7 on schizophrenia. This theme is the importance of *models* in general in biology, including parts and aspects of models such as mechanisms, pathways, and networks or circuits. In chapter 5, a model in the form of a *preferred causal model system* (PCMS) is then shown to be the main tool that biologists use to accomplish partial or creeping reductions in contrast to sweeping, more theoretical reductions.

Chapters 6, 7, and 8 shift back to human behavioral and psychiatric genetics first examined in chapters 1 and 2. Chapter 6 is on personality genetics, but moves also to consider the disorder of depression, as well as some important studies on the interaction between genes and environments pioneered by the Caspi and Moffitt research team as well as quite recent critiques of their approach. These interaction approaches point behavioral and psychiatric genetics both toward environmental studies, in which genetics interacts significantly with the environment over humans' lifetimes, and toward more complex strategies that can still be characterized as reductionistic. Chapter 6 closes with some sobering reflections on the difficulty of obtaining replications in an area such as human personality genetics. Chapter 7 is an in-depth examination of schizophrenia and the extent to which traditional and more modern

(GWAS) genetic techniques can begin to unravel the causes of this devastating disorder. Finally, chapter 8 considers more philosophical implications of the genetic analyses of the first seven chapters. The general theme of this chapter is why genetics seems to generate such strong, even emotional, worries about its implications. A major subtheme revolves around concerns that genetics seems to abrogate the essential human feature of "free will." But another concern among scientists and philosophers is the apparent priority and deterministic aspects that genetic explanations seem to have in contemporary biology. Those two themes appear to interact synergistically, each strengthening the concerns produced by the other. In that chapter I analyze representative contributions from the philosophical, genetic, and neuroscience literatures related to these topics. Chapter 9 is a summary of the main themes of the book and attempts to weave them together in a picture—one that though rather pessimistic about the accomplishments of genetics in the present is optimistic about its future. For those readers who would like to see an extended abstract (and fairly detailed summary) of the themes and results of this book, chapter 9 provides that material.

Reference

Dworkin, Ronald. 1994. *Life's Dominion: An Argument about Abortion, Euthanasia, and Individual Freedom*. New York: Vintage Books.

1

Behaving

Its Nature and Nurture (Part 1)

This is the first of two chapters intended to serve as an overview of the field of "behavioral genetics." Earlier versions of these chapters have proven useful in introducing an interdisciplinary audience to the field and the topics developed in this book. Though my discussion will be focused on a variety of genetic methods and results, I do not mean to suggest that we should think of genetics as the fundamental cause in the analysis and explanation of behavior. My approach to what has traditionally been called the nature-nurture debate sees genetics as part of a larger constellation of causes that crucially includes environments, which work jointly with genetics.

The nature-nurture issue is a perennial one, with its modern roots dating back to Galton's writings in the late nineteenth century. After the rediscovery of Mendel in 1900, the issue and ensuing debates have usually been couched in terms of genes versus environments and their respective influences on the organism. Themes revolving around nature and nurture have been especially contentious when behavioral and mental traits (and disorders) are at issue. ("Psychiatric genetics" is widely viewed as a part of the field of behavioral genetics.) This contentiousness arises not only in our society at large, where the specters of discrimination and eugenics are quickly raised, but also in the social sciences and psychiatry. There have been several waves of vigorous debates about behavioral genetics, often centering on the issue of IQ as well as genetic dispositions to violence and criminality. It is also possible that a very recently published book on genetics and race (Wade 2014) will rekindle this kind of debate yet again. The chapters below touch on those developments in several places, such as where "heritability" arises (which is frequently). IQ is discussed in chapter 2, and a study relating genes and environment to criminality is considered in chapter 6. These contentious topics deserve more space and in-depth analyses than can be provided in this book, however, and another book is currently in process that will provide that analysis (Schaffner, in development).

The past 40 years have seen a shift from an earlier period in which behavioral or psychiatric disorders were seen as primarily environmental, due to poor parenting, for example, to the contemporary view that amalgamates both genetic and environmental influences as major causal determinants. This shift, which has not been without controversy (Rowe and Jacobson 1999), reflects broader shifts in psychosocial studies of nature and nurture (Reiss and Neiderhiser 2000).

In these two initial chapters, I review some of the recent empirical methods and results of studying genes and environments in behavioral (and psychiatric) genetics, and I also touch on a few more general issues including genetic determinism and complexity. Because of the breadth of studies dealing with nature and nurture in behavioral studies and in psychiatric disorders, I focus on several *prototypical* examples. Prototypical examples are concrete exemplars that will help us to grasp the roles of genetics and environment by providing specific instances of how genes and environmental factors are believed to work. Frequently I will employ schizophrenia as a prototypical disorder, but I will also deal with normal mental functions, including "general cognitive ability"—a notion often related to the more contentious terms *intelligence* and *IQ*. I will, in addition, consider the trait of "novelty seeking" and, to some extent, the concept of "impulsivity." I will approach impulsivity primarily in connection with another prototypical disorder, attention deficit hyperactivity disorder (ADHD). In two later chapters, personality genetics and a detailed analysis of schizophrenia will be developed.

These two initial chapters contain three dialogues between a behavioral geneticist and an appeals court judge who wishes to find out more about behavioral genetics. The first dialogue introduces the basic notions of the more traditional "quantitative" (often called "epidemiological" or "classical") approaches to understanding the influence of genes on behavior. The second dialogue (in the next chapter) introduces the newer, "molecular" approaches to understanding the genetics of behavior, including results from more recent genome-wide association studies (GWAS) and analyses of the effects of copy number variants (CNVs). In the third dialogue, I introduce two hypothetical legal cases involving testing for genes implicated in IQ and ADHD, in order to focus much of the discussion from dialogues 1 and 2. Finally, I conclude with a preliminary projection of where the current debates and new methodologies may lead regarding genetics and the understanding of human behavior in a combined nature-nurture perspective.

BASIC TERMS

Though concepts like "gene" and "environment" are widely used both in science and in public discourse, they sometimes have special meanings in behavioral genetics. To begin, we will understand *behavior* in a very broad sense, to include the reactions and interactions of an organism to its environment, including other organisms. Since we will focus on *human* behavior, this broad meaning will also include people's (reported) thoughts and feelings as well as their observable bodily movements (compare BSCS 2000; Plomin et al. 2013). *Genetics* is the science that deals with the inheritance of those characteristics that are physically passed on from parent(s) to offspring, and includes the simpler forms of classical Mendelian genetics of pea plants that some readers may remember from high school. The field of genetics includes, as well, later developments such as the study of chromosomes and the molecular basis of the unit of inheritance, the *gene*, constituted by DNA sequences (or RNA in some viruses such as the AIDS virus, HIV).

Millions of DNA sequences, linked together in a chain and partially covered with special proteins, constitute the chromosomes, of which there are 23 pairs in human females (males have 22 pairs and an X and a Y sex chromosome). Genes come in different forms at their typical position or *locus* along a chromosome, and those forms are called *alleles*. The *genotype* of an organism sometimes refers to all of the genes (or more accurately alleles) in that organism and sometimes refers to all of the alleles that are found at that particular locus. For example, a gene that helps determine whether pea color is yellow or green has two alleles, Y and G. The genotype for a pea plant at the pea color locus thus could be YY, YG, or GG. The observable effects of the genes are called the *phenotype*, for example, the yellow or green color of the aforementioned peas.

The genes that affect a trait may have a *dominant* form, which means that only a single allele at a given locus is needed to produce a particular phenotype. For example, if the Y allele is dominant for yellow pea color, then a YG genotype produces a yellow pea. (A YY genotype would do the same.) The genes that affect a phenotype may also have a *recessive* form, which means that two alleles of the same type must be present at a given locus for a trait to be expressed. If green pea color is recessive, then only a GG genotype will produce a green pea. In a number of cases the alleles are codominant, which means they blend to produce an effect that is midway between the two pure allele forms.

Observable features are also called *traits*. Traits can be virtually any described or measured feature of an organism, including behaviors. Whether those features are unambiguously determined by any single allele or even combinations of alleles, however, must always be discovered through empirical genetic research. A trait that varies continuously, like height, is called a *quantitative trait*, and a gene that helps determine such a trait is termed a *quantitative trait locus* (QTL); the plural is quantitative trait loci (QTLs).

Behavioral geneticists understand the term *environment* very broadly. As Robert Plomin and his coauthors write, in the field of (quantitative) behavioral genetics

> The word *environment* includes all influences other than inheritance, a much broader use of the word than is usual in the behavioral sciences. By this definition, environment includes, for instance, prenatal events, and biological events such as nutrition and illness, not just family socialization factors. (Plomin et al. 2013, 106)

As we will see later, the prenatal, *in utero* environment may exert an extremely important influence on highly prized phenotypic traits.

Behavioral geneticists distinguish environment into two forms, called the *shared environment* and the *nonshared environment*. Intuitively, included in the shared environment are those things that are experienced in common in a family and make family members similar, and included in the nonshared environment are those things that are experienced differently and make individuals different, whether experienced within the family or through outside-the-family interactions. In fact, that is not exactly what behavioral geneticists mean when they use that distinction. Toward the end of this chapter, I will discuss some of the subtle and perhaps odd technical aspects of this distinction that is so important for behavioral genetic studies.

Finally, though environment is broadly conceived in behavioral genetics, there is no "theory of the environment." This stands in stark contrast with genetics, which can appeal to the *framework* of genes, chromosomes, and general knowledge about gene actions and interactions discovered by classical and molecular biology (BSCS 2000). The lack of such a theoretical orientation or environmental framework will prove important at several points later in this chapter and also in later chapters. Toward the close of this chapter, I will return to several research projects that address various aspects of the environment concept, and in the next chapter I will introduce the concept of the *envirome* to describe such projects.

THE FIRST OF TWO MAJOR CLASSES OF METHODS FOR STUDYING THE INFLUENCES OF NATURE AND NURTURE: QUANTITATIVE (OR EPIDEMIOLOGICAL) METHODS

Behavioral geneticists use two broad approaches to study the influences of nature and nurture. The first sort of approach is called *quantitative* (or often *epidemiological*) and the second is called *molecular* (see generally Plomin 2008). As we will see in detail below, epidemiological or quantitative approaches use twin, adoption, and family studies to investigate the influence of genetic and environmental factors. Molecular approaches use the tools of molecular genetics to identify specific genes implicated in a given behavior. Neiderhiser (2001) provides an overview of methods, while Plomin et al. (2013) undertake a systematic analysis.

In this chapter I will introduce some of the quantitative types of studies, and then we will move to an interlude—the first of three dialogues—in which we probe how and why some of the methods work—and do not work. Later on we will return to consider the second class of methods for analyzing behavioral genetics, namely various molecular approaches.

Quantitative methods are used to distinguish nonspecific genetic and environmental contributions to traits or features of individuals or, more accurately, to assess correlations and interactions between genetic and environmental factors that account for *differences* between individuals.[1] In any given study, all individuals have the trait of interest, but they have the trait to different extents. Quantitative approaches study *collections* of individuals or *populations* and typically do not examine individual identified genes, nor individually identified environments. Rather, these methods examine what *proportion* of the phenotypic differences between individuals is due to genetics or environment (or interactions between the two). Behavioral geneticists who employ the quantitative methods use fairly simple equations to infer the effects of genes and environments on the (mostly) quantitative traits. The methods of quantitative behavioral genetics include family, adoption, and twin types of study, and each of these can be coupled together with the other types.

Types of Studies

Family studies look to see if a trait or a disorder "runs in families." Family members with the trait or disorder are identified, and then other family members are screened for the presence of the trait or disorder. Usually "first degree" relatives (parents, siblings, and offspring) are examined, but second-degree relatives (aunts and uncles, cousins, and grandparents) may also be included in a study (Neiderhiser 2001). Family studies can *suggest* that a trait or disorder is genetic, but since both genes and environments are shared by family members, contributions of nature and nurture cannot be disentangled in family studies. Different types of studies are thus needed.

Adoption studies examine genetically related individuals in *different* familial environments, and thus can prima facie disentangle contributions of nature and nurture. If a trait or disorder tends to be present in both adopted-away children and their biological parents, then researchers suspect the existence of an important genetic influence. (At least at first blush it appears to be genetics—not environments—that the biological parent and adopted-away child share.) If, on the other hand, adopted children and their adopting parents tend to share a trait or disorder, then researchers suspect the existence of an important *shared* environmental influence. (After all, the adopted child and adopting parents are genetically quite different.) This nice reasoning depends on several assumptions that can be questioned. For example, it is usually assumed that there is no selective placement that would create a correlation between traits of the biological parents and the adopting parents, such as socioeconomic status or intellectual capacity. If this assumption is violated, the interpretation of the study results needs to be corrected. It is also assumed that the parents and offspring studied represent the population as a whole, which is questionable. Also, the possible effects of the prenatal *environment* need to be considered. If, say, prenatal (and thus preadoption) nutrition might affect an offspring's trait or disorder, this effect might be erroneously ascribed to *genetic* causes. We will discuss this issue specifically in dialogue 3, where we consider IQ and the maternal environment. (For a further discussion of these assumptions and ways of dealing with them, see textbox 6.2 in Plomin et al. 2013, 78–79.)

The first adoption study of schizophrenia conducted by Heston in 1966 resulted in the then startling finding that shared rearing environment has little, if any, effect on the risk of developing schizophrenia. It was also found that being adopted had no effect on the risk of developing schizophrenia (Heston 1966). This study was the first nail in the coffin of the idea of the "schizophrenogenic mother," an environmentalist concept articulated at the time by psychoanalyst Frieda Fromm-Reichmann. These results were confirmed by other later adoption studies, including many by Seymour Kety. Adoption studies, however, are increasingly difficult to conduct in contemporary society. This is because there are fewer adopted children, and also because the confidentiality of adoption records makes it difficult to accurately assess the traits and disorders of biological parents.

Twin studies, the third and final study type, compare identical and fraternal twins, both within the same familial environment and (in adoption studies) in different familial circumstances. Twin studies have been used extensively in behavioral genetics and in psychiatry to understand the extent to which a trait or disorder is influenced by genes or environments. One of the simplest ways to report the results of twin studies uses *concordance rates*. A concordance rate can be thought of as a risk factor for the second twin of having a trait or a disorder *given* a diagnosis of the first twin.[2] As an example, concordance studies of twins reported by Gottesman and his associates over many years, including several new ones, indicate that the risk of developing schizophrenia is ~48% for monozygotic (MZ) twins, 17% for dizygotic (DZ), and 9% for siblings, if the other has been so diagnosed (Cardno and Gottesman 2000; Gottesman and Erlenmeyer-Kimling 2001). (The risk for developing schizophrenia over a lifetime assuming no affected relatives is about 1%.) This concordance pattern suggests that schizophrenia is caused by many genes, which interact in a complex, non-Mendelian fashion; it also suggests that the environment has a major (>50%) effect; details are in chapter 7 below.

Twin studies are also used to obtain *heritability estimates* for behavioral traits. Heritability estimates, which we'll explore in some depth shortly, are meant to indicate how much of the phenotypic *variation* in a population is due to genetic variation. The bigger the heritability estimate, the more researchers think that genetic differences help to explain why people are *different* with respect to a trait. Heritability estimates tend to run about 30%–50% for personality traits, and for most major psychiatric disorders heritabilities range from about 30% (e.g., depression) to about 70% for schizophrenia, though one recent review estimates schizophrenia's heritability at about 85% (Cardno and Gottesman 2000). But it is important to know that approximately 63% of all persons suffering from schizophrenia will have *neither first- nor second-degree* relatives diagnosed with schizophrenia (Gottesman and Erlenmeyer-Kimling 2001), reinforcing the idea that genetic and environmental contributions to the disorder are complexly related.[3]

Readers may wonder how quantitative behavioral geneticists can consider schizophrenia to be a continuous or "quantitative" trait when it looks so "discontinuous," like you either have it or you don't. The key to understanding this is to bear in mind the difference between a continuum of phenotypes and a continuum of liability to express a categorical (present or absent) "trait" like schizophrenia. When behavioral geneticists say that schizophrenia is a quantitative trait, they have not traditionally meant that there is a continuum of phenotypes between, say, no schizophrenia and flagrant schizophrenia as there is a continuum between those who are very short and those are very tall (or those who are prone to have the occasional blues to those who are prone to major depression). They have meant, instead, that there is a continuum of liability or risk for schizophrenia: individuals only exhibit schizophrenia if they pass some threshold of risk (much as a bridge breaks only if it is subject to some threshold of stress). This liability approach is represented in what is called a *threshold model,* first introduced by Falconer (1965), and then developed and applied by Gottesman to schizophrenia in a long series of influential papers as well as in his book (Gottesman and Wolfgram 1991). I will return to further considerations on the threshold model in chapter 7 on schizophrenia.

Because heritability is such a confusing concept, and because the models that estimate it serve as the foundation for quantitative (or epidemiological) genetics, I will introduce that concept and those models through an imagined dialogue between an inquisitive judge and a behavioral geneticist. Judge Jean is an imaginary member of the Ohio State Supreme Court. She expects she may be asked to hear cases in the near future that will appeal to the results of behavioral genetics, both in its traditional quantitative forms and its newer molecular forms.[4] This is the first of three dialogues that are intended to be fairly simple introductions to the basic concepts and methods of behavioral genetics. More details can be found in the notes within the dialogues, and also in the text preceding and after the dialogues, and also in later chapters. Also see Turkheimer (2000).

DIALOGUE 1. BETWEEN A STATE SUPREME COURT JUDGE AND A BEHAVIORAL GENETICIST ON HERITABILITY AND TWIN STUDIES

JUDGE JEAN (JJ): I keep seeing in news stories about behavioral genetics that they have been able to calculate "heritabilities" for all sorts of human behaviors, from personality traits, IQ, and even criminal behavior, to severe psychiatric illnesses, like schizophrenia. I even heard something the other day about "missing heritability." What does this word mean? Why should I care about it?

BEHAVIORAL GENETICIST (BG): *Heritability* is a term behavioral geneticists have adopted from quantitative genetics. The idea had its original applications "down on the farm," as it were, improving crop strains and breeding better farm animals, such as cattle. The general idea is that a heritability estimate tells you how much of the total phenotypic *variation* in a trait (i.e., the variation due to both genetics *and* environment) can be explained by genetic variation alone. Another way to put it would be this: a heritability estimate is supposed to tell you how good a prediction of the phenotype you can make if you know the genotype. I will be more precise about the definition as we talk more. *Quantitative* here means the traits of interest continuously vary, like height and weight do among people. (Geneticists do not estimate heritabilities for traits like the yellow *or* green color of Mendel's peas, because they would be trivially 1, meaning that the phenotypic differences among the plants would depend completely [100%] on the plants' genotypes.) Heritability is symbolized by H^2. It's a squared term for reasons relating to statistics that we need not be concerned with at this point.[5] Heritability is important because for quantitative traits we can estimate how much of the phenotypic variation in a population may be influenced by genetics and how much by the environment. And if a trait appears to be highly genetically influenced in terms of its variation, maybe we can identify the genes that causally influence that trait.

JJ: Wait a minute! You just said H^2 estimates how much the trait *may be* influenced by genetics, and then you said *maybe* we can identify the genes that influence that trait. Why the two *maybes* here?

BG: The first is because usually we have no *direct* evidence of genetic influence on the trait's range. (In this classical heritability approach, there are rare cases where we know what the underlying genotype is that affects the trait, and we are able to genotype the individuals in the population and get *direct* evidence of influence.) But when we don't have *direct* evidence, that is, we don't know the real genotype and cannot measure it, we can still use a wonderful experiment of nature to make a good *indirect* guess at how much of a trait's variation is genetic. We'll talk about that natural experiment a little later.

I used the second *maybe* because it is still very controversial whether high heritabilities indicate that we will be able to find a specific molecular genetic basis for the trait (or a disorder) with that high heritability. For today's discussion we will concentrate on getting a clearer idea about how quantitative genetics defines heritability and obtains heritability estimates. Another day (in dialogue 2) we will look at how molecular methods work. At that point we will consider what the relationship might be between the heritabilities that are given to us by quantitative genetics and what molecular studies tell us, and whether the quantitative results can really assist the molecular research programs.

JJ: OK, let's put aside the molecular approaches for now. But in response to my very first question you told me that heritability estimates give you a sense of the extent to which genetics explains the variation that is observed in a population with respect to a particular trait. Then you told me that "maybe" heritability estimates help you determine the extent of the genetic influence on a trait. Please clarify exactly what is explained by the quantitative approach.

BG: Remember that a quantitative trait, like height or weight, in any collection of individuals we look at usually varies. By the way, a collection of individuals that we examine is usually called a *population*. The scores of the trait of interest, like heights, are plotted along an X axis and the number or percentage of individuals with that score is plotted along the Y axis. Often the figure that results looks like a bell-shaped curve. You've seen curves like this for Scholastic Aptitude Test (SAT) scores and for IQ scores. The figure is also called a *distribution*, since it tells us how the scores are distributed. Every distribution has an average or *mean*, and also a spread around the mean. The spread tells us how much variation there is in the scores, and it can be made precise by the notion of a *variance,* but I won't ask you to learn about the mathematical definition of variance.[6] Heritability is a measure of how much of the *differences among the individuals in a population* is estimated to be genetic and how much is estimated to be environmental.

One has to be very careful with this heritability concept. As three geneticists once put it in their lingo: "heritability does not describe the quantitative contribution of genes to . . . any . . . phenotype of interest; it describes the quantitative contribution of genes to *interindividual differences* in a phenotype studied in a particular population" (Benjamin et al. 2002, 334; my emphasis).[7] That is, a heritability estimate may tell us something about how much of the phenotypic differences between individuals is due to genetic differences among those individuals, but it cannot tell us anything about how much genes affect the phenotype of a given individual. One of the surprising features of the heritability concept is that if, for example, there are no interindividual *differences* in a trait, then the heritability of that trait is zero. Of course that isn't to say that genes don't influence that trait in a given individual! I'll give you some examples of this subtle but critical distinction in a bit.

JJ: I think examples would help, because I'm confused. Can you go over that again? Maybe you can describe individual differences and what's genetic and what's environmental variation in another way?

BG: Sure. There are a few rare cases for which we know exactly what the underlying genotype is and what the phenotype and its distribution look like. Here is a good example based on the writings of the geneticist Richard Lewontin, who used data from a study by Harry Harris. Work with me on it for a bit. Lewontin's example uses some real data involving a human trait, red blood cell acid phosphatase activity, which is governed by three different gene forms, or alleles. But since the function of acid phosphatase is not well understood,[8] though it can be measured for any individual in the lab, I am going to modify the example, just like I do for my students, to try to give you a more intuitive grasp of the concepts. We will actually use the *same data and the same picture and summarizing table* as Lewontin did, following Harris, but now we will interpret the trait as the height of hypothetical ornamental rubber tree plants or, better, a population of hypothetical rubber trees that a tree breeder is studying.[9] These particular rubber trees, like Mendel's pea plants and like humans, are diploid (they have two copies of every chromosome),[10] and, as was found in red cell acid phosphatase, these rubber plants have three alleles (A, B, and C) that are known to affect this trait. Thus any given rubber tree must have one of six possible genotypes (AA, AB, AC, BB, BC, or CC). Now Lewontin and Harris used data on the phosphatase activity phenotype and provided the graph (figure 1.1) of the activity for each genotype (on the horizontal axis) and relative frequencies of the percentage of individuals with that activity (on the vertical axis). For us the graphs represent the heights of the rubber tree plants with the five or six different genotypes (genotype CC was actually too rare to graph) on the horizontal axis, and the vertical axis represents the number of rubber trees that have that height.

Note that each genotype has a nearly bell-shaped curve distribution. The AA genotype has the least height, but it ranges from about 80 up to 160 cm (about 32 inches up to 64 inches), whereas AB runs from about 100 to over 200 (40 to 80 inches), and other genotypes have their own distributions as shown (again, CC is not graphed because too little is known about this very rare genotype to have any data). Each genotype has a range or variance for its height, and of course each of those variances has a mean or average. Since the five or six genotypes are fixed in the plants studies, all of the phenotypic variance (i.e., variation in height) associated with *each* genotype is due to environmental variance. The rubber trees heights are environmentally affected by the climate and the soil in subtle and uncontrollable ways at this stage of our tree breeder's research. The different genotypes have those different means, however, and a part of the phenotypic variance in the population as a whole is because of the variance of those five or six different *genotypic means*. But, again, this genetic variance does not account for all of the population's variance, since within each genotype there is variance that is due to those unknown environmental causes (Lewontin 1995, 66). Thus the population's phenotypic variance is a mix of genotypic variation and environmental variation.

It might be good to emphasize again that our hypothetical height genes here, the As, Bs, and Cs, do not by themselves provide the instructions for constructing a plant of a specific height. Maybe one influences the concentrations of some plant hormone, and another affects water transport into a cell. But it takes thousands of genes working in concert with proteins and chemical environmental factors to grow a rubber tree plant. Each of the genotypes we show has a mean value associated with it, but the genotype does not produce that mean in anything but the most indirect sense.

JJ: OK, I can see there are two different sources of variation, or if you want, variance, in that figure. So what?

BG: Heritability is the ratio of the genetic variance to the total (phenotypic) variance. If you don't mind me writing the formula, the ratio is $V_g/V_p = H^2 \cdot V_p$ is also just equal to $V_g + V_e$. (These variances or measures of the range of individual differences in the heights—the spread-outness—can be read off from the figure, or more accurately calculated from the data that were used to graph the distribution. [Specific values are presented in table 1.1 based on Harris's measurements.] In figure 1.1, the curves labeled with letters show the environmental variation for each genotype, and the unlabeled [top] curve shows the total phenotypic variation.) This ratio (V_g/V_p) is, in words, just the average of the genotypic variance divided by the variance of the total (phenotypic) distribution. Remember: What we're trying to figure out is how much

Table 1.1 RED BLOOD CELL ACTIVITY OF DIFFERENT GENOTYPES OF RED-CELL ACID PHOSPHATASE IN THE ENGLISH POPULATION

Genotype	Mean activity	Variance of activity	Frequency in population
A/A	122.4	282,4	013
A/B	153.9	2293	0.43
B/B	188.3	380.3	0.36
A/C	181.8	392.0	0.03
B/C	212.3	531.6	0.05
C/C	~240		0.002
Grand average	166.0	310.7	
Total distribution	166.0	607.8	

NOTE: Averages are weighted by frequency in population. The "grand average" of 310.7 is the average of the environmental variances within each genotype. The total phenotypic variance is 607.8 and is measured in the population as a whole. The genetic variance is 607.8 – 310.7 = 297.1, since there are only two sources of variation, and genetic variance plus environmental variance = total variance. Thus the heritability of the trait of red cell acid phosphatase activity in this (English) population = V_g/V_p – 297.1/607.8 = 49%. (Compare Lewontin 1995, 67.)

SOURCE: Harris 1980, 194. Reprinted with permission of Elsevier.

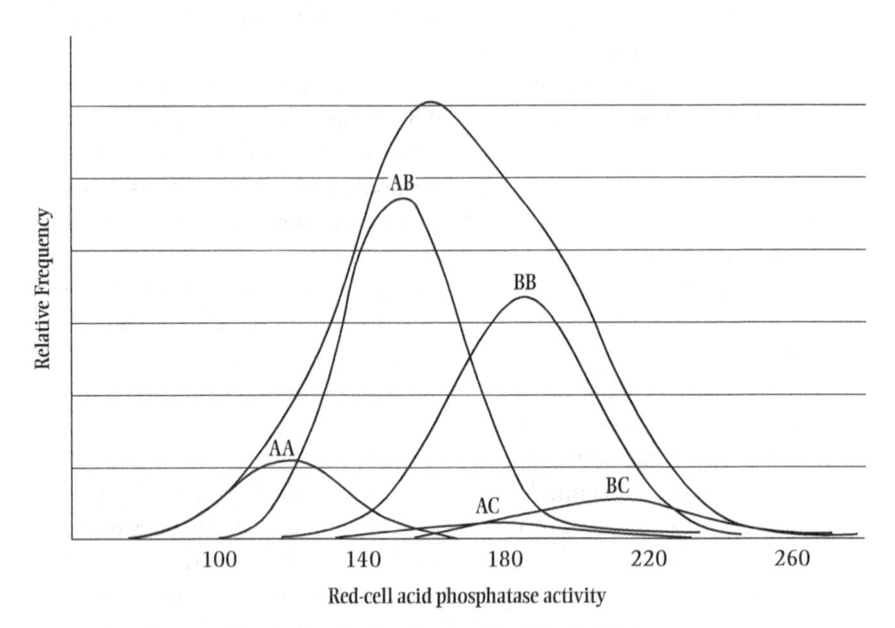

Figure 1.1 The activity of red-cell phosphatase (and the height in centimeters of hypothetical rubber tree plants).
SOURCE: Adapted from Harris 1980, 196. Reprinted with permission of Elsevier.

of the total phenotypic variation is due to genetic variation. In this case we know that for each genotype all of the phenotypic variance is due to environmental variance. Because by definition the total phenotypic variance is equal to the sum of the environmental and genetic variances ($V_p = V_e + V_g$), we can calculate the genotypic variance by subtracting the average of the environmental variances for each genotype from the total phenotypic variance ($V_p - V_e = V_g$). If you want to trust me, I'll tell you that when you do the numbers, it turns out that about 49% (which we can round up to 50% for discussion purposes) of the total phenotypic variation is due to genetic variation. Thus the heritability of rubber tree height in this plant population is said to be ~50%. If you don't trust me, take a look at the table of data (table 1.1) and do the calculations for yourself.[11] Again, this ratio, or heritability estimate, is thought to be a measure of how much of the variation in the height in this rubber tree population in this environment is influenced by genetic variation.

JJ: That's a strange graph. If I knew only about the unlabeled curve value for someone, I could not figure out exactly what genotype that individual had.

BG: You are exactly right. If the phenotypic value for an individual was low, say about 120 cm, that individual probably would have genotype AA, but not necessarily. And for a plant height up around 160, the rubber plant could have *any* of the underlying genotypes.

JJ: Another thing. If the rubber trees (or people in the original phosphate example) in this population were to change, I mean if for some reason there were none of the B or C genes around, it looks like the unlabeled curve would be the same as the AA curve. And if I followed you, there would be no genetic variance at all, since all variance would be environmental, even though the genotype AA is clearly of major influence here. And doesn't that mean a heritability of zero even though the AA genotype is exerting all the influence on the trait?

BG: You got it! But to impress upon you just how tricky this heritability notion can be, let me ask you to think about another example. What would you say the heritability of a human having a brain is?

JJ: Well, if all humans have a brain, or, as you geneticists would say, if there's zero variation among humans with respect to having a brain, then I guess the heritability ratio would be given by zero divided by some environmental variation—maybe due to toxins, for example, that could produce anencephalic infants—so the heritability would be zero. Wouldn't that mean that, even though our genes are crucially involved in the fact that we have brains, the heritability of having a brain is zero?

BG: Again you got it! Variance is a very tricky notion to reason with, and when you are reasoning about genetic and environmental variance, you really have to keep your eye on the ball—or rather on the genes and everything else at once. It is very easy to make the wrong inferences even if you are trying very hard. Part of the problem is due to the fact that the definition of heritability is computed from measurements on a specific population in a particular environment, and strange as it may seem, *if the environment changes, the heritability will almost certainly change.*

Many scientists who work with the heritability concept think it would be clearer if we used a different way to represent the influences of genes and environments on the phenotype that would make this very labile nature of heritability explicit. One popular alternative is called a "norm of reaction." A norm of reaction graphs the value of the phenotype for different genotypes in different environments. A good illustration of norms of reaction is from a widely cited example of another type of plant called *Achillea*. Seven different strains or genotypes of the plant were raised at three different elevations on a mountain, with the results shown figure 1.2. One can even look at the differences in expression of phenotypes over time by introducing a third dimension. In this case the reaction norm becomes a *reaction surface*. We can look at an example and a picture of a reaction surface later on.

JJ: I'm not sure, but I think someone once told me that the biologist you referred to—Lewontin was it?—had shown that heritabilities were no good at all. But it looks like you have just been using Lewontin's work to show that the heritability idea is legitimate. Can you explain?

BG: Richard Lewontin, now an emeritus professor at Harvard, has been a severe critic of behavioral genetics, at least of applications he thinks go beyond very well-founded empirical and theoretical results. He has especially been critical of IQ analyses and published a critical response to Jensen's famous (some would say infamous) long 1969 essay on IQ and race (Lewontin 1970). But probably what you heard about was Lewontin's 1974 paper that sought to undermine any general application of heritability notions to human traits, though Lewontin has written many more critiques of heritability, including the analysis we have been discussing.

Interestingly, the Lewontin 1974 critique has convinced most philosophers, including philosophers of biology, that heritability cannot be used in any way that is scientifically reliable, and not just ideological. Our discussion indicates that though heritability is a difficult concept to grasp correctly and often depends on assumptions that have not been adequately tested in the populations being investigated, it can be an appropriate, if limited, guide. More recently, several philosophers have responded to Lewontin's arguments, including Sesardic (2005) and Tabery (2014), but most philosophers have not read those responses as yet.

JJ: OK. I think I understand, but to get back to the most recent plant example, that amount of variation in the *Achillea* plants due to both genotypes and environments is astounding. I look forward to coming back to reaction norms and surfaces later. But talking just about heritability again, you said the rubber plant—or more accurately the blood phosphatase—case you showed me was rare. Why was it rare again?

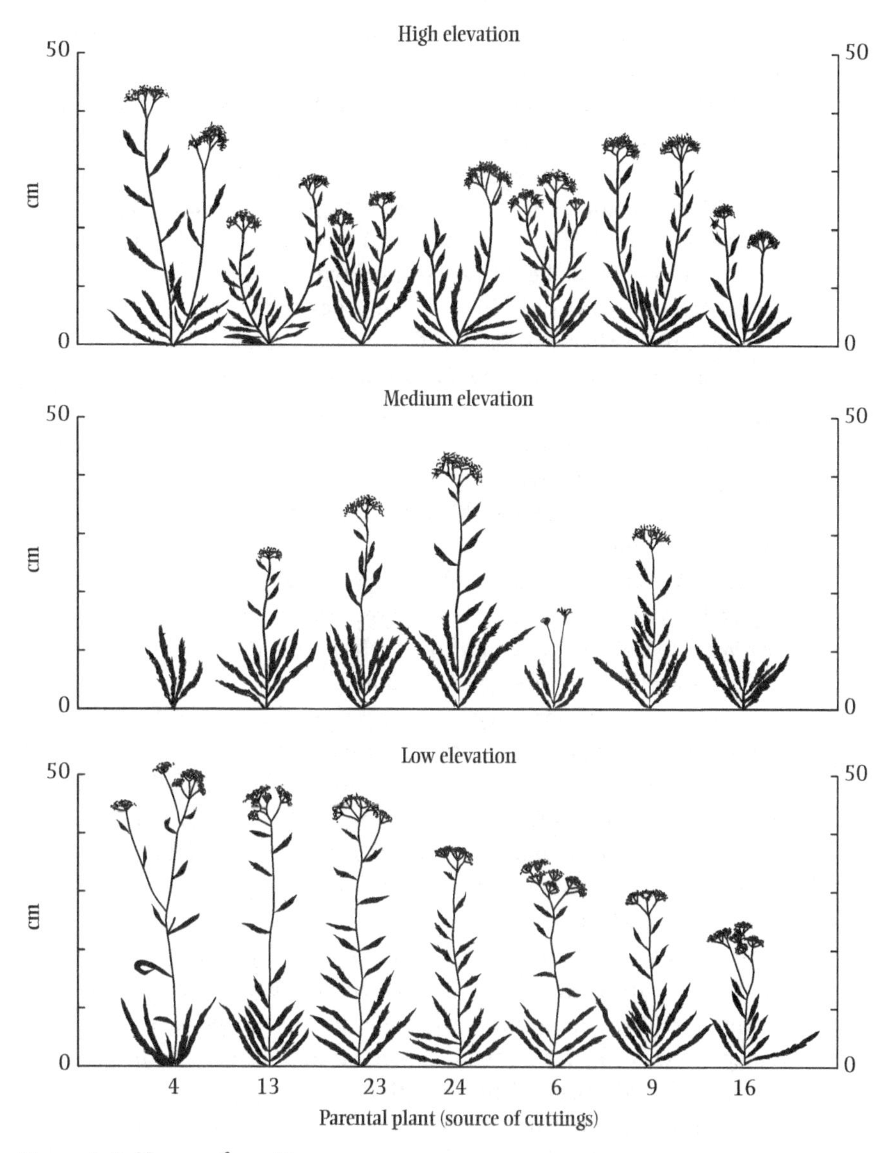

Figure 1.2 Norms of reaction.
SOURCE: Carnegie Institution of Washington. Reprinted with permission.

BG: Because in the cases that behavioral geneticists usually study, we do *not* know what alleles and genotypes underlie the trait, and of course then we do not have distributions of genotypes we can directly test for and plot out, like we did for those black curves that underlie the total population's distribution.

JJ: So what do the behavioral geneticists do then? How do they get information about how much genes might influence a trait—or, more accurately, how genetic variation might influence phenotypic variation?

BG: They use twins. They study both identical twins and fraternal twins, and they look at those types of twins both in the same family and in the different families into which they were adopted. Twins are a wonderful "experiment of nature"—identical twins have identical genotypes, and fraternal twins, though they are born about the same time, have only about 50% of their genes alike.

JJ: Just 50%? Where did you get that number? I once heard Dr. Francis Collins, the director at the National Institutes of Health, say that humans were about 99.9% identical as regards their DNA. How can any random pair of humans be 99.9% identical with respect to their DNA, but fraternal twins have only 50% of their genes alike?

BG: You have to remember this: Francis Collins was saying that at only 1 in a 1,000 points (base pairs) along our chromosomes, we're different. The thing is, small differences at those points make for different alleles (or versions) of the same gene. Behavioral geneticists are interested in the percentage of *alleles* the twins share. It is a little complicated to explain this, but see if you can follow the argument here. If not, or you want to think more about this later, fine, but maybe you can trust me in the meantime. We start from the fact that everyone has two parents, at least biologically. We are not cloning humans yet! And everyone has two copies of a gene (two alleles), and that goes for Mom and for Dad. Mom gives one of her alleles to each sibling and so does Dad.

Some elementary probability arguments show that, on average, (non-identical twin) siblings share the same allele at a given locus 50% of the time and identical twins share the same allele at the same location 100% of the time. Further, it is assumed that, if on average siblings share only one allele at a given location, then, with respect to the relevant trait, they will resemble each other about half as much as twins, who share two alleles at the same location.[12]

Note that we are *assuming* a very simple linearity here; that is, one dose of gene sharing gives one unit of resemblance and two doses give two units. *Additive* genetic variance is that portion of the total genetic variance which can be attributed to the straightforward addition of the effects of single alleles. This "additivity" assumption is important, because without it we can't use the numbers of ½ for fraternal twins versus 1 for identical twins, and those numbers function in absolutely central ways in obtaining heritabilities from twin studies.[13] Heritability estimates based upon this simplifying, additivity assumption are called *narrow sense* heritability estimates, as distinguished from *broad sense* heritability estimates, which allow for more realistic gene-gene interactions. Heritability in the narrow sense is a component of heritability in the broad sense. Some behavioral geneticists say that the narrow heritability component will do just fine and we can confine our focus to narrow heritability, but others say the distinction is important and should not be ignored. We will come back to this distinction several times later.

JJ: This "narrow sense" still seems pretty complicated to me, but even granted it's not as complex and realistic as it all could be, and we can talk about that later maybe, what do behavioral geneticists do with these 100% and 50% figures?

BG: First, they measure quantitative traits in identical twins and in fraternal twins, comparing scores within a family and then aggregating twin pairs from a number of families. It often turns out that the identical twins are more alike on a trait that's measured than are the fraternal twins, like height, and maybe like IQ or disposition to a disease such as heart disease or schizophrenia. If so, that's at least a hint that the trait has an important genetic component. It's only a hint, because the identical twins could be more alike than the fraternal twins because they were treated more alike. But behavioral geneticists assume there is no difference in the way the two types of twins are treated that affects the traits they want to study. This is called the "equal environments assumption" or EEA, and though it is controversial, behavioral geneticists claim they have reasonable empirical evidence for this assumption.[14]

JJ: What's next? How do you get from these measured differences between identical and fraternal twins to the heritability estimates I read about in the *New York Times* and *Washington Post*?

BG: We need to appeal to what's called a correlation coefficient here. Correlation measures how similar the deviations in one variable are to deviations in another variable. I say deviations because remember that we are looking at variances that track together, or covary, often referred to as covariance. (For example, if one twin is five feet, ten inches, then what's the likelihood that her twin will have the same height?) I think I can make the correlation coefficient idea simple. We symbolize this correlation coefficient by r, and it ranges from +1 (perfect correlation or perfect predictability) through 0 (no correlation) to −1 (perfect reverse correlation). A high correlation coefficient (say $r = .75$) can be depicted graphically as a fairly tight cloud of points along a line (like a somewhat diffuse airplane contrail). And that cloud narrows into a straight line as the correlation becomes higher and goes to 1.0. If we are looking at the covariances between phenotype and genotype, such a high coefficient indicates a high heritability. And as an aside, it's worth noting that the correlation coefficient between phenotype and genotype can give us the heritability—all we have to do is square the correlation coefficient! That's because the statisticians tell us that the square of the correlation coefficient gives us a percentage of how much the one variable (the genotype's variation) predicts (or accounts for) the phenotype's variation. (See figure 1.3 for four graphs representing zero, low, medium, and high correlations.)

Getting back to the twins, the correlation coefficient can be used to give a number usually between 0 and 1, which indicates how much the groups of twins are alike, in the sense of how closely they covary, on their measured scores.

JJ: Hmm . . . Maybe that squaring of the correlation coefficient to get a heritability can help me understand that concept better, but let's not move

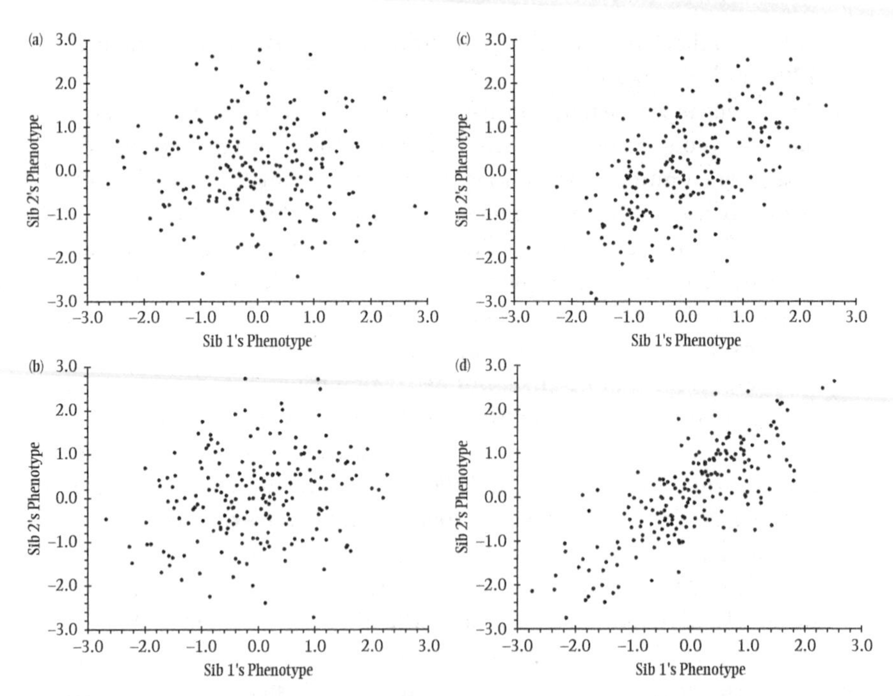

Figure 1.3 Correlation coefficients for values 0, .25, .50, and .75 (for panels a through d, respectively).
SOURCE: Carey 2003, 282–83. Reprinted with permission of Sage Publications.

so fast! I think I might be getting lost in all the math. Let's just talk about twins and what's correlating with what again, and what happened to the two different types of twins?

BG: OK, let's start with the identical twins first—but remember this is a *population* of identical twins, so they have a range of scores, and we look at how closely one twin in a pair resembles the other one in a large collection of twin pairs, maybe hundreds, say half of which are identical twins and half of which are fraternal twins, which we will consider in a moment. Let's assume the identical twins correlate pretty well on a verbal reasoning test score, say about .75, like that last figure in the correlation graphs. Reasonably enough, behavioral geneticists argue that these identical twins resemble each other because of genetics and environment. The genetic or heredity influence (variance) is represented by H^2, as always. The environment influence is represented by c^2 (for common environment). As we will discuss later, the common environment is defined by behavioral geneticists to be that which contributes to the similarity of the trait under consideration (as opposed to the unshared environment, which, by definition, does not contribute to that trait similarity). Then the correlation between identical twins is just the sum of those two influences, H^2 and c^2; in the case of the .75 correlation among the identical twins on the verbal reasoning score, we could write

$.75 = H^2 + c^2$. If you will let me write this as a more general equation, we get:

$$r_{MZ} = H^2 + c^2. \tag{1.1}$$

JJ: OK, so that's how we bring in the identical twins—I guess the MZ stands for monozygotic? But what about the other type of twins?

BG: The dizygotic or DZ twins share only one half their genes—remember you agreed to (partly) trust me on this 50% number—so the joint effects of genetics (or heredity) and common environment that produce the correlation are $h^2/2$ and c^2. (Remember from grammar school fractions: 50% of H^2 is the same as $\frac{1}{2} \times H^2$, which can be represented as $H^2/2$.) We are assuming the common environments (c^2) work *the same* for *both* types of twins, and again this is known as the equal environments assumption (EEA). Let's assume the DZ twins' correlation on the verbal reasoning test score is .50, as in the third correlation graph. They are not as similar as the sets of MZ twins on verbal reasoning scores. Thus $.50 = h^2/2 + c^2$. And the DZ twins general equation is this:

$$r_{DZ} = h^2/2 + c^2. \tag{1.2}$$

JJ: Where is this going? Are you ever going to get to how we can compute a heritability? And don't try to pull any fast ones on me—I saw that you wrote the heritability for the MZ twins as a broad heritability (H) and the heritability for the DZ twins as a narrow heritability (h), and earlier you had pointed out that these two concepts were different.

BG: You are one sharp-eyed judge. What most behavioral geneticists say is that for this kind of simple model, we are just interested in the narrow heritability, so we can assume that $H^2 = h^2$, for the time being. If we can accept this, we are actually where we can compute heritabilities. All we have to do is take those two equations and solve for h^2. I hate to remind you about high school algebra, but you can subtract the second equation from the first after we have rewritten H^2 as h^2 for the MZ twins, and you would get $r_{MZ} - r_{DZ} = h^2 - h^2/2 + c^2 - c^2$, so you can drop the c^2s since they subtract away, and then we have $r_{MZ} - r_{DZ} = h^2/2$, or, multiplying both sides of the equation by 2:

$$h^2 = 2(r_{MZ} - r_{DZ}).$$

So, if we substituted in those sample correlations of .75 and .50 from the verbal reasoning case that we talked about earlier, the heritability for this trait is 50%:

$$2(.75 - .50) = 2(.25) = .5(\text{or } 50\%).$$

JJ: I get it. I think I get it! Generally heritability is twice the difference between the similarity of the identical twins compared to the fraternal twins. What a great equation!

BG: You got it, and it is a frequently used equation. But wait—there's more!

JJ: Why do I need more? I now understand heritability and twin studies.

BG: We need to talk more about the environmental effects. We can get a measure of c^2 from our first equation, just by moving h^2 to the other side and making it a minus, so $c^2 = r_{MZ} - h^2$. But behavioral geneticists are not only interested in the common environment or shared environment. They're also interested in the nonshared environment—what makes people different.[15] We can get this into the picture very easily. The only factors we consider in this simple model are heritability, shared environment, and nonshared environment.[16] This last term we will abbreviate as e^2. We can state this idea as, if you will excuse me, another (third) equation:

$$1 = h^2 + c^2 + e^2. \tag{1.3}$$

This holds because h^2, c^2, and e^2 are each proportions, and proportions must sum to 1. (The genetic influence, plus the influence of the shared environment, plus the influence of the unshared environment is said to equal the total influence on the trait.)

And since we already know how to get h^2 and c^2, we can get e^2 as well from this equation. And all these values come *just from the observed similarities (expressed as correlation coefficients) of the monozygotic and the dizygotic twins!* I will not do the algebra with you, but it's simple enough, and it turns out that

$$e^2 = 1 - r_{MZ}.$$

JJ: All right! But let's go back to the beginning and see where this long argument has led us. Can you summarize?

BG: Twin studies give us a way to infer genetic and environmental contributions to quantitative traits, like heights, weights, and verbal reasoning scores. Typically we do not have any identified genes or genotypes that we can directly assess in the latter cases. We used the amount of shared genetic variance found in identical versus fraternal twins and their scores on some quantitative trait of interest to portion out how much was genetic, and how much was common environment and nonshared environment.

But in the argument we made lots of assumptions. Two big ones were equal environments ($c_{MZ}^2 = c_{DZ}^2$) and additivity of gene effects (both in obtaining the ½ figure and asserting that $H_{MZ}^2 = h_{MZ}^2$). There are other assumptions, too, that we did not fully make explicit, such as denying that there are any correlations between genes and environments, and interactions between genes and environments, but those are complex issues we can largely skip for the moment. But also remember that heritability estimates can change, depending on the populations and environments that are tested. Please don't forget that graph we discussed in connection with norms of reaction studies in plants! And norms of reaction studies for humans have not been done and cannot, for ethical reasons, be done in any well-controlled way. (We can't breed human beings to be genetically identical and then place them in different environments to study the relative importance of genetics and environment for a given trait.) But for the population studied in the specific environments examined, if we have a high heritability, many behavioral geneticists think we may have a good pointer to an underlying genetic architecture. But that is actually still to be proven. We will discuss this later (in dialogue 2).

The twin model is a start, and the version we have discussed is a very oversimplified one at that. Behavioral geneticists can make more complex models that deal with some of these complexities and measure the effects, such as broad and narrow heritabilities. Furthermore, the advent of a new method called genome-wide association analysis (GWAS) has enabled what the researchers call "an assumption-free method" for ascertaining at least a lower-bound estimate of heritability, for traits such as human height and general intelligence. But that's for another day.

JJ: Another day is fine, but before we break, maybe you can quickly tell me if this strange heritability concept is ever used outside the human area, where it seems so hard to study. Is it ever used in animal studies?

BG: The answer is yes, and interestingly, in animal and plant studies, heritability is extensively used in the *narrow* sense that we found we had to move to in the twin-study analysis. (Remember that narrow sense heritability assumes linear or "additive" relationships among genes.) But in the animal and agricultural world, investigators are mainly interested in breeding programs, so as to improve the weight of offspring of cattle, for example. These researchers use narrow heritability as a *breeding value coefficient* to predict how much change they will see in the next generation, since narrow heritability represents the genetic source of the resemblance between parents and offspring. If the narrow heritability is 100% (i.e., if all variation is genetic), then the value of the trait of interest, such as height, in the offspring will be exactly midway between the trait values in the two parents. Thus it is the narrow heritability that is relevant to any evolutionary effects that genes have, and this is an important point to remember in any eugenics discussion.

JJ: Uh-oh! The concept of a "breeding program" related to behavioral genetics scares me, as does the eugenics concept. But do the animal-breeding programs really work on behaviors? Can you give me an example?

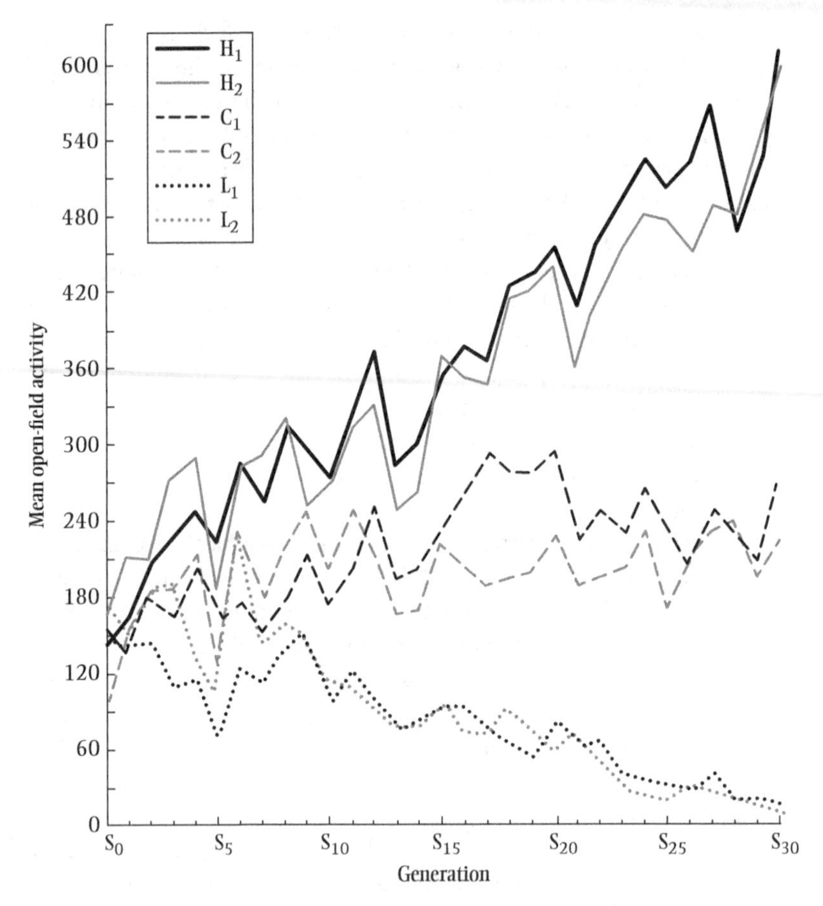

Figure 1.4 Selecting traits. Results of a selection study of open-field activity. Two lines were selected for high open-field activity (H_1 and H_2), two lines were selected for low open-field activity (L_1 and L_2), and two lines were randomly mated within each line to serve as controls (C_1 and C_2).
SOURCE: DeFries, Gervais, and Thomas 1978.

BG: The classic study was done in mice, in which populations were measured for fearfulness, which was tested for by putting the mice in a brightly lit box called an "open field." "Fearful" mice froze and lost excretory control—they were scared—less, as one prominent neuroscientist has indelicately put it.[17] "Brave" mice, on the other hand, actively explored their surroundings. Mice from the brave and the fearful lines were interbred over 30 generations, and as controls the two lines were also crossbred. The results of the breeding program on open-field activity is shown in figure 1.4.

JJ: Interesting! Though the lines zig-zag around a bit, I can imagine roughly straight lines though the three sets of points, so I guess that allows for predictability of what you'll see in each successive generation. But exactly how does heritability come in here?

BG: Maybe it would help to use a simpler example, say one involving our rubber plants again, but now breeding them to increase their height. We will now assume, as in virtually all cases involving quantitative genetics, and especially behavioral genetics, that we do *not* know about those A, B, and C alleles, but only about the heritability of height. (Quantitative geneticists have means of estimating heritabilities from phenotypic resemblance between parents and offspring, and also using resemblances among other relatives.) Look at the Lewontin figure (1.1) again. Note that the plant height varies all the way from a minimum of some around 2½ feet (80 cm) up to the really tall almost 9-foot rubber plants at the other extreme. The average height in this entire collection of plants is about 160 cm (~5 feet). Suppose our plant breeder anticipates orders for a large number of big rubber trees for some new malls. If he can interbreed just very tall parents (which he selects from those that have heights about 260 cm) and if the heritability of height in this population is 50%, then the average height of their offspring will be about 210 cm (about 6′10″), well above the average of the *original* population. (This comes from taking 50% of the *average deviation* of 100 [the difference between the average height of the tall plants (260) and the average height of the original population (160)] and adding this figure of 50 cm to the average height of 160 for the original population.) In case you're interested, the equation that predicts this response is given in standard textbooks of quantitative genetics as

$$R = h^2 S,$$

where the response (R, the increase above the average in the offspring) is equal to the heritability estimate (h^2) times the selection difference value (S, the amount above the average) in the breeding parents.[18]

This easy prediction using the narrow heritability concept, however, assumes that there is *no nongenetic* cause of resemblance between parent and offspring that would affect height, like nutritional or climate differences. (This is assumed as part of the derivation of the equation above as found in genetics texts.) The prediction also assumes that the underlying genes, which are in truth veiled from us now, act to *sum together* in terms of their effects on height (as before, this is called *additive* genetic variation), so that one dose of the allele results in the addition of 1 unit of a trait, two doses of the allele yields 2 units, and so on, just like in our shared resemblance units for human twins. Additive genes are stipulated to not have more complex nonlinear interrelations that would confuse the selection and breeding process, such as the gene dominance effects or gene-gene interaction effects. Though the genes are veiled from us, selection experiments can provide some evidence of additivity of gene contributions through multigeneration breeding experiments like those we saw with the mice.

Finally, you might be interested in knowing that one of the critics of quantitative behavioral genetics, Douglas Wahlsten, wrote in a classic article in 1990 that "the only practical application of a heritability coefficient [h^2] is to predict the results of a program of selective breeding" (1990, 119).

JJ: Well that's very interesting. I can actually grasp intuitively how the heritability estimate works here a bit better—as a partial predictor of a characteristic in the next generation—than I could either looking at the picture of those rubber trees (acid phosphatase activity) or at those twin-study equations. And though I appreciate Wahlsten's viewpoint, after all of our discussion today and all the qualifications about heritability, I imagine many behavioral geneticists see it quite differently. But now we really have to stop, but you never told me about "missing" heritability, so maybe next time.

THE TWIN-STUDY MODEL AND THE ACE PATH DIAGRAM IN BEHAVIORAL GENETICS

Now that we have the rudiments of the heritability concept and understand how shared and nonshared environments can be appealed to, we should note that the twin-study model can be represented graphically in what is called the *ACE model*. The graphical picture may help some readers imagine how twin studies work, and also might assist them in reading other papers in the field. (Readers familiar with ACE diagrams or just in a hurry might skip to the next section.)

This model can be depicted graphically in what is called a path diagram, as with the MZ and DZ twins in figure 1.5. The straight arrows represent causal influence, and the curved arrows correlations.

There are easy and straightforward ways to apply the three equations given above to these diagrams, but the reader must be referred to other sources for these techniques, for example the appendix provided by Plomin et al. (2013). This model uses h^2 to represent heritability, and since most often behavioral

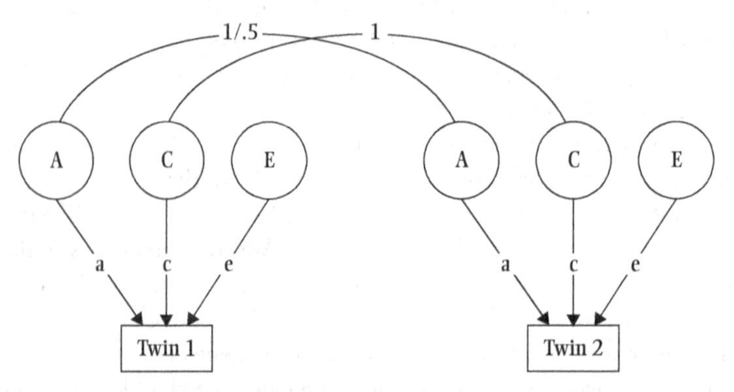

Figure 1.5 ACE paths in a generic simplified format (self-correlations, such as A with A = 1.0, are not shown). See text for definition of variables and discussion.

geneticists assume they are working only with additive genotypic variance, they underscore this additivity by using a^2 in place of h^2, and call this kind of diagram an ACE model, capitalizing the a, c, and e factors that influence a trait. It should be stressed that the circles around A, C, and E indicate that these are theoretical, unobserved, or "latent" variables, in contrast to the observed, empirical, or measured variables represented by the squares for Twin 1 and Twin 2. The long, curved, double-headed arrow labeled with a value of 1.0 (MZ) or 0.5 (DZ) between the As represents the values for shared additive genetic correlation that the behavioral geneticist derived for Judge Jean above. The other long, curved, double-headed arrow labeled with a value of 1.0 between the Cs represents that value which is based on the equal environments assumption (EEA) discussed earlier.

Environmental Complexities and Startling Results from Quantitative Genetics

The c^2 (shared) and e^2 (nonshared) components of the environment that were discussed earlier in this chapter, and that appeared in the dialogue and just now reappeared in the ACE model, represent critically important distinctions that are at root somewhat nebulous. I have noted earlier that many people would probably assume that those things twins experienced in common in a family would be considered part of the *shared* environment, and that those things they did not experience in common would be considered part of the *nonshared* environment. But recall that in the derivation of the equations for the twin model, c^2 was *defined* as all those things that make the twins alike (correlated). Similarly, e^2 is *defined* as all those things that make the twins different. The notions can escape from empirical circularity by being further specified and qualified, but this is where the somewhat nebulous character of the concepts becomes apparent. Ultimately, I will suggest, the distinction between shared and nonshared becomes highly context specific, almost *a term of art*, to be determined in each specific study. That can lead to some problems, such as a reintroduction of circularity, which I note below.

The most explicit discussion of what more needs to be known for an environmental effect to be termed "shared" is by Rowe and Jacobson (1999, 16–17), who note that "four conditions must be satisfied for an environmental effect to count as shared" (all quotations come from these two pages).

Those conditions are that (1) near universals found in a culture from which the subjects are drawn do not count because they would be common to *all* families. For example, the use of the language of English among second-generation Americans is such a near universal. Also, (2) the environmental exposures must be common to *all* siblings. Rowe and Jacobson say that parental divorce would typically count as shared for the involved children, but this is an example we will need to return to again. Another condition (3) requires that "the environmental exposure must have a directional effect on a given trait to be considered an environmental influence on that trait." Directional here seems to mean "plausibly causal and relevant," a meaning that raises a question of circularity (since we are trying to *establish causal properties*).[19] Finally (4), in what seems to be a variant of the third condition, "Environmental effects can be shared environmental influences only to the extent that they reliably change a trait in a *constant* direction." Here the divorce example is returned to, but it is qualified so that if divorce affects siblings differently (say different sibs have distinct emotional/behavioral reactions), then divorce would count as a *nonshared* environmental influence. To add to the potential complexity, Rowe and Jacobson add that as regards this latter type of factor, "the number and variety of potential nonshared environmental influences are nearly limitless." How to reliably specify such nonshared factors and assess their significance turns out to be a *very* difficult problem for behavioral geneticists, as I note in the next subsection.

The bottom line, for this observer, is that the distinction between what counts as shared and what counts as nonshared becomes highly context specific, to be specified only in the individual study. (By "content specific" here I mean that context and often judgment are involved in drawing the distinction, and they may well depend on special data available in particular studies, and not admit of any easily generalizable rules.) But since the individual study is establishing the environmental factors as causal and as either shared or nonshared, circularity may be introduced, which may weaken the empirical force of the study. For such key concepts to have such prima facie vagueness and potential circularity may point to the reasons behind what I'll soon call the "gloomy hypothesis" regarding the nonshared environment.

Startling Results about Shared and Nonshared Environments: A "Gloomy Prospect"?

Twin studies, incorporated within more sophisticated "biometric"[20] path models like the ACE model I mentioned above use the *shared* and *nonshared* environments distinction with the qualifications I just mentioned. Perhaps surprisingly, quantitative studies of normal personality traits (as well as mental disorders) indicate that of all the environmental factors, the *nonshared* ones have the major effect.[21] As noted earlier, it is estimated that about 15% of the variance of the schizophrenia trait is due to the nonshared environment and little, if any, to the shared environment (Cardno and Gottesman 2000). However, an influential meta-analysis of 43 studies by Turkheimer and Waldron (2000) indicated that though nonshared environmental factors were responsible for 50% of the total variation of behavioral outcomes, *identified* nonshared environmental factors accounted for only 2% of the total variance. In this meta-analysis, Turkheimer distinguished two senses of nonshared: objective and effective. An environmental event is objectively nonshared if it is (verifiably, measurably) experienced by *only one* sibling in a family, *irrespective of the consequences* it produces. An event is *effectively* nonshared if it makes siblings *different*, rather than similar (this is how the term is typically *defined*), irrespectively of whether it is experienced by one or both.

The significant claim about the role of the nonshared environment, originally made by Plomin and Daniels in 1987, was that children in the same family are as different as they are because *measurable* differences in their environments made them that way. But, contrary to that claim, the meta-analysis I just mentioned found that, although 50% of the variance in behavioral outcomes was accounted for in the *effective* nonshared sense (the kids *were* different), the median percentage accounted for by objectively definable nonshared event is less than 2% (Turkheimer and Waldron 2000). That is, the nonshared environment may help to explain about 50% of the phenotypic variance, but the specific factors that constitute the nonshared environment typically are not measured, so thus far they only seem to account for 2% of explained variance.

In a related article Turkheimer (2000) infers that these nonshared differences are nonsystematic and largely accidental or random, and thus have been and will continue to be very difficult to study. This possibility had been considered earlier by (Plomin and Daniels 1987) but dismissed as a "gloomy prospect." That "gloomy prospect" now looks more plausible, though Plomin and his colleagues recently have offered a more optimistic gloss, as well as recommendations for further research in this area (Plomin et al. 2013, 100).

The gloomy results regarding specific environmental factors may in part result from the fact that we have no "general theory of the environment," as noted earlier in this chapter. Without some fairly detailed general backdrop view that can suggest how to decompose the environment and measure it reliably, gloomy results may continue. There are only glimmers of hope that this problem will be remedied in the short run.[22] At least one team of investigators in the IQ area declined to explicitly factor in the effects of the nonshared environment and instead treated the nonshared environment as *part of* the *error term*, rather than treating them as distinct terms (Devlin, Daniels, and Roeder 1997). This is a controversial position, and other investigators have proposed that appropriate designs can estimate the amount of error and distinguish it from the nonshared environment term. One way is that, for a given measure, one estimates the total nonshared environment plus error (the E term in an ACE model) and subtracts the error variance from it (obtained as 1.00 minus the reliability of the measure, if using correlations, or that quantity times variance) in order to obtain e^2 proper (personal communication, John Loehlin, 2011).

On the more positive side, quantitative investigations have identified two important ways that genetic variables and variables in the shared and nonshared environment interact to influence phenotypes. The first, genotype-environment *correlation* (usually written as G·E but occasionally as G → E) represents possible effects of an individual's genetics on the environment (e.g., via that individual's evoking different responses from or selecting environments). Such effects were found for both normal and pathological traits studied in the large Nonshared Environmental Adolescent Development (NEAD) study, described in detail in the 2000 book *The Relationship Code* (Reiss et al. 2000). Second, different genotypes have different *sensitivities* to environments, collectively called genotype × environmental *interaction* (G × E). A classic example is phenylketonuria (PKU), which can produce devastating mental retardation unless the diet is controlled for phenylalanine. There were several discoveries made in the years 2002–2005 using molecular methods that involve early environment interacting with genetic dispositions to criminality (Caspi et al. 2002), to depression (Caspi et al. 2003), and to psychosis (Caspi et al. 2005) that have stimulated extensive interest in gene-environment interactions. (For a preliminary discussion of the Caspi et al. articles, see chapter 2 following, with more detail in chapter 6.) Differential sensitivity is important in many models of genetic disorder, including the neurodevelopmental models of schizophrenia genetics mentioned later in chapter 7.

Testing the Models and "Goodness of Fit"

The ACE model I presented was very simple. Contemporary behavioral genetics uses such simple models, but it also employs more complex, multivariate models that can represent common factor genes—genes that can account for correlations between different traits such as verbal and quantitative cognitive abilities. For additional discussion of a genetic contribution underlying several traits, see the following chapter where the "g" factor will be briefly discussed. The more complex multivariate models can also represent additional family relationships that I haven't discussed (e.g., half-sibs and unrelated sibs), and more sophisticated means of assessing family influences over time (i.e., longitudinal models). Given the plethora of conceivable models, assessing the best (or at least better) ones is of major importance. Behavioral geneticists can use multiple models and combine them, and test them jointly using empirical data to determine which models "fit" the data best. "Goodness of fit" tests use highly developed mathematical and computer techniques that are also widely found in other social sciences. An accessible account of how a goodness-of-fit determination is done for some simple ACE models involving MZ twins, related sibs, and unrelated sibs can be found in (Rowe 1994) and in the appendix in Plomin et al.'s (2013) work. Though such goodness-of-fit tests are what Rowe has called the "state of the art" in behavioral genetics, some behavioral geneticists such as Turkheimer (personal communication) have doubts as to whether these empirical tests can adequately control for the mix of additive and nonadditive genetic influences (as well as shared and nonshared environmental influences) involved in human studies. I give an example of a goodness-of-fit approach involving four models in the IQ section of dialogue 3 in the following chapter.

One very useful way to put together the two behavioral genetic research traditions, namely the quantitative tradition in both its simple, heritability form, and its more complex path diagram forms, and the molecular tradition, developed in detail in the next chapter, is to look at them as four levels or approaches. This has been done by Kendler in an article first published in 2005, and modified forms of it are presented in table 1.2 in two slightly different ways.

An excellent, albeit very complex illustration of level 2 involving path diagrams over five stages of a lifetime can be found in figure 1.6. In that study, genetic factors are actually a latent variable constructed from a composite measure of the

Table 1.2 FOUR MAJOR PARADIGMS OF PSYCHIATRIC GENETICS

Paradigm	Samples studied	Method of inquiry	Scientific goals
1. Basic genetic epidemiology	Family, twin, and adoption studies	Statistical	To quantify the degree of familial aggregation and/or heritability
2. Advanced genetic epidemiology	Family, twin, and adoption studies	Statistical	To explore the nature and mode of action of genetic risk factors
3. Gene finding	High-density families, trios, case-control samples	Statistical	To determine the genomic location and identity of suscepti-bility genes
4. Molecular genetics	Individuals	Biological	To identify critical DNA variants and trace the biological path-ways from DNA to disorder

MAJOR PARADIGMS OF BEHAVIORAL AND PSYCHIATRIC GENETICS[a]

General paradigm type	*Subtypes and data samples*	*Scientific aims and goals*
Quantitative/ statistical	Basic—including heritability from family, twin, and adoption studies	Quantitative study of familial aggre-gation and/or heritability
	Advanced—including path analyses of family, twin, and adoption studies	Analysis of genetic and environments and joint risk factors
Molecular	Gene finding involving high density families, trios, and case-control samples; also GWAS searches	Identification of location of genetic risk loci
	DNA, RNA, and protein from individuals and families using gene/protein sequencing	Identification of genes, their regulation and their function(s) in molecular pathways

[a] Distinctions and approaches following Kendler's (2005) earlier analyses of four major paradigms of behavioral and psychiatric genetics. The first two are quantitative, the last two are molecular.

SOURCE: Adapted from Kendler 2005, table 1, p. 4. Reprinted by permission of the *American Journal of Psychiatry.*

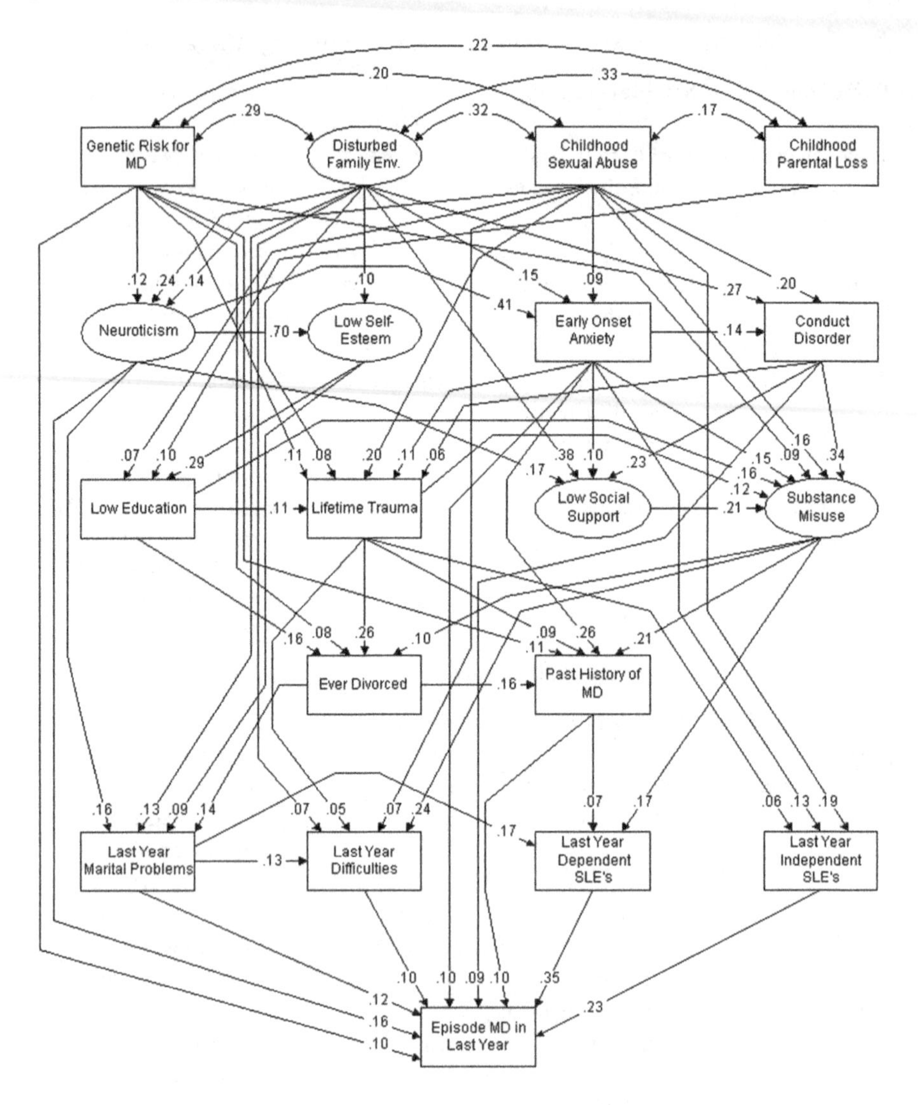

Figure 1.6 Results from an integrative model for the prediction of major depression in the last year in women. Latent variables are in depicted in ovals, observed variables are depicted as rectangles. SLE = stressful life events.
SOURCE: Adapted from Kendler, Gardner, and Prescott 2006. Reprinted by permission of the *American Journal of Psychiatry*. Also based on Kendler and Parnas 2008; reprinted with permission.

lifetime history of major depression in the co-twin as well as in the individual's mother and father, corrected for co-twin status as monozygotic or dizygotic. It needs to be stressed here that specific genes are *not* ascertained in this type of study, which uses familial history data to arrive at a measure of the risk level. It should perhaps be emphasized in addition that what is depicted in the diagram is probabilistic causation, and not deterministic causation. The model has over the years grown quite complicated and best requires color-coding (not possible in these pages) to identify the life stages as well as the large number of causal links among the factors. A figure from the richest of the recent studies by Kendler et al. is given in figure 1.6. This is a male twin study.

I now conclude this introduction to basic concepts and largely traditional, quantitative, or epidemiological behavioral genetics. As I suggested earlier, those traditional approaches produce evidence to suggest that a given trait *may be* importantly influenced by genetics. But to understand whether such a trait *is* specifically influenced by identified genes, researchers have to use newer, molecular approaches. In the following chapter, in a second dialogue between Judge Jean and the behavioral geneticist, we will turn to those molecular approaches. In a third and concluding dialogue with Judge Jean, we will examine the strengths and weaknesses of both the quantitative and the molecular methods by focusing on two specific, though hypothetical, cases of genetic testing for high IQ and for ADHD. That chapter will conclude with some speculations about the future direction of behavioral genetics research in elucidating the relations of nature and nurture.

References

Beckwith, Jonathan. 2006. "Whither human behavioral genetics." In *Wrestling with Behavioral Genetics: Science, Ethics, and Public Conversation*, edited by Erik Parens, Audrey R. Chapman, and Nancy Press, 84–99. Baltimore: Johns Hopkins University Press.

Benjamin, Jonathan, Richard P. Ebstein, and Robert H. Belmaker. 2002. *Molecular Genetics and the Human Personality*. 1st ed. Washington, D.C.: American Psychiatric Publishing.

Biological Sciences Curriculum Study (BSCS). 2000. *Genes, Environment and Human Behavior.* Edited by Joseph D. McInernet. Colorado Springs: BSCS.

Cardno, A. G., and I. I. Gottesman. 2000. "Twin studies of schizophrenia: From bow-and-arrow concordances to Star Wars Mx and functional genomics." *American Journal of Medical Genetics* 97 (1): 12–17.

Carey, G. 2003. *Human Genetics for the Social Sciences*. Thousand Oaks, CA: Sage.

Caspi, A., T. E. Moffitt, M. Cannon, J. McClay, R. Murray, H. Harrington, A. Taylor, et al. 2005. "Moderation of the effect of adolescent-onset cannabis use on adult psychosis by a functional polymorphism in the catechol-O-methyltransferase gene: Longitudinal evidence of a gene X environment interaction." *Biological Psychiatry* 57 (10): 1117–27.

Caspi, A., A. Taylor, T. E. Moffitt, and R. Plomin. 2000. "Neighborhood deprivation affects children's mental health: Environmental risks identified in a genetic design." *Psychological Science* 11 (4): 338–42.

Caspi, Avshalom, Joseph McClay, Terrie E. Moffitt, Jonathan Mill, Judy Martin, Ian W. Craig, Alan Taylor, and Richie Poulton. 2002. "Role of Genotype in the Cycle of Violence in Maltreated Children." *Science* 297 (5582):851–54.

Caspi, A., K. Sugden, T. E. Moffitt, A. Taylor, I. W. Craig, H. Harrington, J. McClay, J. Mill, J. Martin, A. Braithwaite, and R. Poulton. 2003. "Influence of life stress on depression: Moderation by a polymorphism in the 5-HTT gene." *Science* 301 (5631):386–89.

DeFries, J. C., M. C. Gervais, and E. A. Thomas. 1978. "Response to 30 generations of selection for open-field activity in laboratory mice." *Behavior Genetics* 8 (1): 3–13.

Devlin, B., M. Daniels, and K. Roeder. 1997. "The heritability of IQ." *Nature* 388 (6641): 468–71.

Falconer, D. S. 1965. "The inheritance of liability to certain diseases, estimated from the incidence among relatives." *Annals of Human Genetics* 29: 51–76.

Falconer, D. S., and T. Mackay. 1996. *Introduction to Quantitative Genetics*. 3rd ed. Englewood Cliffs, NJ: Prentice Hall.

Gottesman, I. I., and L. Erlenmeyer-Kimling. 2001. "Family and twin strategies as a head start in defining prodromes and endophenotypes for hypothetical early-interventions in schizophrenia." *Schizophrenia Research* 51 (1): 93–102.

Gottesman, Irving I., and Dorothea L. Wolfgram. 1991. *Schizophrenia Genesis: The Origins of Madness*. New York: W.H. Freeman.

Harris, Harry. 1980. *The Principles of Human Biochemical Genetics*. 3rd ed. Amsterdam: Elsevier; New York: North-Holland Biomedical Press.

Heston, L. L. 1966. "Psychiatric disorders in foster home reared children of schizophrenic mothers." *British Journal of Psychiatry* 112 (489): 819–25.

Hettema, J. M., M. C. Neale, and K. S. Kendler. 1995. "Physical similarity and the equal-environment assumption in twin studies of psychiatric disorders." *Behavior Genetics* 25 (4): 327–35.

Jensen, A. R. 1969. "How much can we boost IQ and scholastic achievement?" *Harvard Educational Review* 39: 1–23.

Joseph, J. 2001. "Separated twins and the genetics of personality differences: A critique." *American Journal of Psychology* 114 (1): 1–30.

Kendler, K. S. 2005. "Psychiatric genetics: A methodologic critique." *American Journal of Psychiatry* 162 (1): 3–11.

Kendler, K. S., C. O. Gardner, and C. A. Prescott. 2006. "Toward a comprehensive developmental model for major depression in men." *American Journal of Psychiatry* 163 (1): 115–24.

Kendler, K. S., and J. Parnas, eds. 2008. *Philosophical Issues in Psychiatry*. Baltimore: Johns Hopkins University Press.

Kendler, K. S., M. C. Neale, R. C. Kessler, A. C. Heath, and L. J. Eaves. 1993. "A test of the equal-environment assumption in twin studies of psychiatric illness." *Behavior Genetics* 23 (1): 21–7.

Kendler, K. S., M. C. Neale, R. C. Kessler, A. C. Heath, and L. J. Eaves. 1994. "Parental treatment and the equal environment assumption in twin studies of psychiatric illness." *Psychological Medicine* 24 (3): 579–90.

Kitcher, Philip. 1984. "1953 and all that: A tale of two sciences." *Philosophical Review* 93: 335–73.

LeDoux, Joseph E. 1996. *The Emotional Brain: The Mysterious Underpinnings of Emotional Life*. New York: Simon & Schuster.

Lewontin, R. C. 1970. "Race and intelligence." *Bulletin of the Atomic Scientists* 26 (5): 2–8.

Lewontin, R. C. 1974. "Annotation: The analysis of variance and the analysis of causes." *American Journal of Human Genetics* 26 (3): 400–11.

Lewontin, Richard C. 1995. *Human Diversity*. New York: Scientific American Library, distributed by W.H. Freeman.

Lewontin, Richard C., Steven P. R. Rose, and Leon J. Kamin. 1984. *Not in Our Genes: Biology, Ideology, and Human Nature*. New York: Pantheon.

Neiderhiser, J. M. 2001. "Understanding the roles of genome and envirome: Methods in genetic epidemiology." *British Journal of Psychiatry Supplement* 40:s12–17.

Plomin, Robert. 2008. *Behavioral Genetics*. 5th ed. New York: Worth.

Plomin, R., and D. Daniels. 1987. "Why are children in the same family so different from one another?" *Behavioral and Brain Sciences* 10: 1–60.

Plomin, Robert, John C. DeFries, Valerie S. Knopik, and Jenae M. Neiderhiser. 2013. *Behavioral Genetics*. 6th ed. New York: Worth.

Plomin, Robert, John C. DeFries, Gerald E. McClearn, and Peter McGuffin. 2001. *Behavioral Genetics*. 4th ed. New York: Worth.

Reiss, D., and J. M. Neiderhiser. 2000. "The interplay of genetic influences and social processes in developmental theory: Specific mechanisms are coming into view." *Development and Psychopathology* 12 (3): 357–74.

Reiss, David, Jenae M. Neiderhiser, E. Mavis Hetherington, and Robert Plomin. 2000. *The Relationship Code: Deciphering Genetic and Social Influences on Adolescent Development*. Cambridge, MA: Harvard University Press.

Rowe, David C. 1994. *The Limits of Family Influence: Genes, Experience, and Behavior*. New York: Guilford Press.

Rowe, David C., and Kristin C. Jacobson. 1999. "In the mainstream: Research in behavioral genetics." In *Behavioral Genetics: The Clash of Culture and Biology*, edited by Ronald A. Carson and Mark A. Rothstein, 12–34. Baltimore: Johns Hopkins University Press.

Schaffner, K. F. 1998. "Genes, behavior, and developmental emergentism: One process, indivisible?" *Philosophy of Science* 65 (June): 209–52.

Schaffner, K. F. (in development). *A History of Behavioral Genetics: The IQ and Aggression Debates*.

Schaffner, K. F., and P. D. McGorry. 2001. "Preventing severe mental illnesses: New prospects and ethical challenges." *Schizophrenia Research* 51 (1): 3–15.

Sesardic, Neven. 2005. *Making Sense of Heritability: How Not to Think about Behavior Genetics*. New York: Cambridge University Press.

Tabery, J. 2014. *Beyond Versus: The Struggle to Understand the Interaction of Nature and Nurture*. Cambridge, MA: MIT Press.

Turkheimer, E. 2000. "Three laws of behavior genetics and what they mean." *Current Directions in Psychological Science* 9: 160–1.

Turkheimer, E., and M. Waldron. 2000. "Nonshared environment: A theoretical, methodological, and quantitative review." *Psychological Bulletin* 126 (1): 78–108.

Wade, N. 2014. *A Troublesome Inheritance: Genes, Race and Human History*. New York: Penguin.

Waters, C. Kenneth. 1994. "Genes made molecular." *Philosophy of Science* 61: 163–85.

2

Behaving

Its Nature and Nurture (Part 2)

The previous chapter introduced basic terms, ideas, and methods of behavioral genetics, both discursively in the early sections and then again in the form of a hypothetical dialogue between a behavioral geneticist and Judge Jean. Specifically, I introduced basic concepts at work in the classical, quantitative, and epidemiological approaches to understanding human behavior. This chapter focuses on the more recent molecular approaches of behavioral genetics. (For a concise characterization of the differences between quantitative and molecular genetics, see figure 1.6 from Kendler in chapter 1.) In this chapter, I first introduce the background to the "molecular turn" in behavioral (and psychiatric) genetics, and then I resume the dialogue format between Judge Jean and the behavioral geneticist. In dialogue 2 (dialogue 1 is in the previous chapter), I discuss linkage and association methods, including the newer genome-wide association studies (GWAS) and copy number variant (CNV) results, summarize results of novelty-seeking and Alzheimer's disease studies, and raise some questions about how quantitative and molecular research programs are related. In dialogue 3, I describe two hypothetical cases involving genetic testing for IQ and for attention deficit hyperactivity disorder (ADHD). Judge Jean learns how much is known about both the quantitative and the molecular aspects of IQ and ADHD, and she is introduced to a promising but skeptical vision of the future of behavioral genetics in the context of neuroscientific complexity.

MOLECULAR METHODS

Classical quantitative (or quantitative) studies discussed in the previous chapter can indicate that genes may contribute to complex human behaviors, but they do not identify specific genes or *how* genes contribute to behaviors. At the beginning of the twenty-first century, McGuffin, Riley, and Plomin wrote that the "quantitative approaches can no longer be seen as an end in themselves," and that the field must move to specific genes assisted by the recently completed draft versions of the human genome sequence (McGuffin, Riley, and Plomin 2001, 1232). In fact, my short review of the literature at that time on the nature-nurture issue (Schaffner 2001) indicated that most research in psychiatric genetics had already taken a "molecular turn."

Even 20 or so years ago it was widely acknowledged that most genes playing etiological and pathophysiological roles in behavioral traits and in psychiatric disorders would *not* be single-locus genes of large effect following Mendelian patterns (McGuffin, Riley, and Plomin 2001). Hyman also then noted that mental disorders will typically be heterogeneous and be affected by multiple, partially overlapping sets of genes (Hyman 2000). Mental disorders will thus be "complex traits," technically defined as conforming to *non*-Mendelian inheritance patterns. A classic, though difficult, article on genetics and complex traits appeared in 1994 by Lander and Schork. The article is still essential historical reading for those wanting to familiarize themselves with this still developing notion of a complex trait, which does not deny Mendel's accomplishments, but enriches traditional genetics by moving beyond the study of *monogenic* traits, which result from mutations of a single gene. Subsequently, Lander and his colleagues updated the 1994 account to include GWAS analyses (Altshuler, Daly, and Lander 2008).

Two of the original general molecular methods widely used to search for genes related to behavioral traits and mental disorders are still in use; these methods are *linkage analysis* and *allelic association*. Linkage analysis is the traditional approach to gene identification, but it only works well when genes have reasonably large effects, which does not appear to be the case in behavioral (or psychiatric) genetics.[1] Allelic association studies are more sensitive, but at present they typically require "candidate genes" to examine familial data.[2] To introduce these molecular methods as well as the newer GWAS and CNV, as well as complete sequencing approaches, we return to the dialogue format used earlier and reintroduce the inquisitive Judge Jean.

DIALOGUE 2. A CONTINUATION OF THE DIALOGUE BETWEEN A STATE SUPREME COURT JUDGE AND A BEHAVIORAL GENETICIST: MOLECULAR HOPES AND PROBLEMS

JUDGE JEAN (JJ): Our conversation the other day helped me understand how heritability works and how twin studies can give us some estimates of the roles that genes and environments play. But what I really want to know about are the molecular advances. I know that the sequence of the entire human genome has been completed, and over the last several years I've heard about various genes for novelty seeking and anxiety, as well as about genes that cause Alzheimer's disease and maybe even predispose people to criminality. This molecular area has to be where results about genes could affect my decisions. What can you tell me?

BG: Unfortunately, there has been an awful lot of hype in the molecular area, though there have been some stunning advances as well. The "finished" human genome sequence was completed in 2003 and published in the two leading general scientific journals in October 2004 (Genome

Consortium 2004).[3] But these only identify the sequence of the base pairs of the genomes that the two research groups studied. Information about those DNA sequences is very valuable, but there's still much that we do not understand, such as how much of the DNA codes for genes and how much is nonfunctional, or exactly how many genes there are, and especially how most of the coding sequences (and more recently the noncoding parts, or mainly RNA coding regulatory regions) work to produce phenotypic effects. Work continues at a rapid pace, however, with new discoveries coming every few weeks. Also, it is very hard to tie even general types of behaviors to genes, though some progress has been made, as you mentioned, in the area of Alzheimer's disease. Maybe we should start with that novelty-seeking gene you mentioned and I could explain how it was identified and also some serious problems with it. Then we can talk about the anxiety gene you also mentioned, which seems to have more evidence in its favor since it also has been tied to depression, though those results have also been questioned in very recent times.

JJ: Great! Novelty seeking isn't exactly the sort of "impulsive" behavior that the people who come before me engage in, but maybe that case can give me a sense of how behavioral geneticists try to understand complex behaviors at the level of the gene.

BG: Some researchers did think they discovered a correlation between a particular allele (or form of a gene) and novelty-seeking behavior. That correlation was found by comparing two groups of people. In the first few studies, one group had a *short* form of an allele (called *DRD4*)[4] that codes for a receptor for the neurotransmitter molecule dopamine, which I will explain in a moment. The second group had a *long* form of the same allele. Using a standard psychological questionnaire, the researchers compared how the first group's level of novelty seeking compared with the second group's. The researchers found that in comparison with the first group, the second group exhibited greater novelty-seeking behavior.

That kind of genetic study is called an *association study*, because it looks to associate a variation in the gene's alleles with a variation in the behavioral phenotype (in this case, the level of novelty seeking). The *DRD4* gene was examined because a prominent psychiatrist, Dr. Robert Cloninger, argued that dopamine is one of the important chemicals used by nerves to communicate either among themselves or with muscle cells, especially about novelty-seeking behavior. This kind of chemical is called a neurotransmitter. Cloninger's argument was based on his studies using rats and other animals. In association studies, typically you first need to have a plausible *candidate gene* from some other study, maybe from an animal study, or from a theory about how a certain gene might be related to a behavior. Then you look for the correlations between the alleles and the behavior.

JJ: That seems straightforward enough. So this gene was a good candidate, and it makes more of a chemical that makes the person with this allele seek out novel experiences?

BG: I don't think they thought that precisely about making more of a chemical, but they did expect an effect related to the biochemistry, which might involve gene regulation, which can be complex. And initially there were two different genetic studies that supported an effect for the long form of the allele of *DRD4* on novelty-seeking behavior. But then there were some disconfirming studies, and the results have gone back and forth between positive effects and no effects. It has turned out it is very hard to confirm, or replicate, molecular studies in the behavioral area. There have been several review articles that pooled the various studies together in what is called a meta-analysis. One early meta-analysis by behavioral geneticist Matt McGue concluded that there is a "significant," but quite small effect. ("Significant" here only means not due to chance alone.) The genetic difference at that one locus (the "genetic variation") appears to help explain about 4% of the difference in novelty-seeking activity between the two groups (the "phenotypic variation"). While interesting, such a finding was thought to be of doubtful clinical value (McGue 2001). McGue also had some doubts about the underlying biological explanation of the effect, though remember that Cloninger thinks he has a theory of how this might work. But then there were two more meta-analyses that suggested the net effect of *DRD4* might be even less than McGue estimated, most likely approaching zero.[5]

JJ: Does anyone know why different scientists get such different, even inconsistent, results in this area?

BG: The reasons are partly technical, having partly to do with statistics. It turns out that to get a solid confirmation you need to look at many more subjects than you looked at in the initial discovery study (Lander and Kruglyak 1995). Doing such large studies is hard to arrange since you need to recruit many subjects (or find the information in a database), and it can be very expensive. But part of the problem is that there is a lot of human variation and we don't really understand whether the same allele has different effects in different groups. It's possible, for example, that a *DRD4* allele has one effect in the US and Israeli groups that were studied, and a different effect in, say, the Finnish group that was studied. Also, remember the *DRD4* long-allele effect was pretty small even in populations where it had detectable effects. Cloninger has suggested that standard meta-analysis approaches may incorrectly pool different populations together and mask the *DRD4* effect (personal communication, May 2004). But more importantly, ignoring different environments in such studies can also mask potentially large genetic effects. In a later, but so far unreplicated study in Finland, a group of behavioral geneticists found that several of the alleles of *DRD4* had a pretty large effect on increasing novelty-seeking behavior, but only if the subjects had experienced a "hostile" childhood environment (Keltikangas-Jarvinen et al. 2004). The researchers defined a "hostile" environment as being emotionally distant, with low tolerance of the child's normal activity and with strict discipline. This result is an example of what is called a gene-environment interaction effect, and it's been expected, based on classical quantitative results in some areas, but it now seems to be showing up in some recent molecular studies, including that anxiety gene we mentioned before, though some of those results have also been recently questioned. There is a rising concern that what is called "publication bias" increases the size and number of positive results, since everyone loves a "winner" (Duncan and Keller 2011; Munafo and Flint 2011). That said, the effect of this gene is still studied with positive results reported (Thomson et al. 2013).

JJ: That's very interesting! And I guess reasons to be cautious. But what are those other studies that show this gene-environment interaction?

BG: There are three or four of these that have received a lot of attention (Caspi et al. 2002; Caspi et al. 2003; Caspi et al. 2005; Caspi et al. 2007). And one of them will be of particular interest to you since it relates to aggression and the increased likelihood of having a criminal record. That study, as well as the second one, was done by an English group headed by Avshalom Caspi and Terrie Moffitt, who had access to the records of a New Zealand population going back over 20 years. In their first study, Caspi and Moffitt looked at two alleles of the monoamine oxidase A, or MAOA, gene. Monoamine oxidase A is an enzyme that metabolizes neurotransmitters, and the short allele form does not metabolize as well as the long allele. Individuals who had the short allele were much more likely to develop antisocial behaviors, including a record of criminal convictions, but only if they had been abused in childhood. Those with the long allele did not display this increased antisocial behaviors, even though they also likely suffered childhood abuse.

JJ: Wow! That seems to accord with what you just told me about gene-environment interactions in the *DRD4* case. What about the other results you mentioned?

BG: They too are similar. The second Caspi et al. study involved an increased likelihood of becoming depressed, but only if those becoming depressed experienced an increased number of stressful life events (Caspi et al. 2003). Interestingly, this gene is also the gene that was first identified as being related to anxiety, and then further to a personality trait that psychologists call "neuroticism" (Lesch et al. 1996).[6] This result was also based on the same New Zealand population studied in the MAOA case, but I should emphasize that it is very hard to get such a resource of data that goes back all the way to childhood and to study the environment so as to reveal these genetic effects. The third Caspi et al. study showed that two alleles of a gene called *COMT* has different effects depending on whether the subjects had a significant environmental exposure to marijuana during a vulnerable period of their lives. *COMT* codes for an enzyme that metabolizes several types of neurotransmitters, so it's a plausible candidate gene.

JJ: That's fascinating. So we really need to try to study the environment as well as we study the genetics. But, as you say, that is very hard to do. Are there other examples of genes that have a substantial and consistent effect on behaviors that seem to do so independently of these hard-to-study environmental influences?

BG: You need to understand that these days behavioral geneticists are expecting that almost all genes that relate to behaviors, including some of the major psychiatric disorders, are *likely* to be ones that have *small* effects, and that there are likely to be *many* of them. There are a series of newer types of studies we can talk about a bit later that look for genetic contributions to a trait across all of the genes. These are called genome-wide association studies, abbreviated as GWAS. Also, it's likely that different mutations in different families or population groups might produce the same phenotype, a notion known as "genetic heterogeneity." So, each allele working at many different locations or loci has a small effect on a quantitative trait such as novelty seeking, and these, as I've mentioned in our earlier meeting are called *quantitative trait loci* or QTLs. But some newer studies have revealed the existence of some rare variants, rare meaning maybe only 1 in 5,000, that have some fairly large effects (Walsh et al. 2008; McCarthy et al. 2009). Again, we can talk more about these rare variants later on.

There is also a new methodology that builds on the GWAS approach and the single nucleotide polymorphisms or SNP results that are the basis for GWAS, and a little later we can talk more about these SNPs. This new methodology is known as "GCTA (an abbreviation for Genome-wide Complex Trait Analysis). As the developers' website states, it was originally designed to estimate the proportion of phenotypic variance explained by genome- or chromosome-wide SNPs for complex traits (the GREML method), and has subsequently extended for many other analyses to better understand the genetic architecture of complex traits."[7] GCTA is discussed more further below and again in chapters 6 and 7.

But that all said, there is a QTL that affects Alzheimer's disease (or AD), which has had strong and consistent replications in most though not all ethnic groups and which thus far seems fairly independent of the environment. This is the famous apolipoprotein E4, or epsilon4, allele that we usually abbreviate by *APOE4*. It is probably the *only* example of a broad-based and replicated allelic finding that is related to a "behavioral" phenotype in humans that is not dependent on difficult-to-study related environmental studies, and some other well-confirmed gene effects in AD have been found in addition.[8] Maybe I should also mention that there is a well-confirmed protective effect against alcoholism of some genes related to alcohol metabolism, but that seems restricted to some Asian populations.

JJ: So is this really the only example that seems to hold that broadly? That must be discouraging for behavioral geneticists. But what does this *APOE4* gene do? Does it make a different, fourth type of neurotransmitter receptor?

BG: No it's not related to a neurotransmitter. The *APOE4* gene makes the APOE4 protein. This and other forms of this kind of protein seem to have at least two rather different roles to play in humans, one in cholesterol metabolism that can affect a disposition to heart attacks, and the other in Alzheimer's disease. In AD the protein may bind to some brain cell receptors, but exactly how the protein works is not yet understood. What is known is that humans have three different alleles for this protein, confusingly called *E2, E3,* and *E4*. Since we have a double dose of chromosomes—remember our discussion of the red cell acid phosphatase the other day?—there are six possibilities (3 × 2) for any given individual's genotype. People with a one or more of the *E4* alleles generally have six times the risk of developing AD that people with no *E4* alleles have, who might have two *E3* alleles or an *E2* and *E3* or even two *E2*s.

JJ: So the *E4* alleles act like a risk factor, just like my brother's high cholesterol increases his risk of a heart attack, but do not determine AD?

BG: Right! It's a "risk-factor allele" if you will, though geneticists usually call this a "susceptibility gene," and it is also a QTL because it quantitatively affects AD risk, and it is just one among others that does so.[9]

JJ: Was this very interesting result found by an association study?

BG: Yes, eventually the *APOE4* result *was* produced by an association study. But initially researchers tried to do a *linkage study* in their search for what turned out to be *APOE4*.[10] Even though the linkage approach didn't work in that case, I should take a minute to explain it. After all, you're looking for an introduction to some of the major molecular approaches to understanding human behavior, and linkage analysis has been the traditional one.

Remember, when researchers do an association study, they begin with a "candidate" gene that they have reason to believe is implicated in the behavior under consideration, and then they see if people who carry that allele exhibit the trait to a greater (or lesser) extent than those who carry a different allele at the same location. If you want, I guess you can think of linkage studies as moving in the opposite direction: in linkage approaches, researchers begin with people who exhibit the behavior and then see if they can identify a shared allele. More specifically, in the linkage approaches, researchers look at the history of a family with a particular disease and construct what is called a *pedigree*—actually a number of families' pedigrees—a family tree as it were. Geneticists then look for evidence that some allele called a "marker" that they can test for shows up most of the time in those individuals affected by the disease. They look for what is referred to as *cotransmission* of the disease and a marker allele. That allele may well *not* be *causative* of the disease, but if it tracks well with the disease, that tracking is evidence the allele is genetically *linked to* the causative gene, in the sense of being fairly near it and certainly on the same chromosome. Thanks to advances in human molecular genetics we can now identify lots of these genetic markers along the chromosomes, which make it easier to make better maps of the chromosomes. The better those maps become, the easier it will be to zero in on the gene that is causally related to the disease under investigation. Researchers hope that once they finally locate the causally related gene and make copies of it, they'll be able to see what it does, maybe what kind of protein it makes that differs in normal and diseased persons.

JJ: Sounds great! Have linkage studies produced any clinically useful results?

BG: Linkage studies have produced some clear results in single-gene types of diseases, like cystic fibrosis and Huntington's disease, and some kinds of cancer, and even special forms of Alzheimer's disease called early-onset AD that I briefly alluded to earlier. But if the linkage approach is going to work, the genes to be identified need to have a large effect (probably at least 10% of the total phenotypic variance)—and the genes implicated in most complex traits just won't have such big effects. There the effects are more like 1.2%–1.3%. Another problem with the linkage approach is that it requires researchers to make big assumptions, like about whether the genes of interest act in a dominant or recessive manner. In fact, a linkage study for schizophrenia, which made some of those assumptions, produced false results—results that even got published in the 1980s.

There is a kind of linkage study, however, called *allele-sharing* that can work well with quantitative traits. A special form of allele-sharing known as affected pedigree member, which we need not go into, was in fact employed by Allan Roses's group at Duke University. They used this method to relate late-onset AD to a region of chromosome 19, later determined to be the *APOE4* allele (more accurately the apoliprotein e epsilon-4 allele). Some studies have looked at personality traits using a linkage design, but these have not worked out well.[11]

JJ: OK, if linkage studies aren't so great for identifying alleles involved in common, complex diseases, then what kind of method did give us this fascinating Alzheimer's disease result—that *APOE4* allele? And also please clarify about these early and late-onset forms of AD, if you can.

BG: Let me start with your second question. One form of Alzheimer's disease appears unfortunately early—often in a person's 50s—so it is called early onset. It now seems this kind is caused by any one of three different types of gene mutations. Whereas late-onset AD appears to be affected by many alleles of small effect, early-onset AD (like, for example, Huntington's disease) is quite rare and is caused by highly penetrant, dominant mutations. The more common kind of AD—the late-onset type that struck President Reagan—is the form related to *APOE4*. To distinguish this *l*ate-*o*nset *AD* form, we sometimes use the abbreviation LOAD.

And again, the *APOE4* discovery was first made using an association study design, just like the novelty-seeking *DRD4* discovery. Roses's group at Duke University suspected that the apolipoprotein gene might play an important role in AD, and therefore would make a good "candidate" gene in an association study.[12] And they got a positive result. It was pretty controversial at first, but is now well confirmed, as I suggested earlier.

JJ: This association design now looks very attractive. Can't it be used for other behavioral traits with results as good as were found for AD?

BG: Remember that it was an association study design that was used for novelty seeking and it had those mixed results. But association designs became increasingly attractive to behavioral geneticists, and two geneticists, Neil Risch and Kathleen Merikangas, have long championed association studies as the best research method in behavioral genetics (Risch and Merikangas 1996). Risch and Merikangas also endorse the more recently developed GWAS analyses that we considered earlier (Risch 2009). But there are some problems with the earlier association method, which is still in use. One of the most famous is the *chopsticks problem*. Two other molecular geneticists, Eric Lander and Nicholas Schork, introduced this in that wonderful though very complex paper I mentioned earlier (Lander and Schork 1994). I've got a quotation from that paper here. They write:

> Suppose that a would-be geneticist set out to study the "trait" of ability to eat with chopsticks in the San Francisco population by performing an association study with the HLA complex [a set of immune response genes that frequently vary between ethnic groups. This would-be geneticist suspects that immune response genes, which he believes are involved in the autoimmune disease multiple sclerosis, may also affect manual dexterity in normals]. The allele *HLA-A1* would turn out to be positively associated with ability to use chopsticks—not because immunological determinants play any role in manual dexterity, but simply because the allele *HLA-A1* is more common among Asians than Caucasians. (1994, 2041)

Lander and Schork urge caution regarding the use of association analyses because of this problem, and offer some guidelines about ways that additional controls and tests might help. Good studies use those additional controls.[13]

JJ: Got it! The association between a gene—or better an allele—and a behavioral trait can be just coincidentally true and really due to another common factor like ethnicity. So caution is advised. If, based on an association study, someone claimed a gene "influenced" some trait, I'd certainly want to obtain the services of a good critical methodologist to make sure appropriate controls and tests were in fact used, even if that someone said it made additional sense because it involved one of those neurotransmitters.

BG: That's very wise, especially given the hype that so often surrounds behavioral genetic results.

JJ: So how do these molecular methods differ from the quantitative or epidemiological results we discussed the other day? Do the classical results help identify where the molecular-oriented scientists should look?

BG: Well, the studies we have been talking about today involve specific, identified genes that we could actually test individuals for. In the twin studies we were talking about before, we almost never had had any idea what specific genes might be involved. Remember, in those heritability studies, we were making *inferences* about the percentage of genetic influence on a trait. Now you would *think* that traits with high heritability would be the ones that we would find specific underlying genes or alleles for, but it has *not* yet worked out that way. In fact one prominent geneticist whose work I just mentioned, Neil Risch (Risch 2001, 739), wrote, and I quote him:

> Heritability estimates should . . . not necessarily be viewed as a good predictor of the ease with which molecular genetic analysis can identify the actual susceptibility genes involved. In fact, looking historically, one would draw the conclusion that molecular genetic success is either independent of or negatively correlated with estimated heritability from twin studies.

On the other hand, another prominent psychiatric geneticist has suggested that if a research program is searching for specific genes (alleles) that affect a trait (or disorder), it makes more sense to look at those traits or disorders that have high heritability, since those with low heritability are likely to be mostly environmentally influenced. But a high heritability, as I have mentioned before, might be the result of thousands of genes with tiny effects on many different chromosomes acting in complex ways, and such gene effects would be virtually invisible to any current genetic methods (Kendler, personal communication, April, 2003). This conjecture by Kendler and others has been supported by the large number of loci found in recent GWAS analyses. One author has even suggested that GWAS analyses when completed are likely to show that every chromosome will be implicated in every disorder (see Goldstein 2009).

In the last few years, GWAS technology has transformed genetic research on complex diseases, such that within the first two years of their implementation, more than 250 loci for over 40 common diseases and traits had been identified. Remember from our earlier discussion that, though we are all more than 99% alike genetically, with ~3 billion nucleotides constituting our DNA, this allows millions of potential variants. One type of variant is a single nucleotide polymorphism, or SNP, which is a site in the DNA where different chromosomes differ in the base they have, for example, a T has a C substitute. The difference may not mean much with respect to trait or disease, but these SNPs can be used as markers along the entire DNA sequence (whence the whole genome or genome-wide language), and can be associated with differences between normals and individuals with a disorder, whence the A in GWAS.

These SNPs can be printed on a gene chip—a small piece of glass less than one inch square—by having up to millions of complementary sequences printed as the tiny dots of complementary DNA. Thus hundreds of thousands of SNPs, even millions, can be scanned for quickly. The scanning is done by processing the chip in a laser scanner and automatically interpreting it to reveal the SNP patterns in diseased/disordered individuals versus controls. These GWAS results identify *loci,* but not necessarily genes—though it is suspected relevant genes are close by (or may be identical with) the loci. (The technical term is that the causative genes are believed to be in "linkage disequilibrium" with the identified SNPs.)

The reason that GWAS could not be performed earlier is because the method is dependent on the availability of the HapMap initially completed (first phase) in 2005. The HapMap project is a multi-institutional project identifying millions of common SNPs across various individuals and ethnic groups. Because of the cost of full genome sequencing[14] (still the holy grail in a sense), the HapMap project proposed a shortcut. The HapMap project focuses only on *common* SNPs, those where each allele occurs in at least 1% of the population. With the development of microarrays (gene chips), this technology could be applied to scan for common SNPs by the hundreds of thousands, even millions, very quickly. This is done by having up to millions of complementary sequences printed as the tiny dots of complementary DNA on a gene chip. The chip can then be processed in a laser scanner and automatically interpreted to reveal the SNP patterns in diseased/disordered individuals versus controls. Because of the multiple tests, the *p*-value for significance, a way of trying to reduce errors of commission, or false positives, is set very, very low, at less than 5×10^{-8}, though that number and the idea of a *p*-value may not make that much sense to you.

Even with this powerful new GWAS technology, researchers are only locating small fractions of genes or more accurately loci that account for the genetic influence on traits or diseases. They sometimes refer to this gap between specific genes or loci and general genetic influence as "missing heritability" or "genetic dark matter" (Maher 2008; Visscher et al. 2012), which you said you wanted to learn more about. It is of interest, and important, that much of the discussion of "missing heritability" takes as the gold standard the classical quantitative analyses of that heritability statistic, which as we have discussed (in dialogue 1) can be misleading.[15]

That said, GWAS has also been used to estimate the narrow sense of heritability by a looking only at within-family variance of a trait due to the genes family members have in common, a method that is thought to be free from potential confounds of nature and nurture influences (Visscher et al. 2006). This method is easily generalizable and was applied to the heritability of general intelligence, which we might discuss at another time.

The application of GWAS methods to a trait as noncontroversial and prima facie "simple" as human height, which has a heritability of about 80%, has apparently resulted in a major paradigm shift in the genetics of both "simple" and more complex traits. In 2008, three groups of researchers examined large numbers of humans (the largest study involved 30,000 people) for genetic variants related to height differences. They found about 40 genes. But as Maher reported in the journal *Nature* that year, "There was a problem: the variants had tiny effects. Altogether, they accounted for little more than 5% of height's heritability—just 6 centimetres [out of an expected 27] by the calculations above" (18). This generated the problem we noted earlier of "missing heritability," but more importantly the finding led to the clearer realization that genes *typically* will have tiny effects, and there will be a huge number of them. In a way, this view of genes had been anticipated by Fisher nearly a century ago in his classical paper (Fisher 1918), but Fisher never thought there would be thousands of genes affecting relatively simple traits. One commentator has characterized this realization as a major change in our view of the relation of genetics and traits, saying, "We now live in an era of big data and small effects" (1041).

So the jury is still out about heritabilities and what we will find in the way of molecular entities that account for the heritability. Maybe classical heritability studies with high h^2 values will point the way and serve as a guide for molecular behavioral genetics, but there's no good evidence as yet. Standard accounts take high heritabilities as a kind of threshold requirement, suggesting that a molecular study not be initiated for a trait or a disorder unless it has been shown to have a reasonable if only partial heritability (Cichon et al. 2009, 540). Twin study methodologies, however, might find new ways of application in molecular studies, since twins can be used to control for some complicating circumstances, like different years of birth and familial environments. But those strengths can also mask the differences in effects due to interacting genes and environments, so careful designs and controls are needed here too. In the past three years or so, the field of behavioral and psychiatric genetics has become much more critical of the results accumulated over the past 15 years. Part of that more critical view was occasioned by the GWAS results beginning in 2005, which essentially did not confirm any of the earlier behavioral and psychiatric genetic findings. Those earlier findings were based on the candidate gene design, and currently those discoveries are now seen by many to have not only had statistical flaws, but also to have been subject to considerable positive publication bias, as I mentioned earlier. A recent article that summarizes some important GWAS findings and issues is that of Robinson, Wray, and Visscher (2014).

JJ: That's a very complicated story you just told me, and I will need to think a lot more about it. But I guess that again you are urging me to adopt a *caveat iudex* view, which is fine, and I am often viewed as a "conservative" judge. But before we end, is there some kind of overall perspective you could share with me that ties all of this together?

BG: A prominent psychologist-geneticist, Irving Gottesman, has used the diagram shown in figure 2.1 in several publications (Gottesman 1997) to bring together some of the recent results in cognitive behavioral genetics with that variability of response of a genotype which depends on the environment such as we saw in the diagram that depicted blood acid phosphatase activity.

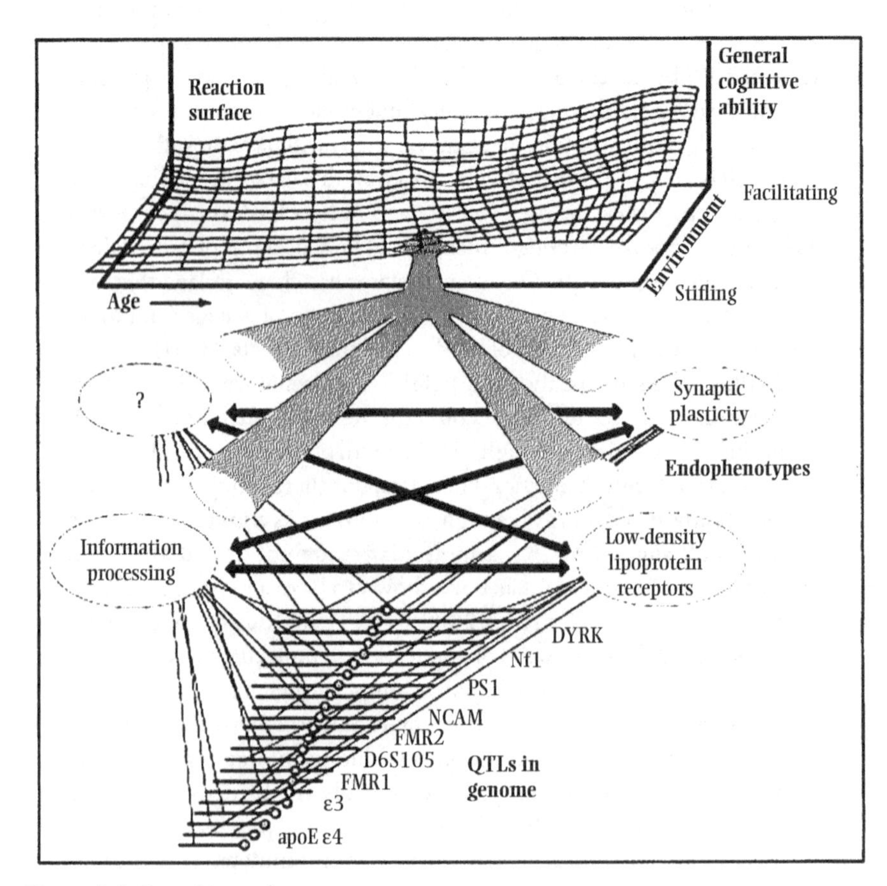

Figure 2.1 Reaction surface.
SOURCE: Gottesman 1997, 1522. Reprinted with permission of the American Association for the Advancement of Science.

Gottesman's diagram looks complicated at first. But it's worth staring at for a minute. The lower portion of the graph depicts QTLs (apoEe4, FMR1, etc.), which are thought to influence "intermediate phenotypes" or "endophenotypes." These endophenotypes are structures or traits (like lipoprotein receptors or synaptic plasticity) that are thought to in turn influence the larger phenotypic trait under consideration (which, in the diagram, is general cognitive ability).

To wrap your mind around this new terminology, maybe it would help to think of it this way. You mentioned that your brother's doctor told him that his good and bad (HDL and LDL) cholesterol counts put him at risk for a heart attack. If his heart-attack risk is the phenotype, then you can think of his cholesterol counts as endophenotypes, or "intermediate" traits that influence his chances of getting a heart attack. Presumably there are also genes of small effect that influence his cholesterols, and these would likely be QTLs. Gottesman's diagram suggests that it might help to think of QTLs as early causes in the chain leading to endophenotypes, with endophenotypes as intermediate causes of the complex phenotypes that are the disorders of interest.[16]

Of course, the premise of behavioral genetics is that how a given trait appears in an individual will be importantly affected by that person's particular complement of genes (or QTLs). But as Gottesman's graph suggests, how that trait appears also will be crucially affected by other variables, such as the age of the individual (a child and adult with the same genome appear differently!) and that individual's environment (orchids don't grow well in the desert).

Take another look at Gottesman's diagram. Do you see in the top portion the two intersecting planes, with the one plane parallel to the page you're reading and the second at a 90-degree angle to the first? Gottesman uses that second plane to represent the age and environmental ranges that affect the phenotype, cognitive ability. Notice that the age range is depicted with the line parallel to the first plane, and the range of environments (from stifling to facilitating) is depicted by the line perpendicular to that plane. Cognitive ability is thus conceived of as a function of one's age, environment, and genetics. The *reaction surface* represents in a three-dimensional way the simple but important idea that different genotypes will have different phenotypes, depending on age and environment.[17] That might not sound very exciting, but it's actually an important step in beginning to appreciate how complex the interactions are, not only among QTLs, but among genes, environment, and time. I should add that though the Gottesman diagram depicts many complexities, it does not depict any downward arrows from the environment affecting which alleles (QTLs) might get turned on and off in response to different environmental conditions. This two-way possibility is, however, acknowledged by Gottesman, and is increasingly stressed by writers who favor what is called a developmentalist perspective (see Moore 2002 and Ridley 2003).

At this point the reader should have a pretty good sense of the general strengths and weaknesses of both quantitative and molecular genetics. But it would be useful to develop the issues a bit more, and do so in a way that the hypothetical judge might find relevant to her legal concerns. To do that, I introduce two more examples, one involving attention deficit hyperactivity disorder (ADHD) and the second involving IQ. Both areas have been extensively researched using quantitative techniques, but ADHD has had far more molecular studies directed against it than has IQ. This difference in the number of studies is probably due to the difficulty of doing molecular studies of a high-level and complex phenotype like IQ or general cognitive capacity (g). Another possible factor is that genetic research on IQ is more socially controversial than would be ADHD research, which might yield pharmacological interventions. Possible pharmacological interventions affecting general cognitive capacities might raise an issue of enhancement in addition to issues of ethnic and racial discrimination—a problem that has periodically plagued genetics and IQ studies, as is noted in the following dialogue.

DIALOGUE 3. IQ AND ADHD GENETIC TESTING

JJ: I called for your advice one more time because I have heard that there are two emerging cases involving genetic testing that I suspect I might see on appeal. While our recent discussions are still fresh in my mind, I'd like to discuss these two case areas with you before actual cases really emerge.

It seems that two families who each have a child in a very "progressive" science-oriented elementary school have sued the school board. One family's child, whose name I have heard through the grapevine is **Ad**am, is described as a bit on the wild side. The school nurse wants him genetically tested for ADHD. If this test and other interviews with Adam's parents and teachers are positive, he will be placed in a special education class—something his parents do not want. The school board supports the nurse as "doing the right thing for the child and other children at the school."

The other family's child, in the same school, is named **Iq**uena. Reportedly she is extremely bright, but before she can qualify for a special gifted students' program, she (like any of her peers who want into that program) has to be tested for the presence of high-IQ genes. The presence of these genes, the school board says, will ensure students are routed into the most appropriate tracks. By the way, Iquena's family happens to be African American; Adam's family is of Eastern European descent.

Both families have retained the services of a prominent Columbus, Ohio, lawyer who has filed suits in both state and federal court on equal protection grounds. The lawyer claims in preliminary papers that both of these students are being discriminated against on the basis of an unverified genetic reductionism and determinism. What can you tell me about the genetics of ADHD and IQ, and these notions of genetic reductionism and determinism?

BG: Well, I think both of these cases are wildly premature. Some might even go so far as to say that, in this context, the science is "junk." But I can give you an overview of these areas if you like.

JJ: Please do. And even if some think this type of testing is "junk science," others clearly seem to disagree. And since I am in a court that follows the Frye rule regarding evidence, and I also worry about the *Daubert* decision,[18] I will need to know about how to decide if this kind of science was legally admissible. (See, we lawyers have pretty elaborate terminologies, too!) Also, there is legal precedent for courts to consider the heritability of IQ (*Johnson v. Calvert*, No X 63 31 90 at 7 (Cal. Super. Ct 1990), so I think I need to be familiar with this kind of thing.[19]

BG: Let's start with the genetics of IQ. The concept is a bit more familiar because everyone has taken an IQ test at some time in her life, or taken tests like the SAT, which produce scores that track IQ test results fairly closely. And, as you just mentioned, at least one court has already considered IQ and its heritability. ADHD is a bit more complex, since it's a disorder and the definition of it is a bit complicated, so we can come back to that later.

I am sure you know that just the *concept* of the "genetics of IQ" is controversial. Partially this is because IQ seems to be important for how one does in life. The idea that such an important aspect of a person is constrained by her genes doesn't sit well with our democratic instincts. Further, some controversial scholars have argued the putative 15-point IQ gap between "blacks" and "whites" is largely "genetic," and that social attempts to close that gap are doomed to fail (Herrnstein and Murray 1994; Jensen 2000). I would not be surprised if Iquena's family was especially sensitive to genetic testing requirements, given this backdrop.

JJ: I remember the controversy about Herrnstein and Murray's *Bell Curve* book in the mid-1990s, and think I recall that this was the second time in the last couple of decades or so that such an IQ and genetics controversy roiled the academy. But sometimes bad news for an individual or a group has a basis in fact. Is IQ genetic?

BG: Remember our discussion about heritability from a few weeks ago? Just as the genetics of schizophrenia that we talked about then has been investigated using twin and adoption studies, so IQ and related cognitive abilities have been extensively examined. Remember our simplified example of the heritability of verbal reasoning? The behavioral geneticists, and many psychologists, prefer to talk not about IQ but rather about "general cognitive ability" or "g." Someday you might want to get into the details of psychological tests and what g means and what its validity is. But for now, since IQ measurements are a good index of g (Chorney et al. 1998), we'll just stick with talking about IQ.

You probably remember that when we used the rubber tree plants' heights (acid phosphatase levels) to explain the heritability concept, we saw that the activity associated with each genotype (and with the sum of those activities) could be depicted by a bell-shaped curve. Well, the distribution of IQ "levels" among humans can also be depicted by such a curve. Figure 2.2 is a diagram from an article by Lubinski that shows IQ distribution and describes what kinds of occupations individuals with different ranges of IQ typically hold. Now this is a phenotypic distribution, and to get to a measure of how much of these individual differences is "genetic" requires that we go back to our earlier discussions about quantitative or epidemiological genetics.

JJ: OK, but are there no molecular genetic results in the IQ area?

BG: In the molecular area, the IQ results have been checkered, and those vagaries may well be important for your case of Iquena, so we will discuss it in a minute. But essentially all the behavioral genetics work on IQ (or g) has actually been of the quantitative, nonmolecular sort. And these quantitative studies have been somewhat controversial themselves. The typical value given for the heritability of IQ is about 50%. But some writers have suggested it may be as high as 80% (Jensen 2000), and others have pinned it at about 30%–35% (Cavalli-Sforza and Feldman 1973; Devlin, Daniels, and Roeder 1997).

JJ: Remind me what heritability means again, and did we not distinguish two types of heritability—the broad and the narrow kinds—when we talked before? Is that not important here?

BG: As you'll recall, heritability is the percentage of the phenotypic variation in a trait that is estimated to be attributable to genetic variation. This is broad heritability. And you remember the distinction correctly.

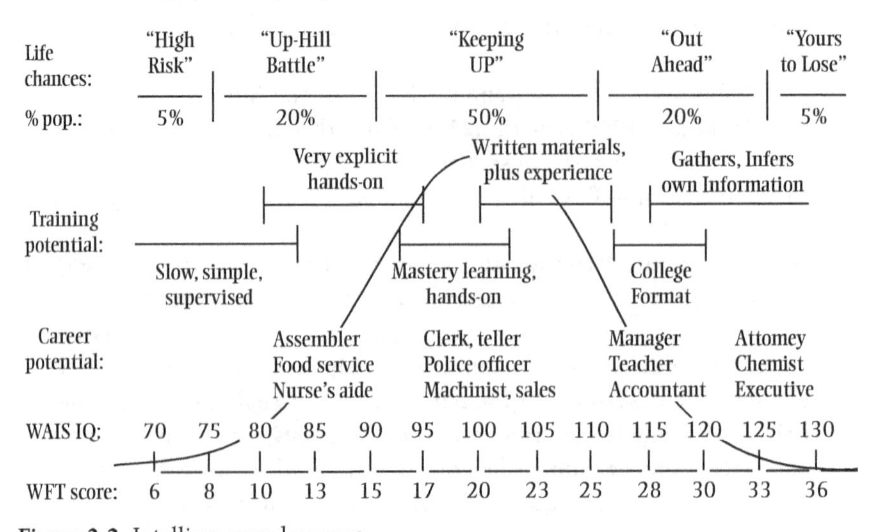

Figure 2.2 Intelligence and success.

SOURCE: Lubinski 2000, 15. Copyright 2000 John Wiley & Sons. Reprinted with permission.

The difference between the two types of heritability—broad and narrow—may well be important for some policy decisions of a general sort, though *not* for testing individuals like Iquena. I'll explain why a bit later. But first let me stress again that we are dealing with *populations* in epidemiological genetics, not specific individuals. Heritability is a characteristic of a group—and only of *that* group in *that* tested range of environments and at *that* time. Thus, inferences to other groups at different times in different circumstances cannot be made using the heritability estimates found for another group at a different time.[20]

JJ: I remember your cautions, but on what kinds of studies does this 50% figure for heritability you cited earlier rest? And do the studies tell us anything really pragmatically useful?

BG: The results are useful but only in a general sort of way—maybe for stimulating further scientific advances—and I'll come back to the notion of usefulness in a moment. The idea of studies of the heritability of IQ really goes back to the founder of modern debates about nature and nurture, Sir Francis Galton, who wrote about the familiality of "genius" way back in 1869. Studies of IQ and its genetic and environmental bases were conducted throughout the twentieth century, though it has been claimed that the data in some of those so-called studies were made out of whole cloth (see Beckwith 2006). But again, if you look at all of the twin, adoption, and family studies aimed at calculating the heritability of IQ, you'll find that the average comes out to about 50%. It's not clear, however, what to make of that 50% figure. Is some ways, it's a nice figure because it's small enough to satisfy many environmentalists and big enough to satisfy many geneticists. The problem is that it's a compromise, which can mask some very deep disagreements.

As I mentioned a little earlier, some researchers think a good heritability-of-IQ estimate for the general, contemporary US population is closer to .8 and others think it's closer to .3 or .4. Matt McGue and his colleagues, for example, examine much of the data collected during the twentieth century and conclude that by the time we are well into our adult years, about 80% of the variation with respect to IQ is due to genetic variation.[21] We don't have time to talk about the details, but let me say this. McGue and his colleagues think that as we get older, what they call "shared environment" becomes less important in explaining IQ differences. That is, they think environmental variables that tend to make siblings in the same family similar earlier in life exert a smaller influence later in life. Plomin thinks that, because individuals seek out environments that reflect their genetic endowments, over the course of a lifetime genetic variables are increasingly important in explaining why people act and appear differently. Put bluntly, according to this theory, individuals who are genetically well endowed seek out environments that stimulate cognitive activity, and individuals who are badly endowed seek out environments that do not stimulate cognitive activity; stimulated abilities tend to flourish and understimulated abilities tend to whither (Plomin et al. 2001).

JJ: Wow, 80%! If you accepted that figure, you might think testing makes a whole lot of sense.

BG: Right, but like I said, some first-rate researchers think that figure is much too high (Devlin, Daniels, and Roeder 1997; Plomin et al. 2001, 175).

They argue that you only get to 80% (or even 50%) by neglecting important environmental factors. In several articles, Feldman has maintained that many twin studies of IQ are flawed because they do not take cultural transmission influences into account and also ignore the special character of the twin relation. (Remember our discussion of the equal environments assumption, which Feldman argues does not hold. And, again, see Beckwith 2006.) Taking those corrections into account, Feldman estimates that the heritability of IQ is about 33% (Feldman and Cavalli-Sforza 1979; Feldman and Lewontin 1975). His analysis suggests that "cultural transmission" also accounts for another 33% of the variation with respect to IQ, with the remaining 33% due to the noncultural environment. Plainly, these studies suggest that the familial and larger cultural environments have a significant effect on IQ.

More recently, Devlin and his colleagues did a meta-analysis of over 200 studies that estimated the heritability of IQ. Those studies included twin, family, and adoption studies, and involved more than 50,000 pairs of relatives. Devlin and his colleagues reached a heritability-of-IQ estimate of 34%, almost identical to the figure arrived at by Feldman and Cloninger. But in the Devlin analysis it was not cultural transmission or special twin effects that were viewed as confounding for the mainstream studies' estimates. Rather, Devlin argued that mainstream studies ignored the maternal, or more accurately the womb, environment as a factor, though Devlin did count the early postnatal period, up until separation of mother and infant, in his analysis.

JJ: Womb environment? How was that studied? Are we still talking about that "simple" twin-study model we explored when we first started our conversations—the one with those three equations?

BG: Yes, we are talking about Devlin and his colleagues' analysis of 200 studies that involved twins. They wanted to see whether the fetus's *environment* had an important effect on the intelligence of the person that fetus would become.

To see whether the in utero environment was an important variable in explaining the phenotypic variation in a population, Devlin and his colleagues created four models, each of which postulated that different factors affected variation in IQ.[22] These sets of factors were just extensions of the ones allowed for by the simple models we discussed earlier. Devlin's models allowed for nonadditive genetic effects (i.e., effects that are dependent on other alleles), as well as for variations in the kinds of environments. The most complex or "richest" of these models, which they happened to call Model IV, allowed for "additive and non-additive genetic effects, twin and singleton maternal effects, and familial environmental effects for twins, siblings and parent-offspring." But this Model IV, which was richest with respect to number of variables it allowed, did not "fit" (or help account for) the data as well as their Model III. The one that "fit" the data best happened to be called Model III. That best-fitting model estimated that the maternal *environment* accounted for 20% of the IQ variance between twins and 5% between (nontwin) siblings.

Since twins share the same womb at the same time, and siblings share it at different times, it was not surprising to find the difference between the 20% and 5% figures. This maternal effect is a quite startling finding, since maternal effects are typically ignored in the traditional IQ analyses. Devlin and his colleagues arguments have tended to be ignored, and Bouchard has posted a strong critique of the approach, but like the Feldman claims of strong cultural effects on IQ heritability, these alternative views are wise to keep in mind.[23]

JJ: That's very interesting. And Devlin's models do seem more nuanced than the models we discussed a couple of meetings ago, though I don't want to ask for the details of how they work. I'll have to trust you on that, for the moment at least. Did using these richer models produce anything else of interest?

BG: As a matter of fact, yes. Devlin and his colleagues suggest that the maternal environment better helps explain IQ variation than does the age-related hypothesis I mentioned earlier, which assumed that the effects of a common environment waned and that there was an interactive snowballing effect. I should emphasize that Devlin estimates that the broad-sense heritability for IQ is about 48%,[24] and that the narrow-sense heritability is only 34%. It is the narrow sense of heritability that is important for any evolutionary or next-generation effects—for reasons that you might remember from our discussion of breeding values. So the Devlin article both identified important maternal effects and estimated the most relevant type of heritability as much less than most behavioral geneticists. It's worth quoting the conclusion of this article to you because of the importance of both points. Devlin, Daniels, and Roeder (1997) wrote:

> These results have two implications: a new model may be required regarding the influence of genes and environment on cognitive function; and interventions aimed at improving the prenatal environment could lead to a significant increase in the population's IQ. Moreover, some of Herrnstein and Murray's conclusions regarding human evolution such as the development of cognitive castes and IQ dysgenics, arise from their belief that IQ heritability is at least 60%, and is probably closer to the 80% values obtained from adoption studies. Our results suggest far smaller heritabilities: broad-sense heritability, which measures the total effect of genes on IQ,

is perhaps 48%; narrow-sense heritability, the relevant quantity for evolutionary arguments because it measures the additive effects of genes, is about 34%. Herrnstein and Murray's evolutionary conclusions are tenuous in light of these heritabilities. (470)

I should add, speaking of Herrnstein and Murray, that those authors make a lot of the 15-point gap in IQ results between United States African Americans and whites, though they do not explicitly attribute the difference to genetics—they only imply that. Jensen, on the other hand, using his "default hypothesis," seems to explicitly attribute a substantial portion of the group difference to heredity (Jensen 2000). But there is a serious question whether the gap even exists, since one large study ignored by both Herrnstein and Murray and by Jensen found the difference only to be four IQ points, and actually within the margin of error of that study when SES corrections were made (Nichols and Anderson 1973). Other studies, one by Flynn whose other work on general IQ increase over time led to it being named the "Flynn effect," has argued that the gap has shrunk a bit, to 10 points (Flynn 1987). There is a lot of misinformation, and misinterpretation, about IQ, especially involving racial differences. Iquena's parents are probably reacting in part to this, though they have other excellent grounds to distrust any genetic test for IQ genes.

JJ: OK, even though now I'll be suspicious of anybody who claims to understand "the genetics of IQ," and I am sensitive to the social dimensions of this issue, tell me how this all affects genetic testing and about any molecular studies of IQ. Why should Iquena's parents distrust molecular studies?

BG: Robert Plomin, whose work I have mentioned before, has had a large ongoing project that has been looking at specific genes of small effect that additively may contribute to high IQ. The rationale for the project is *not* to provide means of testing for such genes, but rather to identify QTLs and then specific genes that may indicate the neurophysiological basis of human cognitive behaviors—genes that make us human, if you will. Several years ago Plomin's project believed it had a sound replication of one gene that related to high IQ, but then that result did not stand up to a further study. But that project continues and other results have been more recently published.

Very recently, Plomin and Deary have summarized the state of IQ and its genetic contributions as well as implications in "five findings." These findings include somewhat complex information on heritability (which varies with age), but noting that "the heritability of traits [including IQ] is caused by many genes of small effect." Though these five findings initially arose from the twin studies methods that we discussed earlier, Plomin and Deary also state that "they are being confirmed by the first new quantitative genetic technique in a century—Genome-wide Complex Trait Analysis (GCTA)—which estimates genetic influence using genome-wide genotypes in large samples of unrelated individuals." You might recall we discussed GCTA in an earlier meeting. Plomin and Deary add further: "Comparing GCTA results to the results of twin studies reveals important insights into the genetic architecture of intelligence that are relevant to attempts to narrow the 'missing heritability' gap" (Plomin and Deary 2015, 98).

JJ: I'm confused. Could you be a little more specific, and with less alphabet soup? What exactly is the situation now?

BG: There have been a number of reports on specific genes related to IQ, but it is not clear that any have panned out. Some of Plomin's initial findings did not replicate. Later, Plomin's work has used pooled DNA along with SNP and microarray technology with some interesting suggestive results (Butcher et al. 2008). Other research groups have gotten positive results, for example Postuma's work with Boomsma's lab on the SNAP-25 gene, located on chromosome 20 (Gosso et al. 2008). However, a very recent comprehensive review of specific IQ genes stated that "despite its high heritability, it is not possible confidently yet to name one genetic locus unequivocally associated with the quantitative trait of intelligence" (Deary, Johnson, and Houlihan 2009, 215). More recently Deary was a coauthor on an article that reiterated the claim that human intelligence is highly heritable, and that a SNP study indicates that at least 50% of the variation in various facets of human intelligence (crystallized and fluid) are accounted for by SNPs (Davies et al. 2011). However these recent studies buy into the thesis that there are many genes of small effects at work, that no specific genes have yet been identified that replicate widely and have any major effects, and that much larger studies will be needed to identify any individual loci that are significant contributors to IQ (again see Plomin and Deary 2015).

JJ: Well, now I understand why you said when we started today that testing for IQ genes was wildly premature, but the science does not seem to be "junk science"—just science in an early stage.

BG: That's what I meant—it's only "junk science" in the context of a genetic testing program directed at Iquena and other students in that progressive school you mentioned. And the results thus far are still very preliminary and nonspecific, but more solid results reporting individual genes may yet come.

JJ: Enough about IQ. Let's talk about Adam's case and the ADHD area. First tell me about this disorder to refresh my memory.

BG: ADHD is a cluster of behavioral symptoms typified by a persistent pattern of inattention or hyperactivity-impulsivity. Persistent means the behavior is seen in more than one setting and also over time. Since everyone is inattentive sometimes and gets a little "hyper" as well, the psychiatrists emphasize that this pattern is more severe and frequent at comparable age levels in those diagnosed with ADHD. Symptoms also need to have begun before the age of seven and have lasted at least six months. By the way, ADHD used to be, and still sometimes is, called attention deficit disorder, or ADD.

ADHD has been reported to occur in many young children in the United States. Sadly, the most recent statistics indicate that just in the United States alone over 5 million kids between the ages of 3 and 17 (nearly 10%) are affected sometime between those ages (Bloom, Jones, and Freeman 2013). The disorder is, unfortunately, overdiagnosed in some populations, typically well-to-do suburban children, and underdiagnosed in poorly health-served areas, including the inner city. Though some behavioral therapies have been effective, especially family-based therapy, the usual treatment is with a stimulant such as Ritalin. And sometimes, in clear cases, Ritalin should be the first-line treatment. The disorder can be debilitating and have negative effects on development. It not infrequently persists into the teen years and can continue into adulthood. About 10% of children diagnosed with ADHD are in special education classes, but if there are learning disabilities associated with the disorder, that number goes up fivefold (see the May 2002 CDC report in Pastor and Reuben 2002 and the update in Bloom, Jones, and Freeman 2013).

JJ: Is genetic testing for ADHD any closer to being useful than testing for IQ?

BG: Studies have tended to suggest high estimates for the heritability (h^2) of ADHD, about 75%, but there is a lot of variation across the studies that have looked at heritability in various subtypes (see Gizer, Ficks, and Waldman 2009; Franke, Neale, and Faraone 2009). The molecular situation in ADHD is somewhat like the state of novelty-seeking gene research that we talked about at our last meeting. In fact, the "novelty-seeking" allele we considered, the 7-repeat allele of the dopamine receptor *DRD4* gene, was one of the alleles implicated in ADHD etiology. But the studies on *DRD4* in ADHD have been inconsistent. Though there is a general view that *DRD4* has an effect on ADHD, it is quite small, probably increasing the risk about 8%, and almost certainly will not be found in all populations.[25] Dopamine genes have been a recurrent player in recent years (Li et al. 2006) as well as a number of other candidate genes (see Gizer, Ficks, and Waldman 2009). However, a 2010 meta-analysis using a GWAS approach found no genome-wide significant association for ADHD (Neale et al. 2010), though there is suggestive evidence that *DRD4* (and a related gene, *DRD5*) may be involved. A very recent GWAS path analysis study "identified 11 candidate SNPs, 6 genes, and 6 pathways, which provided 6 hypothetical biological mechanisms" (Lee and Song 2014, 1189) and additional studies will certainly be developed.

JJ: Well, in spite of the future promise, it does not sound like genetics will help much at this point in ADHD. But getting back to the 7% figure and that huge number of kids that are diagnosed with the disorder, you said something about over- and underdiagnosis? Wouldn't genetics help here?

BG: The genetics, at some point, might actually help make diagnoses more accurate, but a lot more research at a number of levels needs to be done first, and is going on. Some of that research is attempting to define the disorder more specifically and accurately in the clinic, including various subtypes of the disorder. And the genetics *may* help with this at some point. Better definitions of subtypes of ADHD, such as a predominately hyperactive-impulsive or a predominately inattentive type, are being evaluated, as are more refined classifications.

JJ: Okay, I take your point about how genetics might someday help clinicians make better diagnoses. But for now, it looks like the bottom line is that just as in IQ, genetic tests for ADHD are certainly not ready for use in making potentially momentous decisions about what track kids like Adam and Iquena should be put on. I do recognize that Adam might be better placed in a special education situation, but not on any genetic grounds. Given these genetic facts, it's likely that there will be a resolution of this suit even before trial. Nonetheless, I now feel a bit better prepared for this type of case if one materializes. But we have not yet talked about the aspect of the Adam and Iquena cases where the plaintiffs' lawyer is alleging discrimination based on an "unverified genetic reductionism and determinism." Exactly what do behavioral geneticists mean by these terms?

BG: These are not really behavioral genetics terms but are more philosophical ideas, though the terms possibly have legal consequences. Behavioral geneticists do sometimes discuss these kinds of implications of their results. Reductionism means many different things (there is a discussion of about a dozen different senses in Schaffner 1993, chap. 9, and a simplified updated overview of reduction in psychiatric genetics in Schaffner 2013). But the core idea of reduction is simplifying something complex by explaining it in terms of the actions of its smaller parts. Often such a simplifying explanation suggests it is the parts that are the *real things*, and the real causes of what happens. In line with this, genetic reductionism would be attributing all behaviors to the actions of genes. If the genes acted so as to ensure 100% predictability of behavior based on knowledge of the genes, you would have *genetic determinism*—the genes would fully determine the behavior. Even in rare cases like Huntington's disease, where we say that an allele is 100% penetrant, we are not 100% sure that the disease will be expressed, and we certainly do not know exactly when or how the disease will be expressed (Rubinsztein et al. 1996). Behavioral genetics research into common, complex behaviors does not support deterministic explanations. Most human behavior is the result of highly complex interactions among genes and environmental factors. Genes will likely have very small influences and result in small probabilistic changes in risk. Basing a decision such as class tracking for Adam and Iquena mainly on genetic test results is unsound and at best wildly premature.

Actually, there are relatively few legal cases where a defendant's lawyer claimed that the defendant, who had been judged guilty of murder, was less than fully responsible for his actions. One case involved an "aggression gene" found in an extended Dutch family. But that lawyer's request that his client be tested for the so-called aggression gene (in the hope of reducing his client's sentence) ultimately went nowhere.[26] (A related case is discussed below in chapter 8.) The case does raise the perennial philosophical topic of *free will*, but you will have to read elsewhere on that topic, since I am not qualified to discuss it with you.[27] A good place to start reading is in a book that deals with a variety of behavioral sciences impact on the criminal law (Farahany 2009). A recent article in *Science* also indicates how biological factors might affect a judge's decision (Aspinwall, Brown, and Tabery 2012).

JJ: OK, maybe I'll look up those papers on free will and the biological and behavioral sciences sometime. But now that we have discussed many of the claims about genetics and behavior in our last three meetings, I have to say that I feel I have been overinfluenced by "genetics hype." Is anybody trying to battle the hype?

BG: Actually, a lot of different people, from a lot of different angles, are trying to offer subtler, more complex accounts of the role of genetics in human behavior. Some bioethicists and policy wonks and journalists have been concerned about the hype, though to be fair, many of them have engaged in it as well. Some researchers from within behavioral genetics have been concerned about the hype because they can see that the greater the expectations of the science grow, so grows the potential for disappointment. Some other researchers, outside of behavioral genetics—for example, from population genetics—have also been concerned about the hype in general, and the oversimplified reductionism and unwarranted determinism in particular. These other researchers, sometimes joined by sympathetic philosophers, are often referred to as "developmentalists." A number of developmentalists are adversaries of behavioral genetics. (See chapters 3 and 4 for a discussion of developmentalism.)

JJ: I am used to adversarial relationships in the law but did not realize they were prevalent among scientists who study human behavior. That could make it difficult to figure out which expert testimony to trust. It's disappointing that there seem to be so few sound results in an area that has been written about so much and in which so much research has already been done. But maybe we are on the cusp of some real breakthroughs in behavioral genetics?

BG: A lot has been done and my field has come a long way, but it's a journey of a thousand miles, and we have just taken the first steps, to echo an old saying. We all need to remember just how complex an organism a human being is. Genes affect behavior largely by collectively building nerve cells, called neurons, and interacting with the organism's internal and external environments to knit the neurons together into networks, via connections called synapses. This can result in one huge network! And it is not accurate to characterize the genes as simply directing cell and pattern formations, since the causal and informational flow is two-way as genes are turned on and off by proteins and environmental conditions as the network gets built and self-regulates.[28] Much of this building and sculpting happens in the organism's development, but for humans, the nervous system grows until the person is an adult, with extensive remodeling during the teenage years. And there is also evidence of brain changes all the way through human life. Behavioral and psychiatric geneticists and molecular geneticists are both beginning to acknowledge to importance of mapping the environment—it's even been called the *envirome*—so in addition to mapping the genome, some are now calling for substantial enviromics studies (see Anthony 2001). It is startling, and sobering, to compare Dean Hamer's articles (and books) written in the 1990s, which were extremely optimistic regarding human behavioral genetics, with his later comments on the subject.[29] Robert Plomin, whose work we discussed earlier, also became much more cautious about what can be found easily in human behavioral genetics (see Plomin 2003, especially the introduction and conclusion, and his and his coauthors' more recent comments in their forecast chapter in Plomin et al. 2013).

It is difficult to overstate the complexity of the human brain and nervous system. Just the facts give a sense of the overwhelming numbers of cells involved. It has been estimated that the human brain has over 100 billion neurons, which are so connected with each other that there are over 100 trillion synapses (some say up to 1,000 trillion synapses). Groups of the connected neurons constitute circuits, probably each containing hundreds and thousands of neurons. These circuits probably have another thousand-fold number of synapses each, which all work together in largely unknown fashion to produce mental and behavioral processes.[30] The complexity of circuits and the difficulties with learning what is going on in them only began to be appreciated by the geneticists and neuroscientists about ten years ago (see Cowan, Kopnisky, and Hyman 2002, esp. 28–29, and also Hamer 2002,[31] as well as the vision developed by Caspi, Moffitt, and Hariri and their colleagues in Caspi et al. 2010).

JJ: Goodness gracious! What will behavioral geneticists do in the face of that kind of complexity?

BG: Well, we behavioral geneticists seek to find useful, and hopefully robust, simplifications in this vast network, so as to connect the inherited elements, the genes, with specific behavior types. With the help of neuroscientific tools, we need to at least partially decompose the huge network and localize functional parts.[32] Since this is very difficult to do, even with a very simple organism like the common soil worm *C. elegans* (Schaffner 1998, 2000), we don't have any illusions about the difficulties we face. And this worm has only about 1,000 cells, total, with exactly 302 neurons and approximately 5,000 synapses (Schaffner 1998, 2000).

There are new gene chips, known technically as microarrays, that monitor up to a million genes simultaneously that may help. The chips offer a lot of promise in cancer genetics, where they can discriminate the gene expression patterns of different types of cancer (Armstrong et al. 2002). And there were some tantalizing early results in schizophrenia using these chips (Mirnics, Middleton, Stanwood, et al. 2001; Mirnics, Middleton, Lewis, et al. 2001) that actually were replicated (Mirnics and Pevsner 2004). These chips have also been extensively used as SNP chips to pursue the various GWAS projects we talked about earlier. But in basic physiology, development, and behavior, the chips have not yet been able to produce helpful simplifications—so far they have yielded vast amounts of data that are very hard to interpret. Preliminary results applied to behaviors in simple organisms, like this worm, offer promise, but even there the chips suggest that there a lot more genes that seem to be involved, and we do not yet know what most of the genes do, nor exactly how they affect the neurons and the circuits, though progress is being made.[33]

Maybe behavioral geneticists, working with neuroscientists, who can help identify those possibly promising endophenotypes I mentioned earlier, may be able to find genes or alleles of major effect that hold for diverse families and across diverse ethnic groups. In one sense, endophenotypes are at higher levels of aggregation, so they may give us some simplifications that we can understand and use, just like we can fairly easily understand the behaviors of a gas's temperature, pressure, and volume, even though the gas is composed of millions of interacting molecules. Maybe evolution designed us with behavioral modules, with only a few genes affecting them in major ways, which we can dissect out of behavioral patterns with stronger genetic methods and better neuroscientific tools. There are a lot of promising possibilities. But it has been very difficult to get good results in behavioral genetics. And in spite of occasional optimism and frequent hype, and given the complexity of humans just noted, this will be really tough to bring off successfully. Maybe we will just have to get lucky.

SUMMARY AND CONCLUSION

The purpose of this chapter and the preceding one was to introduce the subject of behavioral genetics in a gradual manner, building on clear, albeit somewhat oversimplified, definitions, and then introducing the methods and some results. My intent was to demystify some of the basic ideas for people who are not behavioral geneticists, but who would like to know what the results of behavioral genetics do and do not mean. Some results, such as in Alzheimer's disease research, have stood up under scrutiny. Even though some think that late-onset Alzheimer's isn't squarely within the purview of behavioral genetics, it is a disorder that psychiatrists claim as one of their dementias, and for which they have an entry in their *Diagnostic and Statistical Manual.* Also it is a common, complex disorder, which may suggest some of the possibilities and problems that will be associated with research into other common, complex behaviors. Research on novelty seeking, depression, and schizophrenia (which are squarely within the purview of behavioral and psychiatric genetics) have yielded somewhat controversial results, discussed briefly above but in more detail in chapters 6 and 7 below. But too often it looks like behavioral genetics results are pulled like a rabbit from a magician's hat. One way of looking at these two chapters is that they allow a peek inside the hat. That peek indicates that there are many assumptions that behavioral geneticists make, and also that the magnitude of their findings are often small ones, as well as hard to replicate. Characterizing the limitations in these disciplines and limits in what we can expect to infer from the disciplines' findings is the first step in a wise application of them.

Behavioral genetics will continue to pursue studies, and new and exciting findings can be expected, if not daily, probably at least weekly. These will be hyped in the daily papers and on the evening news, and on the covers of weekly newsmagazines. Hopefully these two chapters will provide part of the information that readers can use to examine critically those frequently all-too-breathless stories. We can look forward to breakthroughs in new understandings of human complexity from these advances. Hopefully, also, both behavioral and pharmacological interventions will emerge that will assist those with behavioral and psychiatric disabilities, and provide some relief from them, with a chance to build a better life. Given the human complexity our behavioral geneticist outlined to Judge Jean in her closing comments, however, we should not expect any magic bullets, but we can anticipate some well-targeted, rationally based interventions, including drugs, in the long run. And we can guard against simplistic genetic determinism and reductionism with an awareness and appreciation of the above-discussed limitations of behavioral genetics.

References

Altshuler, D., M. J. Daly, and E. S. Lander. 2008. "Genetic mapping in human disease." *Science* 322 (5903): 881–88. doi:322/5903/881 [pii] 10.1126/science.1156409.

Annas, G. J. 1991. "Crazy making: Embryos and gestational mothers." *Hastings Center Report* 21 (1): 35–38.

Anthony, J. C. 2001. "The promise of psychiatric enviromics." *British Journal of Psychiatry—Supplementum* 40: s8–s11.

Armstrong, S. A., J. E. Staunton, L. B. Silverman, R. Pieters, M. L. den Boer, M. D. Minden, S. E. Sallan, et al. 2002. "MLL translocations specify a distinct gene expression profile that distinguishes a unique leukemia." *Nature Genetics* 30 (1): 41–47.

Aspinwall, L. G., T. R. Brown, and J. Tabery. 2012. "The double-edged sword: Does biomechanism increase or decrease judges' sentencing of psychopaths?" *Science* 337 (6096): 846–49. doi:10.1126/science.1219569.

Bechtel, William, and Robert C. Richardson. 1993. *Discovering Complexity: Decomposition and Localization as Strategies in Scientific Research*. Princeton, NJ: Princeton University Press.

Bechtel, William, and Robert C. Richardson. 2010. *Discovering Complexity: Decomposition and Localization as Strategies in Scientific Research*. MIT Press ed. Cambridge, MA: MIT Press.

Beckwith, Jonathan. 2006. "Whither human behavioral genetics." In *Wrestling with Behavioral Genetics: Science, Ethics, and Public Conversation*, edited by Erik Parens, Audrey R. Chapman, and Nancy Press, 84–99. Baltimore: Johns Hopkins University Press.

Beecham, G. W., E. R. Martin, Y. J. Li, M. A. Slifer, J. R. Gilbert, J. L. Haines, and M. A. Pericak-Vance. 2009. "Genome-wide association study implicates a chromosome 12 risk locus for late-onset Alzheimer disease." *American Journal of Human Genetics* 84 (1): 35–43. doi:S0002-9297(08)00629-0 [pii] 10.1016/j.ajhg.2008.12.008.

Bertram, L., and R. E. Tanzi. 2001. "Dancing in the dark? The status of late-onset Alzheimer's disease genetics." *Journal of Molecular Neuroscience* 17 (2): 127–36.

Bloom, B., L. I. Jones, and G. Freeman. 2013. "Summary health statistics for U.S. Children: National health interview survey, 2012." *Vital Health Statistics* 10 (258): 1–81.

Butcher, L. M., O. S. Davis, I. W. Craig, and R. Plomin. 2008. "Genome-wide quantitative trait locus association scan of general cognitive ability using pooled DNA and 500K single nucleotide polymorphism microarrays." *Genes, Brain and Behavior* 7 (4): 435–46. doi:GBB368 [pii] 10.1111/j.1601-183X.2007.00368.x.

Caspi, Avshalom, Joseph McClay, Terrie E. Moffitt, Jonathan Mill, Judy Martin, Ian W. Craig, Alan Taylor, and Richie Poulton. 2002. "Role of Genotype in the Cycle of Violence in Maltreated Children." *Science* 297 (5582): 851–54.

Caspi, A., A. R. Hariri, A. Holmes, R. Uher, and T. E. Moffitt. 2010. "Genetic sensitivity to the environment: The case of the serotonin transporter gene and its implications for studying complex diseases and traits." *American Journal of Psychiatry* 167 (5): 509–27. doi:10.1176/appi.ajp.2010.09101452.

Caspi, A., T. E. Moffitt, M. Cannon, J. McClay, R. Murray, H. Harrington, A. Taylor, et al. 2005. "Moderation of the effect of adolescent-onset cannabis use on adult psychosis by a functional polymorphism in the catechol-O-methyltransferase gene: Longitudinal evidence of a gene × environment interaction." *Biological Psychiatry* 57 (10): 1117–27.

Caspi, A., K. Sugden, T. E. Moffitt, A. Taylor, I. W. Craig, H. Harrington, J. McClay, et al. 2003. "Influence of life stress on depression: Moderation by a polymorphism in the 5-HTT gene." *Science* 301 (5631): 386–89.

Caspi, A., B. Williams, J. Kim-Cohen, I. W. Craig, B. J. Milne, R. Poulton, L. C. Schalkwyk, A. Taylor, H. Werts, and T. E. Moffitt. 2007. "Moderation of breastfeeding effects on the IQ by genetic variation in fatty acid metabolism." *Proceedings of the National Academy of Sciences USA* 104 (47): 18860–65. doi:10.1073/pnas.0704292104.

Causton, H. C., B. Ren, S. S. Koh, C. T. Harbison, E. Kanin, E. G. Jennings, T. I. Lee, H. L. True, E. S. Lander, and R. A. Young. 2001. "Remodeling of yeast genome expression in response to environmental changes." *Molecular Biology of the Cell* 12 (2): 323–37.

Cavalli-Sforza, L. L., and M. W. Feldman. 1973. "Cultural versus biological inheritance: Phenotypic transmission from parents to children. (A theory of the effect of parental phenotypes on children's phenotypes)." *American Journal of Human Genetics* 25 (6): 618–37.

Chorney, Michael J., Karen Chorney, Nicole Seese, Michael J. Owen, Johanna Daniels, Peter McGuffin, Lee Ann Thompson, et al. 1998. "A quantitative trait locus associated with cognitive ability in children." *Psychological Science* 9 (3): 159–66.

Cichon, S., N. Craddock, M. Daly, S. V. Faraone, P. V. Gejman, J. Kelsoe, T. Lehner, et al. 2009. "Genome-wide association studies: History, rationale, and prospects for psychiatric disorders." *American Journal of Psychiatry* 166 (5): 540–56. doi:10.1176/appi.ajp.2008.08091354.

Cowan, W. M., K. L. Kopnisky, and S. E. Hyman. 2002. "The Human Genome Project and its impact on psychiatry." *Annual Review of Neuroscience* 25: 1–50.

Daly, M. J., J. D. Rioux, S. F. Schaffner, T. J. Hudson, and E. S. Lander. 2001. "High-resolution haplotype structure in the human genome." *Nature Genetics* 29 (2): 229–32.

Daniels, Michael, Bernie Devlin, and Kathryn Roeder. 1997. "Of Genes and IQ." In *Intelligence, Genes, and Success: Scientists Respond to "The Bell Curve,"* edited by Bernie Devlin, Stephen E. Fienburg, Daniel Phillip Resnick, and Kathryn Roeder, 45–70. New York: Springer.

Davies, G., A. Tenesa, A. Payton, J. Yang, S. E. Harris, D. Liewald, X. Ke, et al. 2011. "Genome-wide association studies establish that human intelligence is highly heritable and polygenic." *Molecular Psychiatry* 16 (10): 996–1005. doi:10.1038/mp.2011.85.

De, S., Y. Zhang, C. A. Wolkow, S. Zou, I. Goldberg, and K. G. Becker. 2013. "Genome-wide modeling of complex phenotypes in *Caenorhabditis elegans* and *Drosophila melanogaster*." *BMC Genomics* 14: 580. doi:10.1186/1471-2164-14-580.

Deary, I. J., W. Johnson, and L. M. Houlihan. 2009. "Genetic foundations of human intelligence." *Human Genetics* 126 (1): 215–32. doi:10.1007/s00439-009-0655-4.

Devlin, B., M. Daniels, and K. Roeder. 1997. "The heritability of IQ." *Nature* 388 (6641): 468–71.

Duncan, L. E., and M. C. Keller. 2011. "A critical review of the first 10 years of candidate gene-by-environment interaction research in psychiatry." *American Journal of Psychiatry* 168 (10): 1041–49. doi:10.1176/appi.ajp.2011.11020191.

Farahany, Nita A. 2009. *The Impact of Behavioral Sciences on Criminal Law.* New York: Oxford University Press.

Faraone, S. V., A. E. Doyle, E. Mick, and J. Biederman. 2001. "Meta-analysis of the association between the 7-repeat allele of the dopamine D(4) receptor gene and attention deficit hyperactivity disorder." *American Journal of Psychiatry* 158 (7): 1052–57.

Feldman, M. W., and L. L. Cavalli-Sforza. 1979. "Aspects of variance and covariance analysis with cultural inheritance." *Theoretical Population Biology* 15 (3): 276–307.

Feldman, M. W., and R. C. Lewontin. 1975. "The heritability hang-up." *Science* 190 (4220): 1163–68.

Fisher, R. A. 1918. "The correlation between relatives on the supposition of Mendelian inheritance." *Transactions of the Royal Society of Edinburgh* 52: 399–433.

Flynn, J. R. 1987. " Massive IQ gains in 14 nations: What IQ tests really measure." *Psychological Bulletin* 101: 171–91.

Franke, B., B. M. Neale, and S. V. Faraone. 2009. "Genome-wide association studies in ADHD." *Human Genetics* 126 (1): 13–50. doi:10.1007/s00439-009-0663-4.

Genome Consortium. 2004. "Finishing the euchromatic sequence of the human genome." *Nature* 431 (7011): 931–45. doi:nature03001 [pii] 10.1038/nature03001.

Gizer, I. R., C. Ficks, and I. D. Waldman. 2009. "Candidate gene studies of ADHD: A meta-analytic review." *Human Genetics* 126 (1): 51–90. doi:10.1007/s00439-009-0694-x.

Goldstein, D. B. 2009. "Common genetic variation and human traits." *New England Journal of Medicine* 360 (17): 1696–98. doi:NEJMp0806284 [pii] 10.1056/NEJMp0806284.

Gosso, M. F., E. J. de Geus, T. J. Polderman, D. I. Boomsma, P. Heutink, and D. Posthuma. 2008. "Common variants underlying cognitive ability: Further evidence for association between the SNAP-25 gene and cognition using a family-based study in two independent Dutch cohorts." *Genes, Brain and Behavior* 7 (3): 355–64. doi:GBB359 [pii] 10.1111/j.1601-183X.2007.00359.x.

Gottesman, I. I. 1997. "Twins: En route to QTLs for cognition." *Science* 276 (5318): 1522–23.

Hamer, D. 2002. "Genetics: Rethinking behavior genetics." *Science* 298 (5591): 71–72.

Herrnstein, Richard J., and Charles A. Murray. 1994. *The Bell Curve: Intelligence and Class Structure in American Life*. New York: Free Press.

Hyman, S. E. 2000. "The genetics of mental illness: Implications for practice." *Bulletin of the World Health Organization* 78 (4): 455–63.

Jensen, A. R. 2000. "The g factor: Psychometrics and biology." *Novartis Foundation Symposium* 233: 37–47; discussion 47–57, 116–21.

Keltikangas-Jarvinen, L., K. Raikkonen, J. Ekelund, and L. Peltonen. 2004. "Nature and nurture in novelty seeking." *Molecular Psychiatry* 9 (3): 308–11.

Kim, D. H., S. H. Yeo, J. M. Park, J. Y. Choi, T. H. Lee, S. Y. Park, M. S. Ock, J. Eo, H. S. Kim, and H. J. Cha. 2014. "Genetic markers for diagnosis and pathogenesis of Alzheimer's disease." *Gene*. doi:10.1016/j.gene.2014.05.031.

Kluger, A. N., Z. Siegfried, and R. P. Ebstein. 2002. "A meta-analysis of the association between DRD4 polymorphism and novelty seeking." *Molecular Psychiatry* 7 (7): 712–17.

Krim, T. M. 1996. "Beyond Baby M: International perspectives on gestational surrogacy and the demise of the unitary biological mother." *Annals of Health Law* (5): 193–226.

Lander, E. S., and L. Kruglyak. 1995. "Genetic dissection of complex traits: Guidelines for interpreting and reporting linkage results." *Nature Genetics* 11 (3): 241–47.

Lander, E. S., and N. J. Schork. 1994. "Genetic dissection of complex traits." *Science* 265 (5181): 2037–48.

Lee, Y. H., and G. G. Song. 2014. "Genome-wide pathway analysis in attention-deficit/hyperactivity disorder." *Neurological Sciences*. doi:10.1007/s10072-014-1671-72.

Lesch, K. P., D. Bengel, A. Heils, S. Z. Sabol, B. D. Greenberg, S. Petri, J. Benjamin, C. R. Muller, D. H. Hamer, and D. L. Murphy. 1996. "Association of anxiety-related traits with a polymorphism in the serotonin transporter gene regulatory region." *Science* 274 (5292): 1527–31.

Li, D., P. C. Sham, M. J. Owen, and L. He. 2006. "Meta-analysis shows significant association between dopamine system genes and attention deficit hyperactivity disorder (ADHD)." *Human Molecular Genetics* 15 (14): 2276–84. doi:10.1093/hmg/ddl152.

Lubinski, D. 2000. "Intelligence: Success and fitness." In *The Nature of Intelligence*, edited by Gregory R. Bock, Jamie A. Goode, and Kate Webb, 6–35. Chichester, UK: Wiley.

Maher, B. 2008. "The search for genome 'dark matter' moves closer." *Nature*, November 17. doi:10.1038/news.2008.1235.

Marcus, Steven, ed. 2002. *Neuroethics: Mapping the Field*. New York: Dana Foundation.

McCarthy, S. E., V. Makarov, G. Kirov, A. M. Addington, J. McClellan, S. Yoon, D. O. Perkins, et al. 2009. "Microduplications of 16p11.2 are associated with schizophrenia." *Nature Genetics* 41 (11): 1223–27. doi:ng.474 [pii] 10.1038/ng.474.

McGue, Matt. 2001. "The genetics of personality." In *Emery and Rimoin's Principles and Practices of Medical Genetics*, edited by David L. Connor, J. M. Rimoin, Reed E. Pyeritz, and Bruce R. Korf, 2791–800. London: Churchill Livingstone.

McGue, M., T. J. Jr. Bouchard, W. G. Iacono, and D. T. Lykken. 1993. "Behavioral genetics of cognitive abilities and disabilities." In *Nature, Nurture and Psychology*, edited by R. Plomin and G. McClearn, 59–76. Washington, DC: American Psychological Association.

McGuffin, P., B. Riley, and R. Plomin. 2001. "Genomics and behavior: Toward behavioral genomics." *Science* 291 (5507): 1232–49.

Melville, S. A., J. Buros, A. R. Parrado, B. Vardarajan, M. W. Logue, L. Shen, S. L. Risacher, et al. 2012. "Multiple loci influencing hippocampal degeneration identified by genome scan." *Annals of Neurology* 72 (1): 65–75. doi:10.1002/ana.23644.

Mirnics, K., F. A. Middleton, D. A. Lewis, and P. Levitt. 2001. "The human genome: Gene expression profiling and schizophrenia." *American Journal of Psychiatry* 158 (9): 1384.

Mirnics, K., F. A. Middleton, G. D. Stanwood, D. A. Lewis, and P. Levitt. 2001. "Disease-specific changes in regulator of G-protein signaling 4 (RGS4) expression in schizophrenia." *Molecular Psychiatry* 6 (3): 293–301.

Mirnics, K., and J. Pevsner. 2004. "Progress in the use of microarray technology to study the neurobiology of disease." *Nature Neuroscience* 7 (5): 434–39.

Moore, David S. 2002. *The Dependent Gene: The Fallacy of Nature/Nurture*. New York: Times Books.

Munafo, M. R., S. M. Brown, and A. R. Hariri. 2008. "Serotonin transporter (5-HTTLPR) genotype and amygdala activation: A meta-analysis." *Biological Psychiatry* 63 (9): 852–57. doi:S0006-3223(07)00824-4 [pii] 10.1016/j.biopsych.2007.08.016.

Munafo, M. R., and J. Flint. 2011. "Dissecting the genetic architecture of human personality." *Trends in Cognitive Sciences* 15 (9): 395–400. doi:10.1016/j.tics.2011.07.007.

Munafo, M. R., B. Yalcin, S. A. Willis-Owen, and J. Flint. 2008. "Association of the dopamine D4 receptor (DRD4) gene and approach-related personality traits: Meta-analysis and new data." *Biological Psychiatry* 63 (2): 197–206. doi:10.1016/j.biopsych.2007.04.006.

Neale, B. M., S. E. Medland, S. Ripke, P. Asherson, B. Franke, K. P. Lesch, S. V. Faraone, et al. 2010. "Meta-analysis of genome-wide association studies of attention-deficit/hyperactivity disorder." *Journal of the American Academy of Child and Adolescent Psychiatry* 49 (9): 884–97. doi:10.1016/j.jaac.2010.06.008.

Nichols, P. L., and V. E. Anderson. 1973. "Intellectual performance, race, and socioeconomic status." *Social Biology* 20 (4): 367–74.

Nussbaum, R. L., and C. E. Ellis. 2003. "Alzheimer's disease and Parkinson's disease." *New England Journal of Medicine* 348 (14): 1356–64.

Pastor, P. N., and C. A. Reuben. 2002. "Attention deficit disorder and learning disability: United States, 1997–98." *Vital Health Statistics* 10: 206.

Plomin, R., and I. J. Deary. 2015. "Genetics and intelligence differences: Five special findings." *Molecular Psychiatry* 20: 98–108.

Plomin, Robert, John C. DeFries, Ian W. Craig, and Peter McGuffin. 2003. *Behavioral Genetics in the Postgenomic Era*. Washington, DC: American Psychological Association.

Plomin, Robert, John C. DeFries, Valerie S. Knopik, and Jenae M Neiderhiser. 2013. *Behavioral Genetics*. 6th ed. New York: Worth Publishers.

Plomin, Robert, John C. DeFries, Gerald E. McClearn, and Peter McGuffin. 2001. *Behavioral Genetics*. 4th ed. New York: Worth Publishers.

Ridley, Matt. 2003. *Nature via Nurture: Genes, Experience, and What Makes Us Human*. New York: HarperCollins.

Risch, N. J. 2000. "Searching for genetic determinants in the new millennium." *Nature* 405 (6788): 847–56.

Risch, N. J. 2001. "The genetic epidemiology of cancer: Interpreting family and twin studies and their implications for molecular genetic approaches." *Cancer Epidemiology, Biomarkers and Prevention* 10 (7): 733–41.

Risch, N. J., R. Herrell, T. Lehner, K. Y. Liang, L. Eaves, J. Hoh, A. Griem, M. Kovacs, J. Ott, and K. R. Merikangas. 2009. "Interaction between the serotonin transporter gene (5-HTTLPR), stressful life events, and risk of depression: A meta-analysis." *JAMA* 301 (23): 2462–71. doi:301/23/2462 [pii] 10.1001/jama.2009.878.

Risch, N. J., and K. Merikangas. 1996. "The future of genetic studies of complex human diseases." *Science* 273 (5281): 1516–17.

Risch, N. J., and K. Merikangas. 2009. "'The future of genetic studies of complex human diseases': Drs. Merikangas and Risch talk about that influential paper 13 years later." *hum-molgen: genetic news*. Accessed August 30, 2015.

Robinson, M. R., N. R. Wray, and P. M. Visscher. 2014. "Explaining additional genetic variation in complex traits." *Trends in Genetics* 30 (4): 124–32. doi:10.1016/j.tig.2014.02.003.

Roses, A. D. 2000. "Pharmacogenetics and the practice of medicine." *Nature* 405 (6788): 857–65.

Rubinsztein, D. C., J. Leggo, R. Coles, E. Almqvist, V. Biancalana, J. J. Cassiman, K. Chotai, et al. 1996. "Phenotypic characterization of individuals with 30-40 CAG repeats in the Huntington disease (HD) gene reveals HD cases with 36 repeats and apparently normal elderly individuals with 36–39 repeats." *American Journal of Human Genetics* 59 (1): 16–22.

Saunders, A. M., W. J. Strittmatter, D. Schmechel, P. H. George-Hyslop, M. A. Pericak-Vance, S. H. Joo, B. L. Rosi, J. F. Gusella, D. R. Crapper-MacLachlan, and M. J. Alberts. 1993. "Association of apolipoprotein E allele epsilon 4 with late-onset familial and sporadic Alzheimer's disease." *Neurology* 43 (8): 1467–72.

Schaffner, K. F. 1993. *Discovery and Explanation in Biology and Medicine*. Chicago: University of Chicago Press.

Schaffner, K. F. 1998. "Genes, behavior, and developmental emergentism: One process, indivisible?" *Philosophy of Science* 65 (June): 209–52.

Schaffner, K. F. 2000. "Behavior at the organismal and molecular levels: The case of *C. elegans*." *Philosophy of Science* 67:s273–78.

Schaffner, K. F. 2001. "Nature and nurture." *Current Opinion in Psychiatry* 14 (September): 486–90.

Schaffner, K. F. 2013. "Reduction and Reductionism in Psychiatry." In *Oxford Handbook of Philosophy and Psychiatry*, edited by K. W. M. Fulford, M. Davies,

R. G. T. Gipps, G. Graham, J. Z. Sadler, G. Stranghellini, and T. Thornton, 1003–22. Oxford: Oxford University Press.

Schinka, J. A., E. A. Letsch, and F. C. Crawford. 2002. "DRD4 and novelty seeking: Results of meta-analyses." *American Journal of Medical Genetics* 114 (6): 643–48.

Sharp, R. R. 2011. "Downsizing genomic medicine: Approaching the ethical complexity of whole-genome sequencing by starting small." *Genetics in Medicine* 13 (3): 191–94. doi:10.1097/GIM.0b013e31820f603f.

Tanzi, R. E. 2012. "The genetics of Alzheimer disease." *Cold Spring Harbor Perspectives in Medicine* 2 (10). doi:10.1101/cshperspect.a006296.

Thomas, A. M., G. Cohen, R. M. Cook-Deegan, J. O'Sullivan, S. G. Post, A. D. Roses, K. F. Schaffner, and R. M. Green. 1998. "Alzheimer testing at Silver Years." *Cambridge Quarterly of Healthcare Ethics* 7 (3): 294–307.

Thomson, C. J., C. W. Hanna, S. R. Carlson, and J. L. Rupert. 2013. "The -521 C/T variant in the dopamine-4-receptor gene (DRD4) is associated with skiing and snowboarding behavior." *Scandinavian Journal of Medicine and Science in Sports* 23 (2): e108–13. doi:10.1111/sms.12031.

Visscher, P. M., M. A. Brown, M. I. McCarthy, and J. Yang. 2012. "Five years of GWAS discovery." *American Journal of Human Genetics* 90 (1): 7–24. doi:10.1016/j.ajhg.2011.11.029.

Visscher, P. M., S. E. Medland, M. A. Ferreira, K. I. Morley, G. Zhu, B. K. Cornes, G. W. Montgomery, and N. G. Martin. 2006. "Assumption-free estimation of heritability from genome-wide identity-by-descent sharing between full siblings." *PLoS Genetics* 2 (3): e41. doi:10.1371/journal.pgen.0020041.

Waldman, Irwin D., and Soo Hyun Rhee. 2002. "Behavioral and molecular genetic studies of ADHD." In *Hyperactivity and Attention Disorders of Childhood*, edited by Seija Sandberg, 290–335. New York: Cambridge University Press.

Walsh, T., J. M. McClellan, S. E. McCarthy, A. M. Addington, S. B. Pierce, G. M. Cooper, A. S. Nord, et al. 2008. "Rare structural variants disrupt multiple genes in neurodevelopmental pathways in schizophrenia." *Science* 320 (5875): 539–43. doi:1155174 [pii] 10.1126/science.1155174.

Zhang, Y., C. Ma, T. Delohery, B. Nasipak, B. C. Foat, A. Bounoutas, H. J. Bussemaker, S. K. Kim, and M. Chalfie. 2002. "Identification of genes expressed in *C. elegans* touch receptor neurons." *Nature* 418 (6895): 331–35. doi:10.1038/nature00891.

3

Genes, Behavior, and the Developmentalist Challenge

One Process, Indivisible?

DECOMPOSING THE WORM

Toward the end of the last chapter, I mentioned that what behavioral geneticists need to do is to seek to find useful and robust simplifications in the vast neural network that constitutes our nervous systems and brain. Such simplifications will be needed in order to connect the variations in the heritable elements, the genes (more accurately the different forms of the gene—the alleles), with specific behavior types. That "simplifying" analysis typically will have to be done with the help of new scientific tools, drawn in no small part from the neurosciences, which will assist in at least partially *decomposing* the huge network and *localizing* functional parts.[1] This, it turns out, is very difficult to do, even with a very simple organism like the common roundworm, *C. elegans*.

In this chapter we will discuss how "the worm," as this organism is affectionately known, has been approached to look at the relation of its genes and its behaviors. We need to start with one of the simplest of organisms that is just complex enough to have both a genome *and* a nervous system. But before we turn to the specifics of the worm, I want to discuss a set of issues that arise in the context of debates between orthodox developmental studies and those I term the "developmentalists" and that pose a series of quite new philosophical problems. Developmentalism, I will argue, is a very useful backdrop in terms of which to discuss the results of the worm's implications for genetics and behavior.

The Developmentalist Challenge

It would be best to start by briefly describing what might be termed the "developmentalist challenge" to a standard view of molecular genetics, and also to behavioral genetics. Though the popular media reinforce an oversimplified single-gene, preformationist, and deterministic picture of behavioral genetics through stories on the fat gene (Zhang, Guo, et al. 2002), happiness genes (Goleman 1996), and genes for sexual orientation (Hamer et al. 1993), behavioral geneticists are more sophisticated.[2] They tend to believe that for any interesting characteristic, like the IQ example discussed in the last chapter, there are many genes each having small joint effects on behavior. Behavioral geneticists are also highly sensitive to the roles of environment and learning. But even this more sophisticated view of behavioral geneticists, ably summarized by Plomin (2008) and Plomin et al. (2013), has been criticized by a number of writers who collectively represent what I term the "developmentalist challenge."

The developmentalist challenge affects a far broader area than behavioral genetics—it has relevance for *any* claims about the separable effects of genes and environment on *any* traits[3]—but it has its greatest force and has been applied most vigorously to behavioral traits (Lewontin 1992, 1995). There, it attacks the traditional "nature-nurture" distinction initially discussed in chapter 1, and also directs some powerful criticisms against the "innate-learned behavior" dichotomy. The term "developmentalist" may not be the best to describe this loosely knit set of criticisms of genetic determinism and DNA primacy. Some who fall into this group prefer the term "interactionist" (e.g., Johnston 1988) or "constructionist" to describe their approach (e.g., Gray 1994, and, I believe, Lewontin 1992), but that latter term carries somewhat misleading connotations of what is known as "social constructionism" with which the developmentalist view should not be conflated.[4] Thus I shall use the term *developmentalist*.[5]

Developmentalists hold to views of differing strength that are critical of the distinctions, such as the nature-versus-nurture dichotomy just mentioned, but I think it fair to say that virtually all (strong) developmentalists appear to accept the following 11 theses. I have divided these up into what developmentalists seem to think are *seven deadly sins* about trait causation, and another *four major mistakes* that classical approaches to the study of nature and nurture have made. Several of these theses also imply a view of how organisms' behavior can be appropriately studied. In each of the 11 items below, I will name the "sin" or the old "mistaken" view using a simple phrase or sentence in italics, and also underline the key concept. Then immediately following this, I indicate what the contrary developmentalist view is. More extensive discussions of the developmentalist view will be found later in the chapter, as well as in chapters 4 and 8. In these analyses I have only been able to partially make use of a very recent book by two prominent developmentalists, Griffiths and Stotz (2013), that does include an excellent chapter titled "The Behavioral Gene," though with a few specific exceptions noted in chapter 4 and 8, I do not think the arguments in this recent text negatively affect my conclusions.

Seven Deadly Sins of Causation

1. *Nature-versus-nature dichotomy.* The nature-nurture distinction is outmoded and needs to be replaced by a seamless unification approach in which genes and environment are "interacting and inseparable shapers [inseparable causes] of development" (Lewontin 1995, 72)
2. *One gene causes one behavior.* The causal relation between genes and organisms is "many-many," and the existence of significant "developmental noise" (chance events or stochasticity during development) precludes both

gene-to-organism trait predictability (including behavioral traits) and organism trait-to-gene causal inferences (Lewontin 1995; Stent 1981). Thus the outcome is emergent (Gottlieb 1995, 135; Lewontin 1995, 27).

3. *Genes are blueprints.* Genes do not "contain" the "information" that is a blueprint for traits; rather, information discernible in maturing organisms *develops*—the information is the *product* of an ontogeny (Oyama [1985] 2000).
4. *DNA sequences contain the essence of behavioral information.* DNA sequences per se have no fixed meaning, but are informational only in a broader causal context of nongenetic interpreting molecules (Lewontin 1992; Oyama [1984] 2000).
5. *Genes cause behavioral traits fairly directly.* Characterizing genes as causes of traits reflects outmoded preformationist thinking (Johnston 1988). Genes do not even make neural structures in any direct way; they produce proteins that affect cell differentiation to yield neurons that become specific types of neurons in specific places with particular connections with other neurons (Gottlieb 1995, 132; Stent 1981).
6. *Genes are the root cause of behavior.* Developmental causation is not just "bottom up," but is also "top down." Genes are not the principal actors that produce traits (including behavioral traits), but are part of a complex system, in which the cytoplasm can influence the genes, extracellular hormones can influence the nucleus, external sensory stimulation can influence the genes, and the hormones can be influenced by the external environment (see Gottlieb 1995, 138; Gray 1994, 180 for references).
7. *A gene produces a single, clear, and specific phenotype.* The most accurate way to describe trait development is to use the "norm of reaction" approach, which for any given genotype "is a list or graph of the correspondence between different possible environments and the phenotypes that would result" (Lewontin 1995, 21), but this does not yield causally deterministic predictions (Gottlieb 1995). Even norms of reactions have to have a temporal developmental dimension added to them (Gray 1994; Gottesman 1997).

Four Major Mistakes of Classical Approaches to Nature and Nurture

1. *Behaviors divide neatly into innate and learned classes.* The classical ethology approach of Lorenz 1965 that distinguished between "learned" and "innate" behavior has to be replaced by an "interactionist," "epigenetic," "ecological," or "life cycle" approach. (See Lehrman 1953; Lehrman 1970; Johnston 1988; Bronfenbrenner and Ceci 1994; Griffiths and Gray 1994.)[6]
2. *Empirical studies can disentangle the effects of heredity and environment into specific percentages.* Classical behavioral genetics is also committed to a false nature-nurture dichotomy that mistakenly believes it can distinguish between the contributions of heredity and environment to behavior (Johnston 1988).

3. *Analysis of variance (ANOVA) methods are powerful tools that reveal developmental effects.* Because classical behavioral genetics is a population-based discipline with its main method being analysis of variance, it can say nothing about the causes of individual development (Gottlieb 1995). An analysis of variance is not the same thing as an analysis of causes (Lewontin 1974; Gottlieb 1995). Classical behavioral genetics thus can only address the question "how much of the variance" is "attributable to heredity and how much to environment," but not *how* hereditary and environment actually produce their effects (Bronfenbrenner and Ceci 1994).
4. *A "heritability" is an excellent summary statistic.* The concept of "heritability" found at the core of classical behavioral genetics is generally useless and misleading (Lewontin 1995, 71–72); the nonadditivity and the environmental dependence of genetic effects will not permit its applicability except in highly specialized artificial circumstances, typically in breeding programs (Layzer 1974; Wahlsten 1990).

In contrast to models of development in which (*a*) genes, or (*b*) environments are determinative Lewontin (1995, 27–28) provides several graphical representations of extreme environmental and genetic determinism (not shown) and of what he characterizes as (*c*) the "correct model of development" that incorporates interactions as well as random "developmental noise." Figure 3.1 represents this correct model.

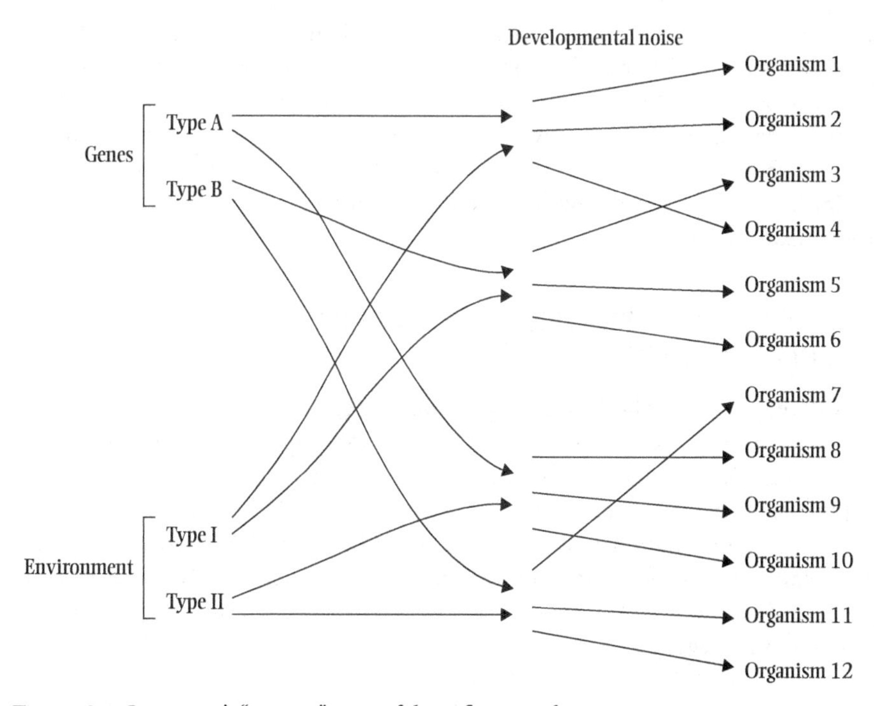

Figure 3.1 Lewontin's "correct" view of the influence of genes on traits.
SOURCE: Lewontin 1995.

The upshot of the developmentalist criticism is to seriously call into question both the methods and the results of the discipline of behavioral genetics. This certainly seems to be the implication of the strongest form of this criticism as we find in Gottlieb (1992, 1995), Gray (1994), and Lewontin (1993, 1995). Others such as Bronfenbrenner and Ceci (1994) accept many of the above developmentalist points, but see ways to employ many of the results and methods of classical behavioral genetics within an expanded approach that these authors term a "bioecological model." Some behavioral geneticists see value in these criticisms but maintain that they are seriously overdone (Turkheimer 1995), and still others are dismissive of the criticisms (Scarr 1995).

The approach in this chapter will initially be bottom-up, in the sense that it will proceed from an account of how a number of contemporary scientists are developing explanations of behavior in simple living systems, frequently called "model organisms." As indicated at the beginning of this chapter, I will start from the simplest model organism that possesses a working nervous system, the round worm *Caenorhabditis elegans* (*C. elegans* for short), and then proceed to a much briefer discussion of that favorite of geneticists since T. H. Morgan's work began in 1910, the fruit fly, *Drosophila melanogaster.* This approach represents a philosophical analog of what is termed the "simple systems" approach, widely adopted in learning and memory studies in "psychology, physiology, biochemistry, genetics, neurobiology, and molecular biology" (Gannon 1995, 205). It is one of the theses of this chapter that only by examining quite recent work at the interface of *molecular* genetics, neuroscience, and behavior can some of the controversies raised by the developmentalist challenge be clarified, and at least partially settled.[7] The conclusions of an account of behavioral explanation in simple systems will then be reassessed in later chapters in connection with more complex organisms, including what these conclusions may tell us about humans. In later chapters we will revisit the claims of the seven deadly sins and the four major mistakes, especially with reference to human studies on personality and on schizophrenia.

C. ELEGANS: THE VERY MODEL OF A MODERN "MODEL ORGANISM"

C. elegans is one of the "model organisms" targeted by the Human Genome Project as a source of potential insight into the working of human genes.[8] Other model organisms include *E. coli*, yeast, *Drosophila*, the honeybee, *Aradopsis*, the zebrafish, the mouse, and the rat.[9]

Bruce Alberts, a noted biologist, former president of the US National Academy of Science, and an author of a highly influential textbook, *The Molecular Biology of the Cell* (Alberts 2008), asked a rhetorical question about *C. elegans* studies nearly 10 years ago:

> Why should one study a worm? This simple creature is one of several "model" organisms that together have provided tremendous insight into how all organisms are put together. It has become increasingly clear over

> the past two decades that knowledge from one organism, even one so simple as a worm, can provide tremendous power when connected with knowledge from other organisms. And because of the experimental accessibility of nematodes, knowledge about worms can come more quickly and cheaply than knowledge about higher organisms. (1997, xii)

Alberts adds:

> We can say with confidence that the fastest and most efficient way of acquiring an understanding of ourselves is to devote an enormous effort trying to understand these and other, relatively "simple" organisms. (1997, xiv)

Historical Overview of the Worm

Though the worm has been closely studied by biologists since the 1870s (see von Ehrenstein 1980 for references), it was the vision of Sydney Brenner that has made *C. elegans* the model organism that it is today. In 1963 Brenner had come to believe, as had some other molecular biologists, including the late Gunther Stent (Stent 1969), that "nearly all of the 'classical' problems of molecular biology" had been solved or soon would be solved, and that it was time to move on to study the more interesting topics of development and the nervous system.[10] Brenner argued that the worm or nematode had a number of valuable properties, such as a short life cycle, small size, relatively few cells, and suitability for genetic analysis, which could make the nematode the *E. coli* of multicellular organisms. By 1967 Brenner had isolated the first behavioral mutants of *C. elegans*, and in 1970 John White began detailed reconstruction of its nervous system (Thomas 1994). In 1974, Brenner published the first major study of the genetics of this organism. In the past 30 years, *C. elegans* has been intensively studied, and landmark collections of essays summarizing the field appeared in 1988 (Wood 1988) and in 1997 (Riddle 1997) known as Worm I and Worm II in the *C. elegans* community. Those volumes, and especially the appendices containing lists of parts, neurons, mutations, and so on, illustrate the power of the "brute force" of the approach taken to this organism.[11] In 1999, the entire genome of *C. elegans* was finally sequenced, and in 2002 the worm won a Nobel Prize. Well, not exactly, but here is what the *New York Times* said in its October 9, 2002, editorial "Ode to a Worm":

> When the Nobel Prize in Physiology or Medicine was awarded to two Britons and an American this week, it should also have gone to a fourth contributor that made their pioneering research possible, the tiny soil worm known as Caenorhabditis elegans, or C. elegans for short. Over the past three decades this unassuming creature has elbowed its way to the forefront of laboratory specimens, joining the legions of rats, mice, fruit flies and bacteria that serve as convenient model organisms for studying the processes of life. The prize was awarded for a series of seminal discoveries by the three scientists about the ways genes regulate organ development and cell death.

> . . . Dr. Brenner, a witty, wide-ranging scientist whom some call the Father of the Worm, was honored for establishing C. elegans as a novel experimental system and using it to show that gene mutations could be linked to specific effects on organ development. Sir John Sulston was honored for developing techniques to map the exact lineage by which one cell of the worm leads to another and for discovering that specific cells always die during the worm's development through a process called "programmed cell death." And Robert Horvitz, of the Massachusetts Institute of Technology, was honored for identifying various genes controlling cell death.

Today the *C. elegans* community is a truly international one, including thousands of researchers who displays extraordinary cooperativity among themselves and with outside researchers as well.[12]

In his pioneering article of 1974, Brenner laid out the rationale and general methodology for studying *C. elegans*. Of related interest to the "simple systems" approach is his comment within this general methodological framework citing the utility of a similar methodology for the study of *Drosophila*. In an enormously important part of that paper that is well worth a long quotation, Brenner wrote:

> In principle, it should be possible to dissect the genetic specification of a nervous system in much the same way as was done for biosynthetic pathways in bacteria or for bacteriophage assembly. However, one surmises that genetical analysis alone would have provided only a very general picture of the organization of those processes. Only when genetics was coupled with methods of analyzing other properties of the mutants, by assays of enzymes or *in vitro* assembly, did the full power of this approach develop. In the same way, the isolation and genetical characterization of mutants with behavioral alterations must be supported by analysis at a level intermediate between the gene and behavior. Behavior is the result of a complex and ill-understood set of computations performed by nervous systems and it seems essential to decompose the problem into two: one concerned with the question of the genetic specification of nervous systems and the other with the way nervous systems work to produce behavior. Both require that we must have some way of analyzing a nervous system.
>
> Much the same philosophy underlies the work initiated by Benzer on behavioral mutants of *Drosophila* (for review, see Benzer, 1971). There can be no doubt that *Drosophila* is a very good model for this work, particularly because of the great wealth of genetical information that already exists for this organism. There is also the elegant method of mosaic analysis which can be powerfully applied to find the anatomical sites of genetic abnormalities of the nervous system. . . .
>
> Some eight years ago, when I embarked on this problem, I decided that what was needed was an experimental organism which was suitable for genetical study and in which one could determine the complete structure

of the nervous system. *Drosophila*, with about 10^5 (= 100,000) neurons, is much too large, and, looking for a simpler organism, my choice eventually settled on the small nematode, *Caenorhabditis elegans*. (Brenner 1974, 72)

Just the (Worm) Facts

We will begin our worm story with a brief summary of the facts about this extraordinary creature. *C. elegans* is a tiny worm, about 1 mm long, that can be found in soil in many parts of the world. It has a transparent body, feeds on bacteria, and has two sexes: hermaphroditic (self-fertilizing) and male. Figure 3.2 from the WormAtlas (Altun 2009) shows a simple diagram of the typical hermaphrodite worm. Its life cycle to the reproductive stage is 3 days with a typical life span of 17 days (Wood 1988). As already noted, the organism has been studied to the point where there is an enormous amount of detail known about its genes, cells, organs, and behavior. The developmental lineage of all cells in the nematode has been traced from the single-celled zygote. The adult hermaphrodite has 959 somatic nuclei and the male 1,031 nuclei; there are about 2,000 germ cell nuclei (Hodgkin et. al 1995). The haploid genome contains about 8×10^7 (= 10,000,000) nucleotide pairs, organized into five autosomal and one sex chromosome (hermaphrodites are XX, males XO), comprising ~19,800 genes. The organism can move itself forward and backward by undulatory movements and responds to touch and a number of chemical stimuli, of both attractive and repulsive forms. More complex behaviors include egg laying and mating between hermaphrodites and males (Wood 1988, 14). The nervous system is the largest organ, being constituted, in the hermaphrodite, by 302 neurons; there are 95 muscle cells to which the neurons can connect. The neurons have been fully described in terms of their location and synaptic connections. The neurons are essentially identical from one individual in a strain to another (Sulston et al. 1983; White et al. 1986) and form approximately 5,000 synapses (nerve to nerve connections), 700 gap junctions (another way cells can communicate), and 2,000 neuromuscular junctions (the contacts between the nerve and muscle cells) (White et al. 1986). The synapses are typically "highly reproducible" from one animal to another, but are not identical.[13]

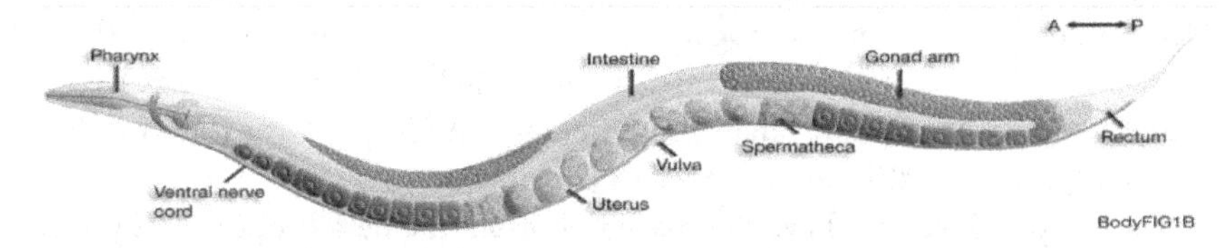

Figure 3.2 Anatomy of an adult hermaphrodite. (Actual length of worm is ~ 1 mm.) Schematic drawing of anatomical structures, left lateral side. Dotted lines and numbers mark the level of each section shown in another figure in Altun and Hall 2009.
SOURCE: Altun and Hall 2009.

In 1988, Wood, echoing Brenner's earlier vision, wrote:

> The simplicity of the *C. elegans* nervous system and the detail with which it has been described offer the opportunity to address fundamental questions of both function and development. With regard to function, it may be possible to correlate the entire behavioral repertoire with the known neuroanatomy. (1988, 14)

Seemingly, *C. elegans* is indeed what Robert Cook-Deegan (1994, 53) called "the reductionist's delight."

C. ELEGANS'S BEHAVIORAL GENETICS: IT'S NOT SO SIMPLE AFTER ALL

Unfortunately there are some limitations that have made this optimistic, reductionistic vision difficult to bring to closure easily. We need to start with a discussion on how neurophysiological and behavioral level studies work in the worm, and then consider the extent to which *developmental* influences involving genes and environmental factors are examined in current research on *C. elegans*.

Neurophysiology and Behavior: Function

METHODOLOGY AND "RULES" RELATING GENES TO BEHAVIOR

In her 1993 review article, Cori Bargmann wrote that "heroic efforts" have resulted in the construction of a wiring diagram for *C. elegans* that has "aided in the interpretation of almost all *C. elegans* neurobiological experiments." But Bargmann went on to say:

> However, neuronal functions cannot yet be predicted purely from the neuroanatomy. The electron micrographs do not indicate whether a synapse is excitatory, inhibitory, or modulatory. Nor do the morphologically defined synapses necessarily represent the complete set of physiologically relevant neuronal connections in this highly compact nervous system. (Bargmann 1993, 49–50)

She added that the neuroanatomy needs to be integrated with other information to determine "how neurons act together to generate coherent behaviors," studies that utilize laser removal (ablations) of individual neurons, genetic analysis, pharmacology, and behavioral analysis (Bargmann 1993, 50).[14] The need for this multilevel approach is still the case over 15 years later.

In this chapter I will not have the space to present the details of the many and varied painstakingly careful studies that have been done comparing behavioral mutants' behaviors with the effects of laser-zapping each nerve (neuronal ablation), in attempting to identify genetic and learning components of *C. elegans*'s behaviors. These include the specifics of Brenner's pioneering work already cited, as well as Avery and Horvitz's work on swallowing and muscle control (see Avery and Horvitz 1989 for references). Rather, I will initially focus on two laboratories' work and say something briefer about two more that provide us with representative studies of the worm's behavior. I will begin with Bargmann and de Bono's work on social and solitary feeding behavior, best thought of as a form of *chemo*-sensory regulated behavior, and then briefly discuss Chalfie and his colleagues' work on the worm's touch response, a type of *mechano*sensory behavior. After that I turn to the electrophysiological investigations of Avery's, Lockery's, Rankin's, and some others' laboratories. After covering the issue of *function* of the genes and neuronal circuits, I then look, in a section on *development*, in more detail at Chalfie's *developmental* investigations of the touch response.

A persistent feature of Sydney Brenner's legacy has been the search for *individual genes* that are strongly related to behaviors. Bargmann's laboratory has investigated several of these genes, particularly those affecting chemosensitivity, susceptibility of the worm to the action of a chemical agent, and chemotaxis, or movement toward or away from a chemical signal. In 1998, she and de Bono published a paper on a genetic explanation for worms' "social" behavior in the prestigious journal *Cell* that attracted wide attention, including a *New York Times* article (Wade 1998). It is to this essay in *Cell* that I turn next. As an interesting contrast to this genocentric perspective, I have elsewhere compared this approach with a paper from Lockery's lab, which virtually ignores gene explainers in favor of a complex neurophysiological account (Schaffner 2000).

An Exceptional Study: A Genetic Basis for Social Behavior

In their 1998 essay, de Bono and Bargmann summarized their results in an abstract (de Bono and Bargmann 1998, 679), which I closely paraphrase here, interpolating just enough in the way of additional information that nonspecialists can follow the near-original language of the published abstract:

> Natural subpopulations of *C. elegans* exhibit solitary or social feeding behavior. Solitary eaters move slowly across a surface rich in the bacteria on which they feed and also disperse on that surface. Social eaters on the other hand move rapidly on the bacteria and bunch up together. A knock-out ("loss of function") mutation in a gene known as *npr-1* causes a solitary strain to take on social behavior. This gene is known to encode a type of protein known as a G-protein coupled receptor, a protein that acts like a switch to open or close ion channels in nerve cells. This NPR-1 protein is similar to a family of proteins called Y receptors that are widely present in the nervous system. Two variants of the NPR-1 protein that differ only in a single amino acid (phenylalanine versus valine) occur naturally in the wild. One variant, termed NPR-1 215F (with phenylalanine, abbreviated as F) is found exclusively in social strains, while the other variant, NPR-1 215V (with valine) is found only in solitary strains. The differences between the F and V variants are due to a single nucleotide difference in the gene's DNA sequence (T versus G). Inserting a gene that produces the V form of the protein can transform wild social strains into solitary ones. Thus these only slightly different proteins generate the two natural variants in *C. elegans*' feeding behavior (figure 3.3).

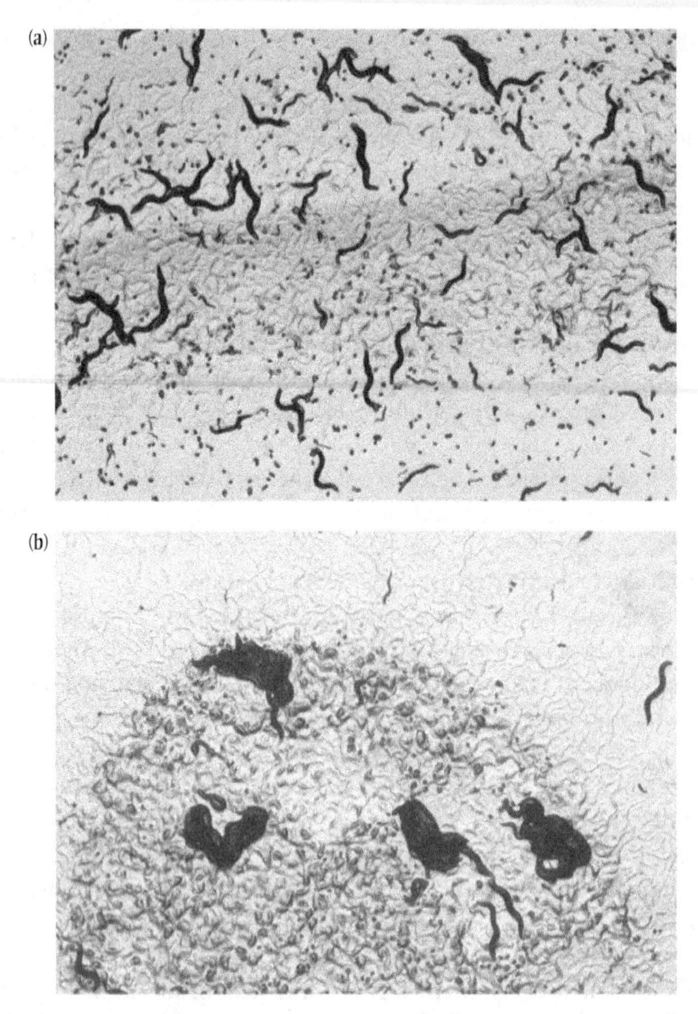

Figure 3.3 Two populations of *C. elegans*, solitary (top) and social (clumped) feeders. SOURCE: de Bono and Bargmann 1998.

This paper by de Bono and Bargmann makes strong claims involving a genetic explanation of behavior. At the end of the introduction to this essay, the authors write that "we show that variation in responses to food and other animals in wild strains of *C. elegans is due to* [my emphasis] natural variation in *npr-1*" (1998, 679). The phenotype difference is actually somewhat complex, and not just related to social or solitary feeding in the presence of sufficient amount of bacterial food supply. As already indicated, the social and solitary strains also differ in their speed of locomotion. Also, the two types differ in their burrowing behavior in the agar jell surface on which the worms are studied in the laboratory. But de Bono and Bargmann contended that "a single gene mutation can give rise to all of the behavioral differences characteristic of wild and solitary strains" (1998, 680).

The authors' strategy went as follows. They obtained, from three other investigators, chemically induced mutants of normally solitary strains that had converted to social feeding behavior. Mapping analysis suggested that all three alleles were due to modifications in a single gene on the X chromosome known as *npr-1*. This gene was then cloned and sequenced, and used to restore ("rescue") the social/hyperactivity phenotype in solitary strains. The difference between social and solitary strains was traced to a single amino acid substitution, phenylalanine for valine, at position 215 in the NPR protein, this due to a single nucleotide G → T mutation. The social behavior mutation is recessive. Sequence comparison with the *C. elegans* database indicated *npr-1* coded for a seven-transmembrane domain receptor, expressed in neurons, that resembled members of the G-protein coupled neuropeptide Y receptor family. This family can be thought of as a set of similar proteins that can control the activity of other molecules, for example, ion channels, second messengers, or enzymes, through a chain of causal influences.

De Bono and Bargmann offer several "different models that could explain the diverse behavioral phenotypes of *npr-1* mutants" (1998, 686). One possibility is that *npr-1* might repress a swarming response by inhibiting sensory neurons. Alternatively food could stimulate production of a molecule that binds to the receptor, and as food decreases, the swarming response might be derepressed. The point to be made here, as the authors write, is that "resolution of these models awaits identification of the cells in which *npr-1* acts, and the cells that are the source of the npr-1 ligands" (1998, 686).

Thus what we have in this account is a genetic difference (one nucleotide) that "causes" the difference between social and solitary behavioral phenotypes. This notion of cause is one that I believe can best be understood as a factor that we can say is "necessary in the circumstances." The "circumstances" here are a huge set of assumptions (what philosophers call a ceteris paribus clause) that assumes organism identity in *all other* relevant aspects. Moreover, the causal pathway(s) relating the gene product (the neuropeptide receptor) and the behavior was not yet known—only speculative models were available. So, in point of fact, the link between gene and behavior *jumps over* not only how the gene produces the tertiary structure of its product (fair enough), but also the nature of the receptor-cell interaction, and cell-to-cell causal influences in a neural net leading to locomotion and clumping. But the claim has been made, not only in the essay per se, but in popular commentary on the paper by a prominent journalist, that here we find a "spectacular" instance of "genes that govern certain behaviors in animals" (Wade 1998). This G-P arcing, or jumping from gene to phenotype, and imputing a causal explanation of the phenotype to a single gene, has been noted before as a problem for reductions or complete molecular explanations (see Kitcher 1984 and also my comments in Schaffner 1998).

This wonderfully "simple" story of one gene that influences one type of behavior in the worm was the case in 1998 as just described. Since then, further work by de Bono and Bargmann has indicated that the story is more complex. I will return to these added complications later in chapter 5. Though this story in its simple form is entrancing—the science reporter for the *New York Times* (Wade 1998) called it "a fascinating beginning of a whole field, molecular biology of social behavior"—the specificity of the gene-behavior connection is extremely unrepresentative, and further research over the course of the following four years showed this to be so (see chapter 5). (One other highly specific connection affecting response of the worm to odors was also found by Sengupta working in Bargmann's laboratory; for details see Sengupta, Colbert, and Bargmann 1994). A more representative picture requires that we understand how the genes work through neurons, as Bargmann and her colleagues will caution us in a moment. It will turn out that the best way to approach this set of issues is to consider how the neurons themselves are connected in simple circuits, which mediate environmental stimuli and the worm's motor responses to those stimuli.

Making and Breaking the Rules

Bargmann and other *C. elegans* investigators (Leon Avery and Horvitz) have suggested some general rules regarding the relation of genes to behavior. Avery, Bargmann, and Horvitz wrote in 1993:

> One way to identify genes that act in the nervous system is by isolating mutants with defective behavior. However, the intrinsic complexity of the nervous system can make the analysis of behavioral mutants difficult. For example, since behaviors are generated by groups of neurons that act in concert, a single genetic defect can affect multiple neurons, a single neuron can affect multiple behaviors and multiple neurons can affect the same behavior. In practice these complexities mean that understanding the effects of a behavioral mutation depends on understanding the neurons that generate and regulate the behavior. (Avery, Bargmann, and Horvitz 1993, 455)

Let us consider these and other *general* principles, which I shall call "rules," governing the relation between genes and behavior that are discernible in the investigations of this extraordinarily well-worked-out simple organism. (I use the term "rules" for these principles, and not laws, because in some cases they admit of exceptions, but hold generally, and I think are default assumptions for all organisms.)[15]

It is probably almost a truism to point out that a single nerve cell (neuron) is the product of many genes, but it is a starting point and might be termed the rule of *many genes—one neuron*. In the quote from Avery et al. immediately above, we encounter several other similar rules. If (1) is the *Many genes → one neuron* rule, then we may term as (2) a *Many neurons → one behavior* rule. Further, it is a generally recognized fact that frequently genes are not specialized to affect just one cell type, but affect many different features and different cell types (Bargmann 1993, 66); this is a widely found phenomenon in genetics termed "pleiotropy." This could be called (3) a *One gene → many neurons* rule. Moreover, in addition to *genetic* pleiotropy, there is the additional fact that any given neuron may play roles in *several different behaviors* (Churchland and Sejnowski call these "multifunctional neurons" [1992, 349]), thus complicating, but not making impossible, an analysis of how behaviors are caused by the neurons. Bargmann cites some minor neurons involved in the chemotaxic response that are also required to regulate the developmental decision between two living states of the worm known as dauer and nondauer development that we need not dwell on in this book (Bargmann

1993, 61).[16] Chalfie et al. (1985) in their investigation of touch circuit neurons (to be discussed later) also point out that one sensory neuron (or one type of sensory neuron) can serve a variety of functions (movement, egg laying, pharyngeal pumping, and possibly the control of other sensory neurons). Similar multifunctional neurons have been identified in the leech and modeled using connectionistic neural nets by Lockery and Sejnowski (1993). This rule might be termed (4) a *One neuron → many behaviors* rule.[17]

There is another consideration raised by Durbin's (1987) observations that apparently strain-identical animals will have somewhat different synaptic connections in their nervous systems. It is not yet clear exactly what is the cause (or causes) of this variation. This may be due to presently hidden genetic differences, an adaptive response to subtly different internal environments in development, or it may possibly be due to partially stochastic processes in development—what Waddington (1957), Stent (1981), and Lewontin (1995) term "developmental noise." For pragmatic reasons, we could aggregate these three processes under the heading of a currently apparent random (more accurately called "stochastic") element, and add this as an additional "rule" that later investigations may further circumscribe. This might best be termed (5) a *Stochastic development → different neural connections* rule. In addition to these five rules, a plasticity or learning/adaptation dimension needs to be considered. Short-term sensory adaptation has been observed to occur in *C. elegans*.

For example, some odor studies on the worm (Sengupta et al. 1993) note that after two hours' exposure to an odorant such as benzaldehyde, the organism loses its ability to be attracted by that substance, though it still is attracted to other odorants. These authors point out that "a more extensive form of behavioral plasticity occurs when animals are starved or crowded. Water soluble chemicals that are strong attractants to naive animals are ignored by crowded, starved animals," and add that "these changes induced by crowding and starvation persist for hours after the worms are separated and fed" (Sengupta et al. 1993, 243; also see Colbert and Bargmann 1995 for additional details). Thus, to the five rules already noted, there is (6) a rule of *Different environments/histories → different behaviors*, which further complicates the predictability of behavior and indicates the impossibility of accounting for behavior from purely genetic information. These six rules are generalizations involving principles of genetic pleiotropy, neuronal multifunctionality, and plasticity (worm plasticity will be revisited again in chapter 5 below). When we turn more explicitly to developmental considerations below, we will have occasion to add to these rules, to represent genetic interactions, but these six will suffice for now. But like virtually any generalization or set of generalizations in biology, they are likely to have exceptions, or near-exceptions. I considered one type of exception involving an almost "one gene—one behavior type" association in the previous section, since it is illustrative of a search for "simplicity" of a sort in behavioral genetics. But the six rules introduced above, I believe, are far more representative of what is found in most studies of genes and behavior in the worm. Rankin seems to disagree with this assessment, writing that "there are pleiotropic mutants in worms as well, but the large number of behavior-specific mutants identified is quite unprecedented, allowing extensive investigation of gene function in behavior" (Rankin 2002, 623).

Finding The Circuit: Connectionist Themes And The Need For Neural Network Modeling In *C. Elegans*

Identification of genes that are "necessary for" specific behaviors, as described in the previous sections, represents one way to indicate the causal role that genes play in generating behaviors. As stressed above, however, genes work in concert and through combinations of neurons synapsing on other neurons and muscle cells to produce those behaviors. To achieve a more *sufficient* explanation of behavior, one that would provide a fuller description of the functional aspect of a nervous system, it is not a *"gene(s) for"* account of behavior that is required; rather it is a *"neural circuit for"* analysis that needs to be provided.[18] *C. elegans* researchers have found a number of such circuits and are in the process of determining additional ones. The chemotaxis circuit, in which Bargmann's and Lockery's laboratories' work will be situated, is one that continues to be further investigated (see Macosko et al. 2009). A fairly detailed neural network for *C. elegans*'s *thermo*taxis behavior was published in 1995 (Mori and Ohshima 1995). Chalfie's and others' work sketched the circuit for *touch sensitivity* in the worm twenty years ago (Chalfie et al. 1985), and though additional features of that circuit have since been identified (see below), much additional work is needed (see Bounoutas and Chalfie 2007 and Chalfie 2009). A closely related (in fact, overlapping) circuit for a tap withdrawal reflex has been characterized in Rankin's laboratory (Wicks and Rankin 1995; Wicks, Roehrig, and Rankin 1996). If this view of the importance of circuits is right, then this suggests we will need a slight modification of our second rule, the one that indicated that *many neurons → one behavior.* The modification is to reflect that the neurons productive of behavior act in an integrated way as part of a well-defined circuit. Thus our rule (2) becomes (2′): many neurons (acting as) *one neural circuit → one type of behavior.*

This chapter cannot present the details of each of these various circuits (but see (Schaffner 2008b for an analysis of a chemosensory circuit). Suffice it for this chapter to show Chalfie et al.'s original touch sensitivity circuit but in a form as recently extended by Goodman. This circuit is somewhat more complex than the original Chalfie et al. (1985) circuit, and contains some neurons and connections that I will not review in detail below (which are available in Goodman 2006 and also in Bounoutas and Chalfie 2007). The circuit diagram (figure 3.4) is, however, more representative of current thinking about mechanosensory connections that produce locomotion in the worm.

It will be helpful to first introduce the identified sensory input neurons and their gross connectivity and placement, and then show in more detail what is known about the more complete circuit including the interneurons and motor neurons that synapse on muscle cells.

The following material on circuits, and the genetics underlying both the development of the circuits and the function of the circuits, is quite technical, and can be skipped on a first reading. The material does underscore a number of

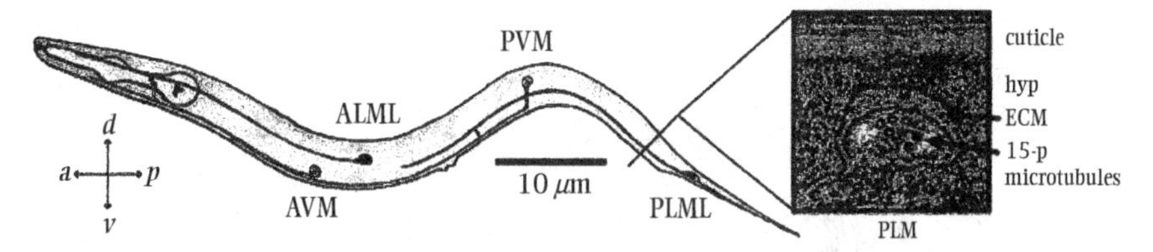

Figure 3.4 Gross and fine morphology of the touch receptor neurons. Position of the cell bodies and neuronal processes of the touch receptor neurons. Only the left side of the animal is shown (left) and electron micrograph of a PLM neuron (right). The electron micrograph is from an L4 animal, courtesy of J. Cueva, Stanford University. SOURCE: Goodman 2006.

the findings expressed as "rules" above and further below, such as pleiotropy and genetic and neuronal multifunctionality, as well as the extraordinary complexity we find in this "simple" organism. The main thrust of these sections is to support a seventh addition to our six rules discussed earlier, one of genetic interaction. That rule is introduced below, where our less technical discussion begins again, and this rule is then followed by a final eighth rule involving environment and learning.

Chalfie's research program examining the worm's touch sensitivity falls into the general area of *sensory mechanotransduction*. Sensory mechanotransduction is defined as "the transformation of mechanical force into electrical signals by specialized cells that communicate with the rest of the nervous system" (Ernstrom and Chalfie 2002, 412). This area includes the sensory modalities of hearing, touch, and balance, with touch itself typically distinguished into nose and head touch, harsh body touch, tap sensitivity, and light or gentle touch—the specific focus of the Chalfie program in the worm.

This neural net, shown in figure 3.5, involves a reflex circuit that generates a movement away from a fine-touch stimulus—typically the stroking of the animal with a thin hair. This circuit has input from identified touch receptor neurons, including ALM, ASH, AVM, PLM, and PVD, then acts through a set of interneurons and motor neurons on muscle cells to generate forward and backward movements (details are in figure 3.5). Goodman (2006) partially describes the general mechanosensitization circuit depicted in figure 3.5 as follows: "A common thread are the four pairs of command interneurons: AVA, AVB, AVD, and PVC. AVA and AVD are needed for backward locomotion, while AVB and PVC appear to mediate forward movement."

It is also necessary to point out that all of these worm circuits are at present strongly underdetermined by any *direct* evidence regarding their polarities and modes of action. As indicated earlier, the electrophysiology of the worm only began to be examined in the mid-1990s, and even in 2006 the neural circuits were noted to be "deduced by integrating the effects of killing individual neurons on touch mediate behaviors with the wiring diagram described in the 'Mind of the Worm' (White's classic work from 1986)" (Goodman 2006). A complete connectivity map at the ganglion level for the worm's nervous system has been constructed and encoded in a computerized database (Achacoso

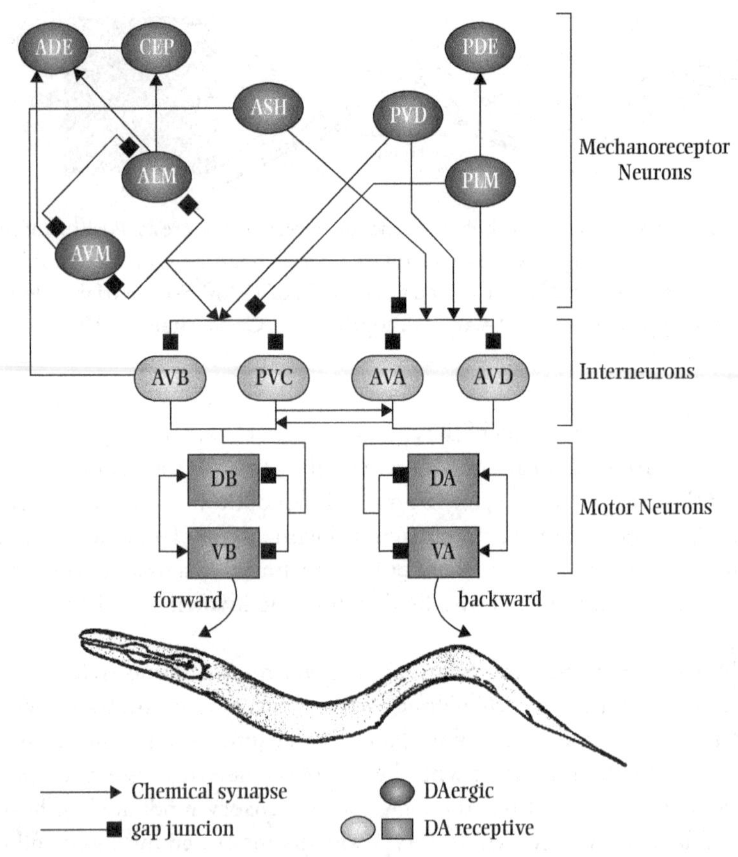

Figure 3.5 Neural circuit that links mechanosensation to locomotion. MRNs are indicated by ovals, interneurons by rounded rectangles, and motor neurons by rectangles. Arrows indicate chemical synapses; squares indicate gap junctions. Dopaminergic MRNs are darker, and cells expressing predicted dopamine receptors have darker borders. Adapted to gray scale.
SOURCE: Goodman 2006.

1992; Cherniak 1994). In 1994, a study done by Raizen and Avery (1994) employed a newly developed method for recording currents, producing what they term "electropharyngeograms," to infer specific neuronal effects. Shawn Lockery's laboratory at the University of Oregon then began a research project to obtain detailed electrophysiological data in *C. elegans.* In the early 1990s, Lockery worked with Terry Sejnowski to develop a sophisticated connectionist model of the bending reflex in the leech (see Churchland and Sejnowski 1992, 339–53, for a general "philosophically-oriented" introduction to this work, and also Lockery and Sejnowski 1993 for an update). Lockery and Sejnowski also constructed what they termed two "preliminary models" to represent chemotaxis in *C. elegans* (see Lockery 1993). In the mid-1990s Lockery developed some special techniques for recording from single neurons in *C. elegans* and, with the assistance of his colleagues Goodman and Ferrée, embarked on a research program to develop and test connectionist models for *C. elegans*. This work continues and has also been further pursued by Goodman at Stanford and Chalfie's laboratory.

In spite of these recent advances noted above, however, current investigations may still need to heavily rely on analogical inferences from the neuronal properties of *C. elegans* cousin *Ascaris* (see Ferrée et al. 1997 and also Wicks et al. 1996). The Wicks et al. paper is from Rankin's laboratory and introduced an intriguing novel methodology for inferring neuronal activity in the tap withdrawal circuit (Wicks et al. 1996) involving a dynamic network simulation.[19] As noted, Lockery's laboratory has both been able to make recordings from individual neurons in *C. elegans* (Lockery and Goodman 1998) and also has modeled those results by utilizing connectionist optimization strategies (specifically a simulated annealing algorithm) that yields predictions in rough agreement with the worm's chemotaxis behavior (Ferrée 1997). Furthermore, the approaches of Rankin's and Lockery's groups both employ general equations for neuronal signal propagation, which though complex, are neuroscientifically well established and introduce powerful physicochemical constraints on the inferred neural networks. Further modeling of *C. elegans* is likely to be done by various investigators involving still higher-fidelity models of neuron compartments and molecular signaling mechanisms. Two promising modeling tools that could be used in such an endeavor are the neural simulation programs NEURON[20] and also GENESIS (Bower 1998).[21] The GENESIS program has evolved into first GENESIS2 and now GENESIS3.[22]

There has been some recent progress in providing more details of neural signaling in the worm. Lockery recently summarized several of these advances, writing: "Solid evidence is beginning to accumulate for synaptic polarities between a small number of key neuron pairs (A → B). Note, however, that all of this evidence concerns only the net effect on B of activating (or inhibiting?) A, which includes all monosynaptic and polysynaptic pathways. Nobody has yet isolated the monosynaptic pathways at any pair of neurons. Most of the evidence is semi-direct, involving inferences from calcium imaging experiments, combined with genetic manipulations. There is now also a small amount of direct evidence obtained from photoactivation of presynaptic neurons combined with patch-clamp recordings from postsynaptic neurons" (Lockery, personal communication, September 20, 2012). Some relevant sources for these recent advances can be found in (Faumont, Lindsay, and Lockery 2012; Lindsay, Thiele, and Lockery 2011; Piggott et al. 2011). Even more recently Lockery added, "I think we can look forward to acceleration in the use of image to infer functional connectivity, now that there is a fairly red-shifted version of channel Rhodopsin, known as Crimson" (Lockery, personal communication, January 15, 2015).

The attentive reader will by now have noted that I have *not* so far said how the circuits are in fact *made*. Accounting for this requires that we look at the *development* of the worm, at least at parts of the developmental sequences that result in specific circuits that account for behaviors.

Making the Circuit: Developmental Neurophysiology and Behavior

GENERATION AND SPECIFICATION: THE LIGHT-TOUCH REFLEX CIRCUIT REDUX

Thus far, I have primarily been characterizing the "functional" aspect of *C. elegans*'s behavioral genetics and have related that to a neural net modeling approach. I now turn to more explicitly developmental considerations, in order to complete the picture, and thus follow Brenner's methodological suggestion that any analysis of behavior also must consider not only the *functioning* of the nervous system, but *also* "the question of the genetic specification of nervous systems." This will also better prepare us to reconsider the developmentalist challenge discussed earlier. Here we look more closely at the light-touch reflex response, a circuit for which was introduced in the previous section.[23] Again, the major original work on this component of the worm's behavioral repertoire was done in Martin Chalfie's lab at Columbia University, where it continues, supplemented by investigations in Goodman's lab at Stanford. Chalfie's group used saturation mutagenesis and identified over 450 touch mutants involving defects in 18 genes, with the mutations being classified into four groups primarily affecting "generation, specification, maintenance, and function" (Chalfie 1995, 179; Zhang, Ma, et al. 2002, 331).

The genes in the groups affecting generation and specification are those centrally involved in the development of the nervous system in *C. elegans*. Three known genes that affect generation of the touch receptors are *lin-32, vab-15,* and *unc-86*. Mutations in these genes result in the cell lineages that normally give rise to the receptors never making the touch receptor cells (Chalfie and Au 1989; Bounoutas and Chalfie 2007). Chalfie summarizes evidence that the protein (UNC-86) that *unc-86* encodes acts as a direct *trans*activator of touch cell differentiation by targeting the *mec-3* gene required for touch cell *specification* (further differentiation) and also *maintenance* (Chalfie 1995, 180). But the process of cell differentiation is probably not a simple linear series of transcription factors successively being activated. Chalfie reported that the downstream *mec-4* and *mec-7* genes—two of the approximately dozen genes that are required for *function*—are under an "accumulative" form of control in which UNC-86 and MEC-3 (the protein that *mec-3* encodes) act in "combinatorial fashion" to activate these downstream genes. (More details concerning the cooperative transcription mechanisms involved in UNC-86 and MEC-3 can be found in Bounoutas and Chalfie 2007).

Differentiation of the touch cells involves some further complications as well. The genes *unc-86* and *mec-3* are expressed in cells that do not become touch receptors, and in seeking to find other regulators, Chalfie and his colleagues have identified seven other genes needed to constrain the number of touch cells to the normal six. These seven include programmed cell death genes *ced-3* and *ced-4* that delete four cells that could become touch receptors. Chalfie Writes, "Together positively and negatively acting genes as well as genes needed for programmed cell death are needed to produce the correct number of touch receptors within the animals (1995, 180). I have mentioned only a few of the *C. elegans* genes that are known to be involved in the differentiation of the touch receptor circuit, and additional ones, their products, and the evidence for them can be found in the articles by Goodman (2006) and Bounoutas and Chalfie (2007). Several more that affect function are described in the following paragraphs, which indicate how highly detailed and fine-structured research on the neurology of worm behavior has become.

As already mentioned, some of the dozen additional genes appear to affect touch cell *function,* in that mutations in these genes do not affect the number or the anatomy of the touch cells, but do result in nonfunctioning touch cells. Two of these (*mec-7* and *12*) encode for proteins found in microtubules specific to the touch cells. Three others (*mec-1, 5,* and *9*) have an effect on the extracellular matrix, perhaps securing a touch receptor to the body wall. Still other genes (*mec-4, 6,* and *10* and relatedly, *mec-2*) seem to be involved in the mechanosensory transduction process, perhaps affecting parts of sodium conductance channels and an adjacent switch mechanism (Chalfie 1995, 181)—more on this in a moment. In 1995 and 1996, Chalfie and his colleagues proposed a model for mechanosensory transduction in the touch receptor neurons that shows how the gene products just mentioned may interact structurally. This original model, which suggested how an ion channel might be made to open, has been replaced by a modified 2007 model in which displacement of the surrounding lipid bilayer produces forces that change the confirmation of the channel (Bounoutas and Chalfie 2007). This follows research by Chalfie's group that has refined the model to provide additional support for the identification of key mechanotransduction *molecules.* A more recent analysis (Chalfie 2009) still favors this model.

Perhaps reflecting a methodological reductionist orientation, such *molecular* identification was referred to as "the 'holy grail' of sensory mechanotransduction" by Ernstrom and Chalfie (2002, 412). It has generally been recognized that the speed at which sensory mechanotransduction occurs requires that the any model of this process—in the worm or other organisms—will involve ion channel complexes. Ion channels and their roles in neuronal cells have been the focus of intense research since Hodgkin and Huxley published their model of action potential generation in the giant squid neuron in the early 1950s, for which they were awarded the Nobel Prize in Physiology or Medicine in 1963. These ion channel models typically "posit a receptor complex containing a channel closely associated with other cellular components that enable it to respond to physical manipulation." (See Ernstrom and Chalfie 2002, 412, for quotation and also references.) These other cellular components are thought to be a critical adjunct to the operation of the channel itself, allowing it to be opened or closed, depending on the circumstances required for normal mechanotransduction. Further work by others on membrane-embedded ion channels has shown that these ion channels have four basic elements or parts: a pore, selectivity filters, sensing modules, and extramembrane domains, which can be combined in various ways to create channels with diverse but distinctive properties (Bass et al. 2002).[24]

Chalfie's lab has used saturation genetic screening in the worm as a means of looking for genetic "lesions"—to use a term suggested by Gunther Stent (1981)—in this case to investigate the roles of elements involved in the worms' ion channel structures. More recent accounts have identified various gain and loss of function mutations in the *mec* degenerin genes (*mec-4, 6,* and *10)* that suggest how the parts of the ion channel may function. For example, *mec-6* codes for a protein that is a single-pass membrane-spanning molecule that can substantially increase the sodium ion currents produced by the MEC-4 and MEC-10 channels. (For still more additional complexity concerning ion channel structure and function in the worm, see Goodman 2006 and Bounoutas and Chalfie 2007, as well as, most recently, Arnadottir et al. 2011.) The take-home lesson here is that various *mec* genes cooperate in forming parts of the degenerin channels, such that mutations in virtually any one will adversely affect ion channel function.

Though considerable progress has been made in identifying the actions of ion channel function genes, much more research will need to be done to clone a few additional involved genes, in order to fully characterize the gene products and modes of action. Chalfie (2009) has discussed the difficulties involved in this research. It is important to realize just how difficult it is to identify the components and interactions of the mechanisms in such entities as ion channels. In spite of the advances to date, the models are based on more elementary systems than eucaryotes (bacteria are intensely studied) and have only been provided for some of the easier-to-study channel types (e.g., potassium). In bacterial crystallography, high enough resolution results can be obtained that the models can be characterized at a very specific molecular level of detail—even at the ion-type level. But for more complex organisms such as the worm, these difficulties and the need to use analogical (homologous) models in this area to infer functions have implications for very strong theses about any simple and sweeping reductionism, which will be the subject of chapter 5.

Also, having suggested several pages ago that Chalfie's approach seemed to reflect a "methodological reductionist orientation," I should note that a most recent publication by Chen and Chalfie (2014) reports data indicating that the worm's "sensory modulation is integrated at multiple levels to produce a single output." The levels are still physicochemical, including environmental vibrations and salt concentration, and produce their effects via "integrin or insulin pathways," yielding a convergence and adapting "touch sensitivity to both mechanical and non-mechanical conditions." This more recent approach is quite consistent with the analysis that I give of partial reduction in chapter 5 below, including the importance of the "pathway" concept.

That all said, the developmental and functional features of the worm considered in this section suggest a seventh addition to our six rules discussed earlier (with the addition of the circuit rule modification added above). This rule notes the frequent effects of genes on other genes, as well as conjoint effects on gene products' function, ranging from possibly simple sequences of activation to the nonlinear combination and complex assemblies noted by Chalfie. Thus we add (7) *One gene → another gene . . . → behavior* (gene interactions, including epistasis and combinatorial and assembly effects).

Environmental Influences: Temperature Sensitivity, Maternal Influence, and the Effects of Social Deprivation on Worm Development

Earlier, the short-term effects of the history and environmental influences on *C. elegans* were briefly mentioned in connection with the worm's behavioral plasticity (learning). Here I consider some of the significant long-term effects of the environment on *C. elegans* that investigators have discovered. I confine my attention to the detection of temperature-sensitive mutants, some of which are seen in "maternal effects" and in the effects of social deprivation in worm development.[25]

Temperature-sensitive mutants are encountered in many species. Briefly, a temperature-sensitive (*ts*) mutant is "an organism or cell carrying an genetically altered protein (or RNA molecule) that performs normally at one temperature but is abnormal at another (usually higher) temperature" (Alberts et al. 1994, G-22). In *C. elegans,* some of the phenotypes are stronger at 25°C than at 15°C. A number of *ts* mutants have been investigated in the worm (they can be found in appendix 4B in Wood 1988). Some of the most interesting are associated with anomalies in sexual differentiation involving work by Hodgkin.

One especially interesting set of temperature-sensitive mutants are those that show a maternal effect. Herman offers the following account of the transforming mutation *tra-3*:

> Homozygous mutants self-progeny of a heterozygous parent are fertile hermaphrodites, but their self-progeny are transformed by *tra-3* to pseudomales. . . . Homozygous *tra-3* hermaphrodites, generated from *tra-3/+* mothers, will produce hermaphrodite progeny, however, when mated with wild-type males; this indicates that zygotic expression of *tra-3 (+)* is sufficient to prevent sexual transformation. (Herman 1988, 27)

Maternal effects in other species, in which the environment apparently has an effect on maternal gene expression in egg production, have been discussed recently (see Pennisi 1996 for references).[26] Similar egg-affecting mutations (but without a demonstrated maternal environmental component) have been investigated in *C. elegans* (see Wood 1988, 256). Maternal, paternal, and parental effects are also discussed more generally by Griffiths and Stotz (2013).

An extensive research program on learning by the worm's tap-withdrawal behavior has been conducted by Catherine Rankin, a prominent *C. elegans* researcher, with a special focus on a simple form of learning known as "habituation" (see Ardiel and Rankin 2008 for a general review). Habituation is the reduction of a behavioral response occurring when a specific stimulus occurs repeatedly.[27]

A perhaps even more dramatic effect of the environment on *C. elegans* is the effect on nervous system structure and behavior of environmental social deprivation during rearing. It is worth providing an extended quotation on this remarkable set of effects. In a review article summarizing the development of the worm's social behavior, Rankin wrote:

> A recent series of experiments examined how behaviour is affected by environmental social deprivation during development. In these experiments, worms were raised either in isolation or in groups from the egg until four days of age. When they were examined on day 4, the isolated worms showed a decreased mechanosensory response to tap, were physically smaller and had delayed development. The tap response uses the tap-touch mechanosensory neurons and the command neurons for forward and backward movement (AVA, AVD, AVB and PVC). Two GFP constructs [green fluorescent protein—reporter molecules signaling gene expression] were used to test whether being raised in isolation led to changes in morphology at the synapse between the sensory neurons and the command interneurons. The constructs allowed visualization of glr-1 receptors on the command interneurons and synaptobrevin—a protein that is associated with synaptic vesicles—on the touch-tap mechanosensory neurons . There was significantly less expression of both GFP constructs in the ventral nerve cords of worms that had been reared in isolation and deprived of mechanosensory stimulation from conspecifics . These results indicate that the amount of synaptobrevin and, by implication, the number of synaptic vesicles in the mechanosensory neuronal terminal, as well as the number of glr-1 glutamate receptors, was determined by the amount of activity in this circuit throughout development. Experience, therefore, can alter both gene expression and the structure of the nervous system. (Rankin 2002, 628)

This information about *C. elegans* temperature sensitivity and social versus solitary rearing effects suggest a final "rule" relating genes and behavior. This is (8) *Environment → gene expression → behavior.* It is of interest to note that *C. elegans* investigators do not use the "norm of reaction" (or "reaction norm") terminology (Bargmann and Chalfie, personal communications), although the *Drosophila* community does so (also see Sarkar 1999, 2003). Nevertheless, the worm community does clearly acknowledge the importance of environmental influences on both genes and developmental processes.

As will be told in more detail (including work on human subjects) in chapter 7 below, the past five years have seen an explosion in studies examining the "epigenetics" of behavior. Epigenetics can be thought of as a set of mechanisms or pathways by which genes are influenced by either the internal or external environments. How best to define the subject of "epigenetics" is still somewhat controversial and will be discussed in chapter 7, but also see the discussion of epigenetics by Griffiths and Stotz (2013). Suffice it to say that the typical mechanisms involve methylation of genes, as well as histones' effects on genes. Two recent articles covering representative studies of the epigenetics of *C. elegans* behaviors can be found in the work of Zheng et al. (2013) and Kelly (2014). These advances are important ones and tilt worm research more in the environmental factors direction, though I believe that they are appropriately covered by Rule 8 introduced above. It is, however, possible that some epigenetic processes are best characterized by Rule 8 in tandem with Rule 7 (on this possibility see Zheng et al. 2013).

A summary of our now eight rules is presented in table 3.1. In my view, these rules, based on empirical investigations in the simplest model organism possessing a nervous system that has been studied in the most detail, should serve as the default assumptions for further studies of the relations of genes and behavior in more complex organisms.[28] These eight rules are generalizations involving principles of genetic pleiotropy, genetic interaction, neuronal multifunctionality, plasticity, and environmental effects, and like virtually any generalization in biology, they are likely to have exceptions, or near-exceptions, but I think these will be rare. That said, neuroscience research continues to advance and disclose ever more complexities and additional forms of plasticity that may require an even more complex set of rules. One intriguing issue is raised by Bargmann and Marder (2013) that involves reprogramming of a neuron's actual neurotransmitters through multiple interactions, suggesting that a rule characterizing the internal dynamics of neurons may need to be considered.

Other Simple Systems

The lessons gleaned above from *C. elegans* and embodied in the eight rules proposed above, also seem to apply to other biological organisms, including *Drosophila*. Earlier I cited Brenner's comparison of *C. elegans* and *Drosophila*, in which Brenner referred to Benzer's early investigation of this organism.

Table 3.1 Some Rules Relating Genes (through Neurons) to Behavior in *C. elegans*

1. *Many genes → one neuron.*
2′. Many neurons (acting as a circuit) → one type of behavior (also there may be overlapping circuits).
3. One gene → many neurons (pleiotropy).
4. One neuron → many behaviors (multifunctional neurons).
5. Stochastic [embryogenetic] development → different neural connections.*
6. Different environments/histories → different behaviors* (learning/plasticity) (short-term environmental influence).
7. One gene → another gene . . . → behavior (gene interactions, including epistasis and combinatorial effects).
8. Environment → gene expression → behavior (long-term environmental influence).

* In prima facie genetically identical (mature) organisms.

Note: The → can be read as "affect(s), cause(s), or lead(s) to."

In the intervening years since Brenner's comment of 1974, Benzer's student, Jeffery C. Hall, and then Hall's own student in turn, Ralph Greenspan, among others, have probed deeply into the cellular and genetic aspects of the behavior of *Drosophila*. Many of these studies have focused on *Drosophila*'s courtship behavior and have resulted in the identification of a number of behavioral mutants, including several types of male mutants termed *fruitless* (*fru*), which court other males as actively as they do females. Hall's investigations into this and other courtship mutants are reviewed in depth in an article in *Science* (Hall 1994), and subsequently Hall collaborated with several other groups in identifying *fru* as "the first gene in a branch of the sex determination hierarchy functioning specifically in the central nervous system (CNS)" (Ryner et al. 1996, 1079). Greenspan's group's work has extended the technique of mosaic creation (Ferveur et al. 1995) and has been summarized by Greenspan (1995) in his still accessible *Scientific American* article.

Greenspan has pursued an extensive research program directed at *Drosophila*'s behavior, including learning, over the course of the past 20 years. Greenspan's book *Fly Pushing* (1997) is a detailed how-to book on the theory and laboratory practice of fly genetics. Greenspan's later, more general text on neurobiology (2007) contains numerous examples of fly biology and behavior, as well as references to other organisms, from jellyfish to honeybees, and Greenspan also notes that "the octopus is famous for its sophistication in solving problems" (145).[29] Finally, the more recent book collaboration with Jonathan Flint and Kenneth Kendler (Flint, Greenspan, and Kendler 2010) on behavioral genetics ranges from simple organisms' behavior to extensive analysis of psychiatric genetics and the common lessons that can be learned regarding the effects of genes on behaviors.

Greenspan's themes resonate well with the eight rules that arise from worm studies. As Greenspan put it, "Behaviors arise from the interactions of vast networks of genes, most of which take part in many different aspects of an organism's biology" (1995, 78; also see his more recent views in Greenspan 2009). To this theme of networks involving multifunctional neurons, Greenspan also added earlier that evidence from *Drosophila's* courtship behavior indicates that both male and female fruit flies "have the ability to modulate their activity in response to one another's reactions," adding:

> In other words, they can learn. Just as the ability to carry out courtship is directed by genes, so too is the ability to learn during the experience. Studies of this phenomena lend further support to the likelihood that behavior is regulated by a myriad of interacting genes, each of which handles diverse responsibilities in the body. (Greenspan 1995, 75–76)

(Drosophila courtship studies continues to be an active field; see Yamamoto, Sato, and Koganezawa 2014 for some additional advances.)

Greenspan has also found additional evidence for multiple and complexly interacting genes in his later work, noting: "One interesting possibility, for which there is increasing evidence, is that the gene networks that subserve any phenotype are much wider ranging than previously suspected. . . . If it is the case that gene networks are wider ranging than we had thought, then it almost certainly must be due to extensive gene interactions" (Greenspan 2009, 5).

If this "network" type of genetic explanation holds for most behaviors, including even more complex organisms than worms and fruit flies, such as mice and humans, it raises barriers both to any simplistic type of genetic explanation, and the prospects of easily achievable medical and psychiatric pharmacological interventions into behaviors. In addition, the eight rules that capture this network perspective may be some of the reasons why it has been so difficult to find single-gene explanations in the area of human behavior. (On this point, also see Schaffner 2008a.) These complexities and the extent they relate to the developmentalist challenge are the topic of the remaining sections of this chapter.

WHICH DEVELOPMENTALIST THEMES DO SIMPLE SYSTEMS SUPPORT?

I have now surveyed a number of aspects of the relations between genes, development, and behavior in simple systems, with a primary focus on *C. elegans*. I now turn to a consideration of the various theses of the developmentalist challenge in the light of this empirical research. For the purposes of this section and the next, it will be useful to distinguish between (1) a set of *five core concepts* found in the developmentalist challenge that apply both to molecular biology and to classical behavioral genetics, and (2) those criticisms more directed at classical behavioral genetics. This section deals with the five core concepts, which I have extracted from the developmentalists' response to the seven deadly sins and four major mistakes of traditional approaches to behavioral genetics. Earlier it was important to cite the *specific sources* of these criticisms, best developed, I think, in those 11 points. Now it is more useful to extract a core of five developmentalist ideas, though the increased specificity of concepts such as heritability will need to be examined later in this book, mainly in chapters 6 and 7.

All of the five core concepts apply to genes, and the last two of these to the relation of genes and environment, in connection with traits or phenotypes. The concepts are those of *parity, nonpreformationism, contextualism, indivisibility,* and *unpredictability.* Basically, *parity* means genes are not special—not "master molecules." *Nonpreformationism* implies that we do not find "traitunculi"—little copies of the traits the genes determine—in the genes. *Contextualism* indicates that genes have little meaning (as "informational molecules") per se, only in context with other genes, and in an environment that is cellular, extracellular, and extra-organismic. *Indivisibility* refers to the thesis that genes and environment cannot be identified by their effects on traits in any separable sense: the effects are a seamless unification—an amalgam. *Unpredictability* means that from total information about genes and environment, we cannot predict an organism's traits: they are, accordingly, emergent. These five concepts seem to capture the core of the 11 points (the sins and mistakes) described above, particularly as they can be addressed in molecular biology and neuroscience. What do successful research programs in the *C. elegans* area tell us about the soundness and applicability of these concepts?

Parity. It would seem that genes *do* have a special set of roles to play in *C. elegans* research (and in biology more generally). Genes are the common bridge between successive generations of organisms, are similar between closely related strains, and display important homolog relations among distantly related species in strongly conserved sequences of DNA. DNA is a *linear* molecule and as such is "one-dimensional and conceptually simple," in contrast to "most other processes in cells [that] result solely from information in the complex three-dimensional surfaces of protein molecules. Perhaps that is why we understand more about genetic mechanisms than about most other biological processes" (Alberts 1994, 223).[30] In an amplification of this point, Morgan points out that if we had a reasonable "metric" for protein tertiary structure, such as we do in terms of DNA nucleotide sequences (A, T, G, and C) then perhaps the developmentalist would have a stronger argument for parity of proteins (Morgan 2001). In addition, there is a general consensus in biology that a simple form of the "central dogma of protein synthesis" is correct. This form of the "central dogma" holds that information flow is *from* DNA (or RNA) *to* protein, and thus that DNA (or RNA) has a special *informational* priority. (I also view this form of the central dogma as a material implementation of the denial of the inheritance of acquired characteristics, though it depends on the fact that the environment, including proteins, does not *systematically* restructure DNA.)[31] (The environment, however, does have significant—if often random—effects on DNA through *epigenetic* interactions with DNA, a topic that I discuss in some detail in chapter 7 below.) Both Oyama (2000) and Lewontin (1993, 1995) would disagree with this informational interpretation, though I hasten to add that neither Oyama nor Lewontin, in disputing an informational priority for DNA, would subscribe to a thesis of the inheritance of acquired characteristics. Genes are also seen as special, because methods have been developed to screen for mutants, map "genes for" traits (as a first approximation), localize those genes, clone them, and test their role as "necessary" elements for a trait using sophisticated molecular deletion and rescue techniques.

No *C. elegans* investigator ever thinks genes act alone—they all recognize the need for the cellular and extracellular supporting environments, and also look for environmental effects on the organism (Rule 6) and on the genes' expression (Rule 8). Naked DNA (or RNA) is not sufficient to produce interesting biological traits, in spite of the significant cell-free systems, and origin of life, experiments that can be accomplished with polynucleotides. Thus, *causally*, genes have parity with other molecules as severally necessary and jointly sufficient conditions (to produce traits), but *epistemically* and *heuristically*, genes do seem to have a primus inter pares status, even in an increasingly "epigenetic" age.[32]

One of the best discussion of the special role that genetics plays in behavior appears in Stent's (1981) analysis, and in his distinction between what he terms the "ideological" and the "instrumental" views of genetics in neuroscience. The *ideological* view represents a complete genetic determinism: sufficient information is in the genes to determine a neural circuit that determines behavior.

(Stent cites early articles by Benzer and Hall and Greenspan that hold this position.) Stent argues against this approach in favor of an *instrumental* analysis, in which "the genetic approach appears as the study of the *differences* in neurological phenotype between animals of various genotypes, without any particular interest (other than *methodologic*) in the concept of genetic specification" (1981, 162; my emphasis).[33] Stent quotes from statements by several authors supporting the instrumentalist approach. Genetic mutations are "an exciting and unique way of *lesioning the system* at the level of cell interactions" (and genetic mutations are "models for inherited disease" both in mice and in humans (Mullen and Herrup 1979, 174; my emphasis). The mutant approach may in addition "provide convenient experimental preparations to which other techniques can be applied" (Pak and Pinto 1976, 398). Though some *C. elegans* investigators *may* hold the stronger "ideological" thesis, the weaker "instrumental" approach clearly is supported by the accounts of work in *C. elegans* described in the present chapter and summarized in Rules 1–8. It should be stressed, however, that this sense of "instrumental" is not to be identified with the philosophical sense of denying realism (see Rosenberg 1994); rather it sees genetics and mutational analysis as powerful heuristics that could point the way toward, but are not equivalent to, a complete realism-based explanation of behavior.[34]

This issue of parity has in recent years taken on a life of its own and been largely disentangled from arguments pro and con developmental systems theory, though not completely, as the frequent reference to parity by Griffiths and Stotz (2013) shows. I will return to some of that parity-related literature in chapter 8 when I consider the issue of the special character of genes more broadly, and where I consider Water's influential defense of the special role that DNA plays as an actual specific difference maker. There I also refer to my agreements and differences with the critique of Waters by Griffiths and Stotz (2013).

Nonpreformationism. No *C. elegans* investigator known to this author seems to think of a DNA sequence as representing a behavioral "trait": the nature of a sequence of DNA nucleotides, that sequence's relation to other sequences or genes (Rule 7), and its relation to protein synthesis machinery, as well as developing heuristics for ascertaining the functional role certain types of protein sequences may play in cells, are part of the training of competent researchers in this domain.

Contextualism. The previous paragraph would seem to indicate that contextualism is accepted by the *C. elegans* community; see again Rule 7, as well as Rules 6 and 8.

Indivisibility. C. elegans researchers do distinguish the causal effects of DNA sequences, operating through protein synthesis and protein folding and assembly, from the effects of other molecules (e.g., pheromones) and conditions (e.g., heat/temperature). The causal schema is a complex web or network, but not an indivisible one from the point of view of analysis. Again, no *C. elegans* investigator ever thinks genes act alone—researchers all recognize the need for the cellular and extracellular supporting environments. It will also be useful here to recall Brenner's perspicacious methodological comment (the full quote is above) that "behavior is the result of a complex and ill-understood set of computations performed by nervous systems and it seems essential to decompose the problem into two: one concerned with the question of the genetic specification of nervous systems and the other with the way nervous systems work to produce behavior" (Brenner 1974, 72). Investigators such as Bargmann's, Lockery's, and Rankin's groups pursue primarily functional investigations: examining the adult neural circuits and components of them to ascertain what the parts are that are necessary components of behaviors, as well as how the parts are connected. Chalfie's group has investigated both functional and developmental issues in connection with touch circuit neurons. Thus these two aspects of a simple organism's behavior can be conceptually teased apart and investigated.

Unpredictability. C. elegans investigators deny a strong unpredictability thesis, but seem to accept the likelihood of some stochasticity, both in development and in analysis of function (Rule 5). They also accept the many-many thesis (see Lewontin on what I termed the second of seven deadly sins of causation above) that follows from Rules 1–4 above. The stochasticity, however, is apparently filterable out for most researchers using populations and averaging, and occasionally employing standard statistical methods (for an example and additional references, see Bargmann et al. 1993, 525–26). *C. elegans* researchers also deny any strong emergentist claim, but that is an issue that deserves its own treatment and a clarification of various senses of emergence, which I present in chapter 5 below.

Thus the assessment of the developmentalist challenge's core concepts is a mixed one. Two of the developmentalists' core concepts, contextualism and nonpreformationism, are accepted as warranted and are utilized. Three are denied: parity (but only heuristically and epistemically), indivisibility, and unpredictability (and a fortiori, emergence). (And again, for a discussion of several senses of "emergence," see chapter 5 below.)

SUMMARY AND CONCLUSION

In this chapter I have examined a simple organism, *C. elegans*, for the light that it can throw on the contentious area of behavioral genetics. Some of the debate about the roles that genes play in phenotypic trait production centered on what I characterized as the developmentalist challenge. In examining that simple system that we know the most about concerning the relations between genes and behavior, it became evident, I believe, that there is no simple type of genetic explanation for behavior: a tangled network with all of the complexities summarized in the eight rules given in table 3.1 is the *default* vision, even in this simplest of model organisms. But several of the claims of the developmentalist challenge were questioned, among them the *indivisibility* and *unpredictability* theses. In addition, a thesis of *emergence*—at least in any strong and "mysterious" sense—was also viewed as not supported. The *parity* thesis was given a somewhat complex reading: genes are special, but are at best "necessary condition explainers," and genes, through the analysis of mutations, offer powerful tools for investigating behavior.

What I have not said much about so far is whether there are good reasons to expect that these findings will generalize to other organisms, including ourselves.

That is an issue I take up in the next chapter, on what makes model organisms *models* for other organisms—an issue I relate to a notion of 'deep homology."

References

Achacoso, Theodore B., and William S. Yamamoto. 1992. *AY's Neuroanatomy of* C. elegans *for Computation*. Boca Raton, FL: CRC Press.

Alberts, Bruce. 1997. "Preface." In *C. elegans II*, edited by Donald L. Riddle, et al., xi–xiv. Plainview, NY: Cold Spring Harbor Laboratory Press.

Alberts, Bruce. 2008. *Molecular Biology of the Cell*. 5th ed. New York: Garland Science.

Altun, Z. F., and D. H. Hall. 2009. Introduction to *WormAtlas*. Accessed November 10, 2012. doi:10.3908/wormatlas.1.1.

Ankeny, R. 1997. "The conqueror worm: An historical and philosophical examination of the use of the nematode *C. elegans* as a model organism." PhD dissertation, History and Philosophy of Science, University of Pittsburgh.

Ardiel, E. L., and C. H. Rankin. 2008. "Behavioral plasticity in the *C. elegans* mechanosensory circuit." *Journal of Neurogenetics* 15: 1–18. doi:905566269 [pii] 10.1080/01677060802298509.

Arnadottir, J., R. O'Hagan, Y. Chen, M. B. Goodman, and M. Chalfie. 2011. "The DEG/ENaC protein MEC-10 regulates the transduction channel complex in *Caenorhabditis elegans* touch receptor neurons." *Journal of Neuroscience* 31 (35): 12695–704. doi:10.1523/JNEUROSCI.4580-10.2011.

Avery, L., C. I. Bargmann, and H. R. Horvitz. 1993. "The *Caenorhabditis elegans* unc-31 gene affects multiple nervous system-controlled functions." *Genetics* 134 (2): 455–64.

Avery, L., and H. R. Horvitz. 1989. "Pharyngeal pumping continues after laser killing of the pharyngeal nervous system of *C. elegans*." *Neuron* 3 (4): 473–85. doi:0896-6273(89)90206-7 [pii].

Bargmann, C. I. 1993. "Genetic and cellular analysis of behavior in *C. elegans*." *Annual Review of Neuroscience* 16: 47–71. doi:10.1146/annurev.ne.16.030193.000403.

Bargmann, C. I., and E. Marder. 2013. "From the connectome to brain function." *Nature Methods* 10 (6): 483–90.

Bass, R. B., P. Strop, M. Barclay, and D. C. Rees. 2002. "Crystal structure of *Escherichia coli* MscS, a voltage-modulated and mechanosensitive channel." *Science* 298 (5598): 1582–87. doi:10.1126/science.1077945 298/5598/1582 [pii].

Bechtel, William, and Robert C. Richardson. 1993. *Discovering Complexity: Decomposition and Localization as Strategies in Scientific Research*. Princeton, NJ: Princeton University Press.

Bechtel, William, and Robert C. Richardson. 2010. *Discovering Complexity: Decomposition and Localization as Strategies in Scientific Research*. MIT Press ed. Cambridge, MA: MIT Press.

Beckner, Morton. 1959. *The Biological Way of Thought*. New York: Columbia University Press.

Bounoutas, A., and M. Chalfie. 2007. "Touch sensitivity in *Caenorhabditis elegans*." *Pflugers Archiv: European Journal of Physiology* 454 (5): 691–702. doi:10.1007/s00424-006-0187-x.

Bower, James M., and David Beeman. 1998. *The Book of GENESIS: Exploring Realistic Neural Models with the GEneral NEural SImulation System*. New York: Springer-Verlag.

Brenner, S. 1974. "The Genetics of *Caenorhabditis elegans*." *Genetics* 77: 71–94.

Brenner, S. 1988. Foreword. In *The Nematode* Caenorhabditis elegans, edited by William Barry Wood, ix–xiii. Cold Spring Harbor, NY: Cold Spring Harbor Laboratory.

Bronfenbrenner, U., and S. J. Ceci. 1994. "Nature-nurture reconceptualized in developmental perspective: A bioecological model." *Psychological Review* 101 (4): 568–86.

Brown, Andrew. 2003. *In the Beginning was the Worm: Finding the Secrets of Life in a Tiny Hermaphrodite*. New York: Columbia University Press.

Chalfie, M. 2009. "Neurosensory mechanotransduction." *Nature Reviews: Molecular Cell Biology* 10 (1): 44–52. doi:nrm2595 [pii] 10.1038/nrm2595.

Chalfie, M., J. E. Sulston, J. G. White, E. Southgate, J. N. Thomson, and S. Brenner. 1985. "The neural circuit for touch sensitivity in *Caenorhabditis elegans*." *Journal of Neuroscience* 5 (4): 956–64.

Chalfie, Martin, and John White. 1988. "The nervous system." In *The Nematode* Caenorhabditis elegans, edited by William Barry Wood, 337–91. Cold Spring Harbor, NY: Cold Spring Harbor Laboratory.

Chen, X., and M. Chalfie. 2014. "Modulation of *C. elegans* touch sensitivity is integrated at multiple levels." *Journal of Neuroscience* 34 (19): 6522–36. doi:10.1523/JNEUROSCI.0022-14.2014.

Cherniak, C. 1994. "Component placement optimization in the brain." *Journal of Neuroscience* 14 (4): 2418–27.

Churchland, Patricia Smith, and Terrence J. Sejnowski. 1992. *The Computational Brain: Computational Neuroscience*. Cambridge, MA: MIT Press.

Colbert, H. A., and C. I. Bargmann. 1995. "Odorant-specific adaptation pathways generate olfactory plasticity in *C. elegans*." *Neuron* 14 (4): 803–12. doi:0896-6273(95)90224-4 [pii].

Cook-Deegan, Robert M. 1994. *The Gene Wars: Science, Politics, and the Human Genome*. New York: Norton.

de Bono, M., and C. I. Bargmann. 1998. "Natural variation in a neuropeptide Y receptor homolog modifies social behavior and food response in *C. elegans*." *Cell* 94 (5): 679–89.

Durbin, R. M. 1987. "Studies on the development and organisation of the nervous system of *Caenorhabditis elegans*." PhD dissertation, Cambridge University.

Ernstrom, G. G., and M. Chalfie. 2002. "Genetics of sensory mechanotransduction." *Annual Review of Genetics* 36: 411–53. doi:10.1146/annurev.genet.36.061802.101708 36/1/411 [pii].

Faumont, S., T. H. Lindsay, and S. R. Lockery. 2012. "Neuronal microcircuits for decision making in *C. elegans*." *Current Opinion in Neurobiology* 22 (4): 580–91. doi:10.1016/j.conb.2012.05.005.

Ferrée, T. C., B. A. Marcotte, and S. R. Lockery. 1997. "Neural network models of chemotaxis in the nematode *Caenorhabditis elegans*." *Advances in Neural Information Processing Systems* 9: 55–61.

Flint, Jonathan, Ralph J. Greenspan, and Kenneth S. Kendler. 2010. *How Genes Influence Behavior*. New York: Oxford University Press.

Fodor, J. 1975. *The Language of Thought*. New York: Crowell.

Gannon, T., and C. Rankin. 1995. "Methods of studying behavioral plasticity in *Caenorhabditis elegans*." In Caenorhabditis elegans: *Modern Biological Analysis of an Organism*, edited by H. F. Epstein and D. C. Shakes, 205–23. San Diego: Academic Press.

Goldhaber, D. 2012. *The Nature-Nurture Debates: Bridging the Gap*. Cambridge: Cambridge University Press.

Goleman, D. 1996. "A set point for happiness." *New York Times*, July 21.

Goodman, M. B. 2006. Mechanosensation. In *WormBook*, edited by C. elegans Research Community. doi:10.1895/wormbook.1.62.1.

Gottesman, I. I. 1997. "Twins: En route to QTLs for cognition." *Science* 276 (5318): 1522–23.

Gottlieb, G. J. 1995. "Some conceptual deficiencies in developmental behavior genetics." *Human Development* 38: 131–41.

Gottlieb, Gilbert. 1992. *Individual Development and Evolution: The Genesis of Novel Behavior.* New York: Oxford University Press.

Gray, Russell. 1994. "Death of the gene: Developmental systems strike back." In *Trees of Life*, edited by P. E. Griffiths, 165–209. Dordrecht: Kluwer.

Greenspan, Ralph J. 1995. "Understanding the genetic construction of behavior." *Scientific American* 272 (4): 72–78.

Greenspan, Ralph J. 1997. *Fly Pushing: The Theory and Practice of* Drosophila *Genetics.* Plainville, NY: Cold Spring Harbor Press.

Greenspan, Ralph J. 2007. *An Introduction to Nervous Systems.* Cold Spring Harbor, NY: Cold Spring Harbor Laboratory Press.

Greenspan, Ralph J. 2009. "Selection, gene interaction, and flexible gene networks." *Cold Spring Harbor Symposium on Quantitative Biology* 74: 131–38. doi:sqb.2009.74.029 [pii] 10.1101/sqb.2009.74.029.

Griffiths, P. E., and K. Stotz. 2013. *Genetics and Philosophy.* New York: Cambridge University Press.

Griffiths, P. E., and R. D. Gray. 1994. "Developmental systems and evolutionary explanation." *Journal of Philosophy* 91: 277–304.

Hacking, Ian. 1999. *The Social Construction of What?* Cambridge, MA: Harvard University Press.

Halder, G., P. Callaerts, and W. J. Gehring. 1995. "Induction of ectopic eyes by targeted expression of the eyeless gene in *Drosophila*." *Science* 267 (5205): 1788–92.

Hamer, D. H., S. Hu, V. L. Magnuson, N. Hu, and A. M. Pattatucci. 1993. "A linkage between DNA markers on the X chromosome and male sexual orientation." *Science* 261 (5119): 321–27.

Hodgkin, J., R. H. Plasterk, and R. H. Waterston. 1995. "The nematode Caenorhabditis elegans and its genome." *Science* 270 (5235): 410–14.

Hull, David L. 1974. *Philosophy of Biological Science.* Englewood Cliffs, NJ: Prentice-Hall.

Johnston, T. 1988. "Developmental explanation and ontogeny of birdsong: Nature/nurture redux." *Behavioral and Brain Science* 11: 617–63.

Kelly, W. G. 2014. "Transgenerational epigenetics in the germline cycle of *Caenorhabditis elegans*." *Epigenetics Chromatin* 7 (1): 6. doi:10.1186/1756-8935-7-6.

Kitcher, Philip. 1984. "1953 and all that: A tale of two sciences." *Philosophical Review* 93: 335–73.

Layzer, D. 1974. "Heritability analyses of IQ scores: Science or numerology?" *Science* 183 (131): 1259–66.

Lehrman, D. 1953. "Critique of Konrad Lorenz's theory of instinctive behavior." *Quarterly Review of Biology* 28 (4): 337–63.

Lehrman, D. 1970. "Semantic and conceptual issues in the nature-nurture problem." In *Development and Evolution of Behavior*, edited by L. Tobach, E. Aronson, D. Lehrman, and J. Rosenblat, 17–50. San Francisco: W.H. Freeman.

Lewontin, Richard C. 1974. "Annotation: The analysis of variance and the analysis of causes." *American Journal of Human Genetics* 26 (3): 400–411.

Lewontin, Richard C. 1992. *Biology as Ideology: The Doctrine of DNA*. New York: HarperPerennial.

Lewontin, Richard C. 1995. *Human Diversity*. New York: Scientific American Library, distributed by W.H. Freeman.

Lindsay, T. H., T. R. Thiele, and S. R. Lockery. 2011. "Optogenetic analysis of synaptic transmission in the central nervous system of the nematode *Caenorhabditis elegans*." *Nature Communications* 2: 306. doi:10.1038/ncomms1304.

Lockery, S. R., and M. B. Goodman. 1998. "Tight-seal whole-cell patch clamping of *Caenorhabditis elegans* neurons." *Methods in Enzymology* 293: 201–17.

Lockery, S. R., and T. J. Sejnowski. 1993. "The computational leech." *Trends in Neurosciences* 16 (7): 283–90.

Lockery, S. R., S. J. Nowlan, and T. J. Sejnowski. 1993. "Modeling chemotaxis in the nematode *C. elegans*." In *Computation and Neural Systems*, edited by F. H. Bower and J. M. Eeckman, 249–54. Boston: Kluwer.

Macosko, E. Z., N. Pokala, E. H. Feinberg, S. H. Chalasani, R. A. Butcher, J. Clardy, and C. I. Bargmann. 2009. "A hub-and-spoke circuit drives pheromone attraction and social behaviour in *C. elegans*." *Nature* 458 (7242): 1171–75. doi:nature07886 [pii] 10.1038/nature07886.

Morgan, G. J. 2001. "Bacteriophage biology and Kenneth Schaffner's rendition of developmentalism." *Biology and Philosophy* 16 (1): 85–92.

Mori, I., and Y. Ohshima. 1995. "Neural regulation of thermotaxis in *Caenorhabditis elegans*." *Nature* 376 (6538): 344–48. doi:10.1038/376344a0.

Mullen, R., and K. Herrup. 1979. "Chimeric analysis of mouse cerebellar mutants." In *Neurogenetics*, edited by Breakefield, 173–96. New York: Elsevier.

Nelkin, Dorothy, and M. Susan Lindee. 2004. *The DNA Mystique: The Gene as a Cultural Icon*. Ann Arbor: University of Michigan Press.

Oyama, Susan. [1985] 2000. *The Ontogeny of Information: Developmental Systems and Evolution*. 2nd ed. Durham, NC: Duke University Press.

Pak, W. L., and L. H. Pinto. 1976. "Genetic approach to the study of the nervous system." *Annual Review of Biophysics and Bioengineering* 5: 397–448. doi:10.1146/annurev.bb.05.060176.002145.

Piggott, B. J., J. Liu, Z. Feng, S. A. Wescott, and X. Z. Xu. 2011. "The neural circuits and synaptic mechanisms underlying motor initiation in *C. elegans*." *Cell* 147 (4): 922–33. doi:10.1016/j.cell.2011.08.053.

Plomin, Robert. 2008. *Behavioral Genetics*. 5th ed. New York: Worth Publishers.

Plomin, Robert, John C. DeFries, Valerie S. Knopik, and Jenae M. Neiderhiser. 2013. *Behavioral Genetics*. 6th ed. New York: Worth Publishers.

Raizen, D. M., and L. Avery. 1994. "Electrical activity and behavior in the pharynx of *Caenorhabditis elegans*." *Neuron* 12 (3): 483–95. doi:0896-6273(94)90207-0 [pii].

Rankin, C. H. 2002. "From gene to identified neuron to behaviour in *Caenorhabditis elegans*." *Nature Reviews Genetics* 3 (8): 622–30.

Riddle, Donald L., et al., eds. 1997. *C. elegans II*. Plainview, NY: Cold Spring Harbor Laboratory Press.

Rosenberg, Alexander. 1985. *The Structure of Biological Science*. New York: Cambridge University Press.

Rosenberg, Alexander. 1994. *Instrumental Biology, or, The Disunity of Science*. Chicago: University of Chicago Press.

Scarr, S. 1995. "Commentary [on Gottlieb]." *Human Development* 38: 154–58.

Schaffner, K. F. 1998. "Genes, behavior, and developmental emergentism: One process, indivisible?" *Philosophy of Science* 65 (June): 209–52.

Schaffner, K. F. 2000. "Behavior at the organismal and molecular levels: The case of *C. elegans*." *Philosophy of Science* 67: s273–s278.

Schaffner, K. F. 2008a. "Etiological models in psychiatry: Reductive and nonreductive." In *Philosophical Issues in Psychiatry*, edited by K. S. Kendler and J. Parnas, 48–90. Baltimore: Johns Hopkins University Press.

Schaffner, K. F. 2008b. "Theories, Models, and Equations in Biology: The Heuristic Search for Emergent Simplifications in Neurobiology." *Philosophy of Science* 75: 1008–21.

Schaffner, K. F, and K. C. Tabb. 2014. "Varieties of Social Constructionism and the Problem of Progress in Psychiatry." In *Philosophical Issues in Psychiatry III*, edited by K. S. Kendler and J. Parnas, 83–115. New York: Oxford University Press.

Sengupta, P., H. A. Colbert, and C. I. Bargmann. 1994. "The *C. elegans* gene odr-7 encodes an olfactory-specific member of the nuclear receptor superfamily." *Cell* 79 (6): 971–80. doi:0092-8674(94)90028-0 [pii].

Sengupta, Piali, Heather Colbert, Bruce Kimmel, Noelle Dwyer, and Cornelia I. Bargmann. 1993. "The cellular and genetic basis of olfactory responses in *Caenorhabditis elegans*." In *The Molecular Basis of Smell and Taste Transduction*, edited by Derek Chadwick, Joan Marsh, and Jamie Goode, 235–50. Chichester, UK: Wiley.

Stent, Gunther S. 1981. "Strength and weakness of the genetic approach to the development of the nervous system." In *Studies in Developmental Neurobiology*, edited by Maxwell Cowan, 288–321. New York: Oxford University Press.

Stent, Gunther S. 1969. *The Coming of the Golden Age: A View of the End of Progress*. Garden City, NY: Published for the American Museum of Natural History by the Natural History Press.

Sulston, J. E., E. Schierenberg, J. G. White, and J. N. Thomson. 1983. "The embryonic cell lineage of the nematode *Caenorhabditis elegans*." *Developmental Biology* 100 (1): 64–119. doi:0012-1606(83)90201-4 [pii].

Thomas, J. H. 1994. "The mind of a worm." *Science* 264 (5166): 1698–99.

Turkheimer, E., H. Goldsmith, and I. I. Gottesman. 1995. "Commentary [on Gottlieb]." *Human Development* 38: 142–53.

van der Weele, Cor. 1995. *Images of Development*. The Hague: CIP—Gegevens Koninklijke Bibliotheek.

von Ehrenstein, Gunter, and Einhard Schierenberg. 1980. "Cell lineages and development of *Caenorhabditis elegans* and other nematodes." In *Nematodes as Biological Models*, vol. 1, *Behavioral and Developmental Models*, edited by Bert Zuckerman, 1–71. New York: Academic Press.

Waddington, C. H. 1957. *The Strategy of the Genes: A Discussion of Some Aspects of Theoretical Biology*. London: Allen & Unwin.

Wade, N. 1998. "Can social behavior of man be glimpsed in a lowly worm?" *New York Times*, September 8.

Wahlsten, D. 1990. "Insensitivity of the analysis of variance to heredity-environment interaction." *Behavioral and Brain Sciences* 13: 109–61.

Waters, C. Kenneth. 1994. "Genes made molecular." *Philosophy of Science* 61: 163–85.

White, J. G., E. Southgate, J. N. Thomson, and S. Brenner. 1986. "The structure of the nervous system of the nematode *Caenorhabditis elegans*." *Philosophical Transactions of the Royal Society of London*, Ser. B, 314: 1–340.

Wicks, S. R., and C. H. Rankin. 1995. "Integration of mechanosensory stimuli in Caenorhabditis elegans." *Journal of Neuroscience* 15 (3, Pt. 2): 2434–44.

Wicks, S. R., C. J. Roehrig, and C. H. Rankin. 1996. "A dynamic network simulation of the nematode tap withdrawal circuit: Predictions concerning synaptic function using behavioral criteria." *Journal of Neuroscience* 16 (12): 4017–31.

Wood, William Barry. 1988. *The Nematode* Caenorhabditis elegans. Cold Spring Harbor, NY: Cold Spring Harbor Laboratory.

Yamamoto, D., K. Sato, and M. Koganezawa. 2014. "Neuroethology of male courtship in *Drosophila*: From the gene to behavior." *Journal of Comparative Physiology A: Neuroethology, Sensory, Neural, and Behavioral Physiology* 200 (4): 251–64. doi:10.1007/s00359-014-0891-5.

Zhang, Y., K. Y. Guo, P. A. Diaz, M. Heo, and R. L. Leibel. 2002. "Determinants of leptin gene expression in fat depots of lean mice." *American Journal of Physiology, Regulatory, Integrative and Comparative Physiology* 282 (1): R226–R234. doi:10.1152/ajpregu.00392.2001.

Zhang, Y., C. Ma, T. Delohery, B. Nasipak, B. C. Foat, A. Bounoutas, H. J. Bussemaker, S. K. Kim, and M. Chalfie. 2002. "Identification of genes expressed in *C. elegans* touch receptor neurons." *Nature* 418 (6895): 331–35. doi:10.1038/nature00891 nature00891 [pii].

Zheng, C., S. Karimzadegan, V. Chiang, and M. Chalfie. 2013. "Histone methylation restrains the expression of subtype-specific genes during terminal neuronal differentiation in *Caenorhabditis elegans*." *PLoS Genetics* 9 (12): e1004017. doi:10.1371/journal.pgen.1004017.

4

What's a Worm Got to Do with It?

Model Organisms and Deep Homology

This chapter both asks and answers the set of issues raised in the chapter title. The views expressed in this chapter were provoked by three sets of comments generated by my original article against the developmentalist challenge, which appeared in 1998 (Gilbert 1998; Griffiths 1998; Wimsatt 1998), but the chapter introduces some newer philosophical themes that will be helpful as we look at more complex genetic systems, including humans. One theme is the way in which biological knowledge in general is structured. The contrast is with physics, where the main "explainers" are theories viewed as collections of a small number of interrelated universal statements (e.g., Newton's three or four laws, Maxwell's four or six equations, or the three-axiom version of quantum mechanics), a notion I have called the "Euclidean Ideal" (Schaffner 1986). In biology, with a few (important) exceptions, biological knowledge and biological explanations seem to be framed around a few exemplar subsystems in specific organisms, and perhaps even in specific strains. Examples include the Jacob-Monod *lac* operon in *E. coli* K12, Mendel's pea "factors," Morgan's white-eyed male mutant in *Drosophila*, and Kandel's *Aplysia* model for learning in neurobiology. These exemplar subsystems are used as (interlevel) *prototypes* to organize information about other similar (overlapping) models to which they are related more by analogical reasoning than by deductive elaboration. *C. elegans*, as presented in the previous chapter, is a source of a number of these prototypes, and the social versus solitary feeding exemplar from 1998 will be shown to have become more complex and relevant for a discussion of partial reduction in chapter 5.

Another recurrent theme of this chapter can be reintroduced by recalling the views of Bruce Alberts from chapter 3. Alberts is a noted biologist, former editor in chief of *Science*, former president of the US National Academy of Science, and first author of a highly influential textbook, *The Molecular Biology of the Cell* (Alberts 2008). Alberts wrote: "We can say with confidence that the fastest and most efficient way of acquiring an understanding of ourselves is to devote an enormous effort trying to understand these and other, relatively "simple" [model] organisms (1997, xiv). The idea here is that we can generalize from some specific and simpler prototypes, but the next obvious question is, on what grounds? The answer is to appeal to homologies and conserved genetic features, as is developed further below.

Looking at the issues raised by Gilbert (1998), Griffiths (1998), and Wimsatt (1998) leads to important questions not only within the philosophy of biology, but also for current biomedical research programs more generally. Two of those papers (by Gilbert and Jorgensen and Wimsatt) questioned the biological utility of a focus on *C. elegans* and other "simple systems" or "model organisms." This criticism presumably also applies to highly directed research by biomedical scientists on *Drosophila* and the mouse (*Mus*), and probably also to investigations on *E. coli*, yeast, the plant, *Arabidopsis*, the zebrafish, and primates, and even more recently, the chicken and the honeybee.[1] Griffiths and Knight's original paper (1998) did not dispute my essay's focus on *C. elegans*, but did question one of my main conclusions regarding the heuristic priority of DNA-based analyses. This issue will be raised again in a more general context in chapter 8 below, but it should again be noted (as in chapter 3) that in 2013 Griffiths and Stotz published an extensive analysis of the parity issue.

In what follows, I will first discuss two general strategies of research in the biomedical sciences and provide a twin philosophical rationale for the simple-systems approach. More specific replies to the comments are then offered.

TWO STRATEGIES OF RESEARCH IN THE BIOMEDICAL SCIENCES

To a first approximation, inquiry in the biological sciences can take two contrasting approaches, that we might term (1) narrow but deep (*ND*), and (2) broad but shallow (*BS*). The ideal approach would be both deep *and* broad, but for practical reasons, such an ideal is most likely to be a "long run" strategy. The first type, *ND* analysis, mobilizes the biomedical community's resources around a few prototype organisms, some of which are "simple" (phage λ, *E. coli*, and *C. elegans*) and some that are more complex (*Mus* and various primates). *BS* inquiries, on the other hand, highlight biological variation and diversity, both within and among organisms and environments, and urge we attend to many different species simultaneously in biological investigations. Both Gilbert and Jorgensen and Wimsatt in their original comments *seemed* to strongly favor the *BS* approach, arguing (or implying) that the *ND* approach is misleading and gives a false picture of biological complexity, particularly as involves organism and species plasticity. In my original article, and in the last chapter, I argued for a *ND* "model system" approach, both in biology and in the philosophy of biology, but I did not discuss a background rationale, and that now seems useful to do in the light of the thrust of many of the provocative commentaries. In chapter 3 and again earlier in this chapter I referred to the views of Bruce Alberts. I think Albert's views on model organisms and his way of thinking are representative of a broad consensus among contemporary biological researchers, but it will be useful to examine *why* a "model organism" approach is so widely accepted in contemporary biology.[2]

First, I want to point out that the structure of biological knowledge, from both epistemic and logic-of-explanation perspectives, is organized differently from what we find in standard accounts of the physical sciences.[3] Details of this view are developed in detail in chapter 3 of my 1993 book, but the following brief summary will provide the gist of the analysis as it relates to model organisms. This account will also serve as important background to the discussion of reduction in chapter 5.

As noted at the beginning of this chapter, in physics the main "explainers" are theories viewed as collections of a small number of interrelated universal statements (e.g., Newton's three or four laws, Maxwell's four or six equations, or the three-axiom version of quantum mechanics), whereas in biology the surrogate for such theories is a collection of molecular models functioning as prototypes or exemplars. I would speculate, with that view as a backdrop, that the very existence of confusing *diversity and variation* in biological organisms and processes forces a focus toward simplifying prototypes that can be used to convey information, and laboratory techniques, in a less bewildering way. On such a view—one that mirrors Alberts's account, but more epistemically and as related to the logic of explanation—model systems are a powerful heuristic for biological research.

However, such prototypes of necessity *need* to be *representative*—to connect analogically to other prototypes—if they are to do their job(s) as surrogates for what theories do in other sciences, since they putatively function as the (nearly) common element relating a variety of organisms or biological processes. Though the organisms are typically chosen for partly idiosyncratic historical reasons, there are some *general* reasons for "model" organism choice, including short life cycle, ease of stock maintenance, and experimental tractability (see Ankeny 1997; Bolker 1997 and more recently Ankeny 2011). The *hope*, of course, is that such chosen organisms and subsystems, when probed deeply and broadly enough, will disclose "widely conserved" mechanisms of general applicability, sometimes called "high connectivity models" or "deep homologies." (Scientists also hope that the mechanisms that are revealed are *"simple"*—in the sense that they can be understood by humans [and not just computers], reasoned about, rigorously tested, and usefully applied.)[4] Interestingly, the test for the conservation is initially in *other model organisms*, with *Homo sapiens* being the ultimate pragmatic application, typically in the medical context.[5]

The terms "high connectivity" and "deep homology" are worth some brief additional consideration. In their overview of the utility of *C. elegans* as a model organism, Riddle et al. (1997) write that *C. elegans*, in addition to yeast, *Drosophila*, and a few other model systems, is a "high connectivity model," using a term initially introduced by Morowitz's 1985 report on models in biomedical research. In such models, "knowledge gained in one area of research ultimately 'connects' with research in other areas. This connectivity both expands and reinforces understanding and speeds research progress" (1997, 6). Riddle et al. cite "parallels between the development of the body plan in nematodes, flies, and mice" (6) and also the similarity of proteins used for programmed cell death in both nematodes and humans.

"Deep homology" identifies the same set of near "universals" but adds the beginnings of an explanatory dimension. The term "deep homology" refers to widespread conservation, by descent, of gene sequences, together with identification of *functional similarity* across many organism types (see Fitch and Thomas 1997, 830). In the roughly 15 years since the ancestor of the present chapter was written, the interest in and the number of articles on this subject has grown extensively, and has even resulted a method for systematic discovery of human disease models using orthologous phenotypes (McGary et al. 2010). Molecular biologists frequently search gene databanks for homologous genes as part of their fundamental inquiries into gene function (see the discussion in chapter 3). Though the concept of "homology" admits of a number of different senses (see Hall 1994), the core idea seems to involve some intuition of "sameness."[6] Also compare Brigandt's view (2003), who analyzes a number of different senses depending on the subfield in biology. Also see the special issue of the journal *Biology and Philosophy* on homology, and the introduction and overview of the articles in that issue (Brigandt and Griffiths 2007). Hillis argues that "molecular biologists may have done more to confound the meaning of the term homology than have any other group of scientists," adding that "in many circles of molecular biologists, homology has come to mean 'similarity': a simple quantifiable relationship, for which the word similarity adequately suffices" (1994, 339–40). Finally, for a very recent in-depth analysis of homology, see the book-length contribution by Wagner (2014).

My general view is what all this tells us is that molecular biology has *extended* the original concept of homology to include elements of its initial contrast concept, *analogy, because* of the power of widely conserved genes to identify similar *functions*. Whether this account of deep homology is a useful one is in part an empirical matter, which I address further below.

What is wanted is a further *explanation* of the power of model systems including the features of high connectivity and deep homology. I think that it is here that Wimsatt's concept of "generative entrenchment" can, as he suggests, play an important role. Other like concepts are Waddington's "canalization" (Waddington 1957), Kauffman's self-organizing properties (Kauffman 1993), and Wagner's "generative" and "morphological" constraints (Wagner 1994). All of these appear to seek ways in which genetic and epigenetic factors restrict variation and make some nearly universal mechanisms more likely.

There is a separate but related issue regarding model systems raised by Gilbert and Jorgensen's comment on my earlier version of the last chapter: "The very richness of life that the Developmentalist Challenge claims has been hunted down and eliminated from *C. elegans* research" (1998, 259). Actually, if this was the intent of the "hunters," they failed. It is ironic that such a highly inbred and simple system exemplifies many of the developmentalist systems theorist's (or DST) principles—a result that, as Wimsatt mentions (1998, 273), was surprising to me. But I take a different message from the *C. elegans* community's attempt to restrict variation in the organism—a message that does not relate to DST at all. To me, models are not only intended to be *representative* prototypes, but also to be "idealized" in the sense of sharpened and more clearly delineated. The value of sharpened, simplified idealizations is a lesson that the physical sciences can still teach us, and it is evident in the idealizations found in simple subsystems in biology as well, such as in the original operon model for *E. coli* K12 (Jacob and Monod 1961). Once simple prototypes are preliminarily identified (for example, the so-called wild type of subsystem and some key mutants), *then* variations (often in the form of a *spectrum* of mutants) are sought (or re-examined) to elucidate the operation of simple mechanisms. (For examples of this strategy in the operon area see my 1993, 76–82.) In point of fact, Griffiths and Knight (1998) themselves, as spokespersons for DST, do not in their comments object to my focus on the worm.

The bottom line it seems to me, after these philosophical preliminaries have established a rationale and a context, is whether the model systems approach can be supported by empirical facts, and it is to this issue that I now turn.

EMPIRICAL SUPPORT FOR THE *ND* APPROACH IN *C. ELEGANS* THAT POINTS TOWARD VARIATION AND RICHNESS

Phylogeny and Strain and Species Variation

In his comments on my original article, Wimsatt cited Bolker and Raff's 1997 criticism of the model systems approach as subscribing to a "great chain of being" myth. But in actuality, the place of the worm in phylogeny is the subject of *empirical* investigation, as well as some controversy.[7] In 1997, Fitch and Thomas offered three cladistic possibilities as represented in figure 4.1 (a, b, c), and argued that the data available to them supported 1a. If so, this would license the use of the worm as a predictor for humans (Fitch and Thomas 1997, 817), in agreement with Wimsatt's suggestion that model systems proponents need to take *evolution* into account. But about that time, Aguinaldo et al. (1997) argued that their data analysis of ribosomal 18S rDNA supported something more like figure 1b, and stated that "it had been assumed that developmental mechanisms common to *Caenorhabditis* and to *Drosophila* originated before the protostome-deuterostome divergence and hence should also be found in *Homo sapiens*. Our results imply that mechanisms found in both nematodes and fruit flies will not necessarily be found in humans" (1997, 492).[8] At that time, Fitch (personal communication) believed that 18s rDNA evolved too rapidly in nematodes to be of much use as a phylogenetic instrument, particularly as regards the deep divergences to which it has been applied by Aguinaldo et al. (1997).

This debate has continued, with the two main views now having been named the "Coelomata" and "Ecdysozoa" positions. The former holds to the old 1a view, and the latter to the 1b figure. ("Coelomata" derives from importance attributed to the coelum, a body cavity, of which *C. elegans* has only a primitive variant. "Ecdysozoa" refers to molting and classifies the nematodes together with the arthopods in its cladistic account.) Though the evidence is still not fully in, the Coelomata view does seem to be better supported by the preponderance of

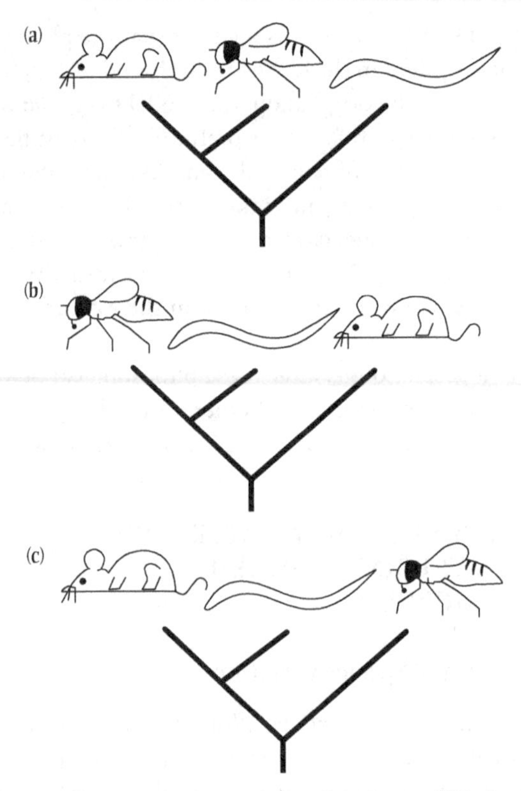

Figure 4.1 Three possibilities of relationships of *C. elegans* ("the" nematode) to *Drosophila melanogaster* ("the" arthropod) and *Mus musculus* ("the" vertebrate): (a) nematodes as an outgroup taxon to vertebrates and arthropods; (b) nematodes more closely related to arthropods than to vertebrates; (c) nematodes more closely related to vertebrates than to arthropods. Obviously, these hypotheses (like the model systems themselves) are overly simplistic representations for enormously diverse phylogenetic groups. Although present data favor a or b, robustly distinguishing which hypothesis is most likely depends on the accumulation of much more data.
SOURCE: Legend from Fitch and Thomas 1997, 818.

the evidence in more recent times, but there are dissenters. A sense of where this contested issue is at present is summarized in a portion of the abstract from Holton and Pisani:

> Solving the phylogeny of the animals with bilateral symmetry has proven difficult. Morphological studies have suggested a variety of alternative hypotheses, of which, Hyman's Coelomata hypothesis has become the most established. Studies based on 18S rRNA have failed to endorse Coelomata, supporting instead the rearrangement of the protostomes into two new clades: the Lophotrochozoa (including, e.g., the molluscs and the annelids) and the Ecdysozoa (including the Panarthropoda and most pseudocoelomates, such as the nematodes and priapulids). Support for this new animal

> phylogeny has been attained from expressed sequence tag studies, although these generally have a limited gene sampling. In contrast, deep genomic-scale analyses have often supported Coelomata. However, these studies are problematic due to their limited taxonomic sampling, which could exacerbate tree reconstruction artifacts. (2010, 310)

The relation of *C. elegans* mechanisms to developmental mechanisms in the mouse, which argues against the Ecdysozoa position, will be cited below. But the general point to be made here is that the relationships among model systems is not viewed in terms of some philosophical great chain of being—although these investigations are termed "tree of life" studies—but is a matter that is in part an empirical investigation, in which awareness of alternative possibilities is actively pursued.

Another point at which empirical results can help us sort out the value of the *ND* and *BS* approaches is in connection with experimental testing of putative "deep homologies," such as mentioned in the previous section. Perhaps the most interesting area in *C. elegans* deep homology involves developmental and pattern-forming mechanisms. There is evidence that *C. elegans* uses the same mechanism as do *Drosophila* and other more complex organisms for cell-fate specification (Fitch and Thomas 1997, 830; Maduro 2010). Fitch and Thomas suggest that a "tool kit" of "basic regulatory mechanisms" is used by all species evolutionarily proximal to *C. elegans*. They also state that "the most striking evidence for the conservation of function is the ability of a molecule from one species to function in a different species as its endogenous homologue," and cite the ability of human *blc-2* to regulate cell death in *C. elegans*, and also the interspecies substitutability of the *hox* genes between worms and fruit flies (1997, 831). Maduro cites support for a similar view using the newer molecular biological techniques developed and further refined in the twenty-first century, including RNAi knockdown.

Thus, in contrast to Wimsatt's (1998) (and Gilbert and Jorgensen's [1998]) originally expressed concerns about the nonrepresentativeness of *C. elegans*, these results actually suggest *C. elegans* could *help* Wimsatt to confirm his belief that the *hox* gene example is paradigmatic of generative entrenchment.

Finally, readers should be aware of the preliminary inquiries currently underway to investigate strain and environmental variations in *C. elegans*, and similar investigations in the nematodes in general. In 1997 Fitch and Thomas pointed out that there were 17 *Caenorhabditis* species then known, though only 4 were currently easily available, and that initial comparisons suggest considerable genetic variation between species in spite of morphological similarity. About that time, Fitch (1997) also issued a request to the *C. elegans* community for information about available non-*elegans* species so as to share them more broadly. By 1997, more than 22 different *C. elegans* species had been identified, 10 of which were in culture and were under further investigation to examine genetic diversity and phylogeny (see originally Fitch and Thomas 1997, 825–30). The species number had grown to 24 as of June 2010 (Fitch and Felix, conference presentation). There are 65 wild isolates kept at the *Caenorhabditis* stock center (see Kiontke and Sudhaus 2006). The number of strains created in various laboratories is estimated by Maduro (personal communication, July 28, 2010) to be huge, perhaps hundreds of thousands created by mutation or transgenesis over the years. Thus *C. elegans* researchers are not solely interested in hunting down and eliminating variation, but wish to use the information generated by detailed investigation of the N2 Bristol strain for comparison with other organisms, both reasonably closely related, as well as phylogenetically more distant classes.

Genes, Neural Plasticity, and Behavior

A related concern about the elimination of variation in model systems, but in studying the relations of genetics to behavior, is expressed at several points by Gilbert and Jorgensen's original set of comments (1998, 261, 263). I disagree with Gilbert and Jorgensen's claim that worm behavioral geneticists can only study traits that are present and absent (1998, 261), and thus will miss any subtle variation. Behavioral assays can identify a number of fairly complex behaviors (see chapter 3 above and more recently Ardiel and Rankin 2008, 2010). Furthermore, subtly different chemotactic and thermotactic behaviors in the worm are modified by experience, and these effects have been extensively studied and related to a variety of genes in *C. elegans* (Bargmann and Mori 1997). Also, various protocols to examine associative learning have and continue to be explored in *C. elegans*, as Gilbert and Jorgensen know well (see Jorgensen and Rankin 1997, 787–90). Interestingly, Jorgensen *himself*, writing with Rankin on neural plasticity in the worm, states that that "once well-defined learning paradigms become established in *C. elegans*, genetic analysis of this organism may resolve several long-standing issues in our studies of learning and memory" (Jorgensen and Rankin 1997, 790). Jorgensen and Rankin also write that "in the future, genetic analyses of the mechanisms involved in the long- and short-term memory phases of habituation should lead to additional insights into the similarities and differences between memory processes in this simple nervous system [*C. elegans*] and in more complex organisms such as *Drosophila, Aplysia,* and mammals" (787).

In general I read Gilbert and Jorgensen's comments as largely in agreement with the main messages of my chapter 3 and the original 1998 essay, though we occasionally use somewhat different language in our formulations of the issues.[9] In addition to issues about model organisms in general already reviewed above, Gilbert and Jorgensen (1998) asked whether worm research can say anything useful about interesting research on *human* cognition, and argued that it essentially cannot. My answer to this question, as indicated in my original paper (also see chapter 3), was more positive. Part of the answer is contained in the comments about deep homology made earlier, but part of the value of *C. elegans* studies is also *methodological*. Worm studies will not tell us anything about consciousness or intention or agency (Gilbert and Jorgensen 1998, §6), for complexities exist in humans not found in simpler organisms. And some of these complexities will include the interaction of linguistic and sociocultural factors with biological developmental processes (see Deacon 1997 for an elaboration of such a view). But some fundamental mechanisms, including simplified analogues of real biological neural nets, are emerging in *C. elegans* studies (see my references to Lockery's research program in my previous chapter and also Wicks, Roehrig, and Rankin 1996). Furthermore, it appears that the molecules and mechanisms of neurogenesis are phylogenetically conserved among worms, flies, and vertebrates" (see Fitch and Thomas 1997, 831, for older supporting references, and also Maduro 2010 for more recent information and references). For even more recent support of this view, two articles appeared in *Science* magazine reporting on a functional vasopressin- and oxytocin-like signaling system in the worm (Beets et al. 2012; Garrison et al. 2012). In an analysis of these articles, the commentary author notes that "even at the highest levels of coordinating fundamental and complex behaviors, the same neural mechanisms are at work in worms and humans" (Emmons 2012, 475).

The types of influences that genes can have on human behavior have been previewed in chapters 1 and 2 above and will be taken up again in later chapters in this book. Where Gilbert and Jorgensen and I may disagree on some useful extrapolations to humans lies in the area of psychiatric genetics. There was a useful homologue available in *C. elegans* that relates to Alzheimer's disease (Levitan and Greenwald 1995), and more recently a transgenic *C. elegans* Alzheimer's disease model has been investigated (Hassan et al. 2009). And the model system and transgenic approaches are only in the early stages of application to such other serious diseases as schizophrenia, where a potentially suggestive mouse model was identified (the *dishevelled* gene) back in 1997 that also had some ties back to *Drosophila* (Lijam et al. 1997). A recent report of a knockdown of a schizophrenia susceptibility gene (*DISC 1*) in mice (Niwa et al. 2010) suggests these model systems will tell us more about human mental illness (*DISC 1* is mentioned again briefly in chapter 7 below). A worm is not a human, but worm studies, as well as research on other animals, may offer important lessons about human psychopathology if yoked to other model systems.

DEVELOPMENTALISM: PROS AND CONS

Interestingly, Griffiths and Knight closed their comments on the earlier version of the last chapter by anointing me as a developmentalist (258). It should be clear from my previous chapter that I was at the time of our exchanges (and still am) sympathetic to *some* of those themes. In particular, I think that the *DS* theorists have provided important criticisms of philosophers'—and biologists'—overly restricted attention to genetics and the role(s) of DNA. Griffiths and Knight did criticize me for one of my exceptions to the developmentalist creed, however, namely my views about the heuristic and epistemic priority of DNA and its informational content. In response, I would reiterate my comments in my previous chapter, which I think are powerful arguments, but would also add that I have already acknowledged the importance of epigenetic inheritance factors, albeit briefly, both where I touch on the importance of "maternal effects" as well as on when reporting on recent (2013–2014) studies of epigenetics in the worm by Chalfie and by Kelly. (In addition, in chapter 7 below, I will discuss in extenso some recent epigenetic research programs in schizophrenia.)

The sense of "information" I intend is related to the question what molecule types account best for individual and species *differences*.[10] Though it is possible that we may find extensive variations in molecules other than DNA that constitute the set of severally necessary and jointly sufficient conditions for embryogenetic differences, at present the *variations* among individuals and species seem largely resident in the DNA, though this will ultimately be an empirical matter as further progress is made on maternal enzymes and in epigenetics. Of course, as Griffiths and Knight suggested, this may be because this is where we have *looked for* causes—there is (as yet) no large-scale Human *Phenome* Project that might focus on molecules other than DNA and on higher-order properties, as well as on the environment.[11] Sarkar, whom Griffiths and Knight cite as critical of the informational concept of the gene, also seems to be struggling to capture this sense of information, though he does not find any plausible, developed account that characterizes it (see especially Sarkar 1996, 222–23, for his possible alternatives to DNA-based biology). (In recent years there has been an increased interest in the notion of biological information more generally—on this topic see Godfrey-Smith 2008 and also Bogen 2011 and Griffiths and Stotz 2013.)

Developmental systems theory may be one of those possible alternatives (though not one that Sarkar mentions or seems to favor). The danger of DST in its present form, as I see it, is that it gives too much to "context" (see Griffiths and Knight 1998, 256–57) and needs to formulate its categories of interactions more clearly, but that is not a point I can elaborate in this chapter. It is not helpful to assert that everything interacts with everything else, but that could be a problem for DST unless it provides us with some form of prioritized ontology.[12] It would be interesting to see what a current NIH-ish DST research grant program announcement and request for proposals would look like, and I originally encouraged Griffiths and Knight, and other developmentalists, to consider proposing one, and to an extent they did so (Oyama, Griffiths, and Gray 2001). Such a program might end up emphasizing epigenetics, or perhaps even more biological system theoretic approaches. (For more recent comments on the gene concept with DST as a backdrop by leaders in the DST area, see Griffiths and Stotz 2013).

CONCLUSIONS

This set of excellent comments discussed above by some leading biologists and philosophers of biology originally pushed me to consider a number of foundational issues that were only implicit in the original version of chapter 3. There is still much more to be done. For example, Gilbert and Jorgensen asked, "Why do genes sell?" They also asked, to whom do I need to tell this story about DST and the extent to which it is supported by *C. elegans* results? More colorfully, they asked on whose door I should nail these eight theses, suggesting behavioral geneticists and journalists need to know. I would agree that those groups are prime audiences, and to an extent, this book may address some of those groups. Other papers and projects that were completed in the past 10 years will likely be the source of messages to these and other policymaking groups.[13]

My previous chapter, however, and this rejoinder to the set of three original comments on the ancestor of chapter 3, are still mainly directed at the philosophy-of-science community, and at those scientists who have an interest in philosophical issues. It is my hope that this interaction will raise the discussion of genetics and behavior to a new level in terms of philosophical clarity married to scientific detail. Certainly the commentators on my original paper accomplished that, and I hope I have in this revised chapter as well.

References

Alberts, Bruce. 1997. "Preface." In *C. elegans II*, edited by Donald L. Riddle, et al., xi–xiv. Plainview, NY: Cold Spring Harbor Laboratory Press.

Alberts, Bruce. 2008. *Molecular Biology of the Cell*. 5th ed. New York: Garland Science.

Ankeny, R. A. 1997. "The conqueror worm: An historical and philosophical examination of the use of the nematode *C. elegans* as a model organism." PhD dissertation, History and Philosophy of Science, University of Pittsburgh.

Ankeny, R. A., and S. Leonelli. 2011. "What's so special about model organisms?" *Studies in History and Philosophy of Science* 42 (2): 313–23.

Ardiel, E. L., and C. H. Rankin. 2008. "Behavioral plasticity in the *C. elegans* mechanosensory circuit." *Journal of Neurogenetics* 15: 1–18. doi:905566269 [pii] 10.1080/01677060802298509.

Ardiel, E. L., and C. H. Rankin. 2010. "An elegant mind: Learning and memory in *Caenorhabditis elegans*." *Learning and Memory* 17 (4): 191–201. doi:17/4/191 [pii] 10.1101/lm.960510.

Bargmann, C. I., and I. Mori. 1997. "C. elegans II." In *Cold Spring Harbor Monograph Series*, edited by D. et al. Riddle, 717–38. Plainview, NY: Cold Spring Harbor Laboratory Press.

Beets, I., T. Janssen, E. Meelkop, L. Temmerman, N. Suetens, S. Rademakers, G. Jansen, and L. Schoofs. 2012. "Vasopressin/oxytocin-related signaling regulates gustatory associative learning in *C. elegans*." *Science* 338 (6106): 543–45. doi:10.1126/science.1226860.

Bilder, R. M., F. W. Sabb, T. D. Cannon, E. D. London, J. D. Jentsch, D. S. Parker, R. A. Poldrack, C. Evans, and N. B. Freimer. 2009. "Phenomics: The systematic study of phenotypes on a genome-wide scale." *Neuroscience* 164 (1): 30–42. doi:S0306-4522(09)00048-7 [pii] 10.1016/j.neuroscience.2009.01.027.

Bogen, J., and P. Machamer. 2011. "Mechanistic information and causal continuity." In *Causality in the Sciences*, edited by P. K. Illari, F. Russo, and J. Williamson, 845–64. New York: Oxford University Press.

Bolker, J. A., and R. A. Raff. 1997. "Beyond worms, flies, and mice: It's time to widen the scope of developmental biology." *Journal of NIH Research* 9: 35–39.

Brigandt, Ingo. 2003. "Homology in comparative, molecular, and evolutionary developmental biology: The radiation of a concept." *Journal of Experimental Zoology B: Molecular and Developmental Evolution* 299 (1): 9–17. doi:10.1002/jez.b.36.

Brigandt, Ingo, and Paul Edmund Griffiths. 2007. "The importance of homology for biology and philosophy." *Biology and Philosophy* 22 (5): 633–41.

Deacon, Terrence William. 1997. *The Symbolic Species: The Co-evolution of Language and the Brain*. New York: Norton.

Emmons, S. W. 2012. "Neuroscience. The mood of a worm." *Science* 338 (6106): 475–76. doi:10.1126/science.1230251.

Fitch, D. H. A., and W. K. Thomas. 1997. "Evolution." In C. elegans *II*, edited by Donald L. Riddle, Thomas Blumenthal, Barbara J. Meyer, and James R. Priess, 815–50. Plainview, NY: Cold Spring Harbor Laboratory Press.

Freimer, N., and C. Sabatti. 2003. "The human phenome project." *Nature Genetics* 34 (1): 15–21. doi:10.1038/ng0503-15.

Garrison, J. L., E. Z. Macosko, S. Bernstein, N. Pokala, D. R. Albrecht, and C. I. Bargmann. 2012. "Oxytocin/vasopressin-related peptides have an ancient role in reproductive behavior." *Science* 338 (6106): 540–43. doi:10.1126/science.1226201.

Gilbert, S. F., and E. M. Jorgensen. 1998. "Wormholes: A commentary on K. F. Schaffner's 'Genes, Behavior, and Developmental Emergentism.'" *Philosophy of Science* 65: 259–66.

Godfrey-Smith, P., and K. Sterelny. 2008. "Biological information." In *The Stanford Encyclopedia of Philosophy*, edited by Edward N. Zalta. Fall 2008 ed.

Griffiths, P. E., and R. D. Knight. 1998. "What is the developmentalist challenge?" *Philosophy of Science* 65: 253–58.

Griffiths, P. E., and K. Stotz. 2013. *Genetics and Philosophy*. New York: Cambridge University Press.

Hall, Brian Keith. 1994. *Homology: The Hierarchical Basis of Comparative Biology*. San Diego: Academic Press.

Hassan, W. M., D. A. Merin, V. Fonte, and C. D. Link. 2009. "AIP-1 ameliorates beta-amyloid peptide toxicity in a *Caenorhabditis elegans* Alzheimer's disease model." *Human Molecular Genetics* 18 (15): 2739–47. doi:ddp209 [pii] 10.1093/hmg/ddp209.

Hedges, S. B. 2002. "The origin and evolution of model organisms." *Nature Reviews Genetics* 3 (11): 838–49. doi:10.1038/nrg929 nrg929 [pii].

Holton, T. A., and D. Pisani. 2010. "Deep genomic-scale analyses of the metazoa reject Coelomata: Evidence from single- and multigene families analyzed under a supertree and supermatrix paradigm." *Genome Biology and Evolution* 2: 310–24. doi:evq016 [pii] 10.1093/gbe/evq016.

Jacob, F., and J. Monod. 1961. "Genetic regulatory mechanisms in the synthesis of proteins." *Journal of Molecular Biology* 3: 318–56.

Jorgensen, E. M., and C. Rankin. 1997. "Neural plasticity." In C. elegans *II*, edited by Donald L. Riddle, Thomas Blumenthal, Barbara J. Meyer, and James R. Priess, 769–90. Plainview, NY: Cold Spring Harbor Laboratory Press.

Kauffman, Stuart A. 1993. *The Origins of Order: Self-Organization and Selection in Evolution*. New York: Oxford University Press.

Kiontke, K., and W. Sudhaus. 2006. Ecology of *Caenorhabditis* species. January 9. In *WormBook*, edited by C. elegans Research Community. doi:doi/10.1895/wormbook.1.37.1.

Lijam, N., R. Paylor, M. P. McDonald, J. N. Crawley, C. X. Deng, K. Herrup, K. E. Stevens, et al. 1997. "Social interaction and sensorimotor gating abnormalities in mice lacking Dvl1." *Cell* 90 (5): 895–905. doi:S0092-8674(00)80354-2 [pii].

Longino, Helen E. 2013. *Studying Human Behavior: How Scientists Investigate Aggression and Sexuality*. Chicago: University of Chicago Press.

Maduro, M. F. 2010. "Cell fate specification in the *C. elegans* embryo." *Developmental Dynamics* 239 (5): 1315–29. doi:10.1002/dvdy.22233.

McGary, K. L., T. J. Park, J. O. Woods, H. J. Cha, J. B. Wallingford, and E. M. Marcotte. 2010. "Systematic discovery of nonobvious human disease models through orthologous phenotypes." *Proceedings of the National Academy of Sciences USA* 107 (14): 6544–49. doi:10.1073/pnas.0910200107.

Morowitz, H., and National Research Council Committee on Models for Biomedical Research. 1985. *Models for Biomedical Research: A New Perspective*. Washington, DC: National Academy of Sciences Press.

Niwa, M., A. Kamiya, R. Murai, K. Kubo, A. J. Gruber, K. Tomita, L. Lu, et al. 2010. "Knockdown of DISC1 by in utero gene transfer disturbs postnatal dopaminergic maturation in the frontal cortex and leads to adult behavioral deficits." *Neuron* 65 (4): 480–89. doi:S0896-6273(10)00045-0 [pii] 10.1016/j.neuron.2010.01.019.

Oetting, W. S., P. N. Robinson, M. S. Greenblatt, R. G. Cotton, T. Beck, J. C. Carey, S. C. Doelken, et al. 2013. "Getting ready for the Human Phenome Project: The 2012 forum of the Human Variome Project." *Human Mutation* 34 (4): 661–66. doi:10.1002/humu.22293.

Oyama, Susan, Paul E. Griffiths, and Russell D. Gray. 2001. *Cycles of Contingency: Developmental Systems and Evolution, Life and Mind*. Cambridge, MA: MIT Press.

Riddle, Donald L., T. Blumenthal, B. J. Meyer, and J. R. Priess, eds. 1997. C. elegans *II*, Plainview, NY: Cold Spring Harbor Laboratory Press.

Sarkar, Sahotra. 1996. *The Philosophy and History of Molecular Biology: New Perspectives*. Boston: Kluwer Academic.

Schaffner, K. F. 1980. "Theory structure in the biomedical sciences." *The Journal of Medicine and Philosophy* 5: 57–97.

Schaffner, K. F. 1986. "Exemplar reasoning about biological models and diseases: A relation between the philosophy of medicine and philosophy of science." *Journal of Medicine and Philosophy* 11 (1): 63–80.

Schaffner, K. F. 1993. *Discovery and Explanation in Biology and Medicine*. Chicago: University of Chicago Press.

Schaffner, K. F. 2006. "Behaving: Its nature and nurture—part 1 and part 2." In *Wrestling with Behavioral Genetics: Science, Ethics and Public Conversation*, edited by Erik Parens, Audrey R. Chapman, and Nancy Press, 3–73. Washington, DC: Georgetown University Press.

Schaffner, K. F. 2008. "Etiological models in psychiatry: Reductive and nonreductive." In *Philosophical Issues in Psychiatry*, edited by K. S. and Kendler, J. Parnas, 48–90. Baltimore: Johns Hopkins University Press.

Waddington, C. H. 1957. *The Strategy of the Genes: A Discussion of Some Aspects of Theoretical Biology*. London: Allen & Unwin.

Wagner, G. P. 1994. "Homology and the mechanisms of development." In *Homology: The Hierarchical Basis of Comparative Biology*, edited by Brian Keith Hall, 274–99. Chichester, UK: Wiley.

Wagner, G. P. 2014. *Homology, Genes, and Evolutionary Innovation*. Princeton, NJ: Princeton University Press.

Wicks, S. R., C. J. Roehrig, and C. H. Rankin. 1996. "A dynamic network simulation of the nematode tap withdrawal circuit: Predictions concerning synaptic function using behavioral criteria." *Journal of Neuroscience* 16 (12): 4017–31.

Wimsatt, W. 1998. "Simple systems and phylogenetic diversity." *Philosophy of Science* 65: 267–75.

5

Reduction

The Cheshire Cat Problem and a Return to Roots

> "Well! I've often seen a cat without a grin," thought Alice; "but a grin without a cat! It's the most curious thing I ever saw in all my life!"
>
> —Lewis Carroll, *Alice in Wonderland*

TWO THESES ABOUT REDUCTION

We have now examined some general features of behavioral genetics as it applies to humans, and then looked at how behaviors are explained in some much simpler organisms, focusing on the worm *C. elegans* as an example where the investigators working with this tiny organism have been able to drill down to the level of one-nucleotide change in the genes and to genes affecting the pores that allow ion flows that affect motion. The question we will consider in this chapter is the apparent gulf between what we know, and can fairly easily know, about such simple organisms and what can be said about the biochemical mechanisms and pathways that govern human behavior.

To start, I want to clear away some potential distractions by making some distinctions. One distinction is between two kinds of reductionism. The first is what I call "sweeping reductionism," where we have a sort of "theory of everything" and there is *nothing but* those basic elements—for example, a very powerful biological theory that explains all of psychology and psychiatry. The second kind is "creeping reductionism," where bit by bit we get fragmentary explanations using interlevel mechanisms. In neuroscience, this might involve calcium ions, dopamine molecules, and neuronal cell activity, among other things, of the sort we encountered in the explanation of the worm's reaction to light touch stimulation in chapter 3.

Sweeping reductionism, I think, is probably nonexistent except as a metaphysical claim. There is, however, some scientific bite in trying to do something like this in terms, say, of reducing thermodynamics to statistical mechanics, or, as I will describe below, reducing optics to electrodynamics, but even these sweeping reductions tend to fail somewhat at the margins. So I don't think that sweeping reductionism really has much in the way of cash value in the biological and psychological (and social sciences). It's a scientific dream for some, and for others, a scientific nightmare.

Creeping reductionism, on the other hand, can be thought of as involving partial reductions—reductions that work on a patch of science. Creeping reductionism is what neuroscientists do when they make models and propose mechanisms. Creeping reductions do not typically commit to a nothing-but approach as part of an explanatory process. Rather, they seem to tolerate a kind of pragmatic parallelism, or emergence, working at several levels of aggregation and discourse at once. And creeping reductions are consistent with a coevolutionary approach that works on many levels simultaneously, with cross-fertilization.

Clearing away another distraction requires distinguishing between two kinds of determinism. "Sweeping determinism," regarding a powerful theory we think might be fundamental and universal, states that given a set of initial conditions for any system, all subsequent states of the system are predictable and determined. This is what some Newtonians believed. And quantum mechanics, though it's indeterministic in the small, is essentially deterministic for bodies, like cells and organisms, that are medium-sized and larger. In the genetics area, where we focus on the presence of powerful genes (alleles) related to disorders and traits, this kind of determinism is called genetic determinism. I referred to an example of this in chapter 2 where I briefly mentioned Huntington's disease.

But sweeping genetic determinism has so far failed to be instantiated empirically, and sweeping neuroscientific determinisms are not yet even close for humans, as noted in chapter 2, and in chapter 3 we found incomplete reduction even in narrowly focused worm behaviors. What we have seen, then, is "creeping," or partial, neuroscience, with determinism that may be coming. As mechanisms get elaborated, neuroscientists will get roughly deterministic explanations for some types of behavior in some people. And I say "roughly" because here too there will be some problems at the margins.

Claims of sweeping determinism worry philosophers and the philosophically inclined, but a mechanical determinism of a sweeping sort has never had any ethical or especially legal relevance, so far as I know. Nobody ever brought somebody into court and said that he was mechanically determined by Newton's theory of motion. We will revisit this issue again later in a concluding chapter 8 when I consider "free will."

In this chapter I want to elaborate on these distinctions and propose two theses, and then examine what the consequences of those theses might be for discussions of reduction and emergence as philosophers have discussed these notions. The first thesis is that what have traditionally been seen as robust reductions of one theory or one branch of science by another more fundamental one are largely a myth outside of some examples in the physical sciences. In the biological sciences, these prima facie sweeping reductions tend to fade away, like the body of the famous Cheshire cat, leaving only a grin, whence the epigraph to this chapter. The second thesis is that the grins, maybe better to say "smiles," that remain are fragmentary patchy explanations, and though patchy and fragmentary, they are *very important,* potentially Nobel Prize-winning advances. To get the best grasp of them, I want to argue, we need to return to the roots of discussions and analyses of scientific explanation more generally, and not focus mainly on reduction models.

I did not always think that the first thesis was true. Particularly in the physical sciences, it appeared that we had strong reductions that were constituent parts of actual science—and not mere philosophical quests for unified science. When I studied physics in the 1950s and 1960s, thermodynamics was taught as a separate course in physics departments, but everyone knew that statistical mechanics was the science underlying thermodynamics. Similarly there were courses offered in optics, but the nature of light was known to be an electromagnetic wave (at least to a good first approximation), and Maxwell's equations could be mathematically manipulated to generate a wave equation, which in turn could be used to explain various laws of optics, such as Snell's law of refraction, and even more complex relations such as Fresnel's formulas for the relative amplitudes of reflected and incident polarized light waves (see Schaffner 1972, 18).

Closer inspection of the explanatory process, however, revealed difficulties.[1] Though one can get Snell's law by derivation from Maxwell's electromagnetic theory, one does not obtain the entire range of Fresnel's theory of physical optics (actually theor*ies* is more accurate, since there were several models employed by Fresnel to cover all of optics—again see Schaffner 1972). Furthermore, to get an explanation of optical dispersion, one has to go beyond Maxwell's theory per se to Lorentz's electron theory. But even Lorentz's theory was not enough to account for all of physical optics, since to get an explanation of the photoelectric effect, one has to go beyond it to Einsteinian elementary quantum mechanics, and an explanation of the optics of moving bodies requires special relativity. To an extent, what we see in these theory shifts from Maxwell to Lorentz to Einstein represents a partial—a creeping rather than sweeping—approach to reduction of optics, but these shifts can be overemphasized. In fact, Feyerabend, and especially Kuhn, seized on these shifts and interpreted them as representing not reduction but rather *replacement* of the earlier, higher-level theory of physical optics. That strikes me as too extreme a view, since optics can by and large be preserved at its level, and a very powerful general reduction accomplished by an amalgam of Maxwell, Lorentz, and Einstein, suitably corrected—more in the next paragraph on this point. This view of theory corrections, later termed theory coevolution by Churchland, was sketched in my own essays (Schaffner 1967, 1968).

These types of reductions are, as noted, rare. But they can be extraordinarily powerful, as in the explication of optics by electromagnetic theory. The logical features and explanatory strategies of that example come quite close to fulfilling classical Nagelian reduction conditions (more about these later). Skeptics of this kind of powerful general reduction need to examine a detailed account of exactly how that reduction works, as well as to identify where departures from classical theory are needed. The basis for this inquiry can, in point of fact, be found in two back-to-back books by the distinguished physicist Arnold Sommerfeld. Sommerfeld published six advanced textbooks in the 1940s covering all of physics, which were based on his extensive lectures on the topics delivered in the 1930s. Volume 3 was entitled *Electrodynamics*, and volume 4, *Optics* (Sommerfeld 1950a, 1950b). The optics in volume 4 is developed reductionistically from Maxwell's theory as delineated in volume 3, and the two texts represent an in-depth extended exemplar of a sweeping reduction. This reduction is formulated in the Euclidean-Newtonian mode of entire fields being mathematically derived from a small number of integrated universal physical laws supplemented with simple connections between the fundamental terms in the reduced and reducing theories. The reduction also provides electromagnetic theory with a privileged position ontologically and causally. The light vector and the light wave are not fundamental after this successful reduction, but rather the electric vector in the electromagnetic wave.[2]

But such a comprehensive, sweeping, deductively elaboratable account as found in this reduction seems to be dependent on some rather *stringent* requirements. Both reduced and reducing fields need to be representable in terms of a small number of principles or laws. Also, the connections between the two fields need to be straightforward and relatively simple, though far from obvious. (It is a simple and general statement that that the electric vector *is* the light vector but it is *not obvious*—the identification required experimental and theoretical arguments—see Sommerfeld 1950b, 56–57.) Nor it is at all obvious that light is an electromagnetic wave—Hertz's magnificent experiments, wonderfully recounted in his introduction to his collected papers, *Electric Waves*, show this clearly (Hertz [1893] 1962). Both of these stringent conditions, simple axiomatizability and simple connectability, fail in significant ways in more complex sciences such as molecular genetics and neuroscience, though that they do fail, or would fail, was not necessarily foreseeable at the beginning of the Watson-Crick era in genetics in the 1950s and 1960s.

That one encounters creeping rather than sweeping reductions in biology, moreover, can be illustrated by Kandel's classical explanations of learning in the sea snail *Aplysia* in neuroscience. The standard accounts by Kandel provide explanations of some simple learning behaviors in *Aplysia*, but not *all* of *Aplysia*'s behaviors are explained. (For example, *Aplysia californicum* engages in a kind of California-style sex involving multiple partners, but I have not seen any molecular cartoon describing and explaining this complex behavior.) Additionally, those Kandel models are only partial neural nets and partial molecular cartoons that describe what happens to strengthen synapse connection (Kandel, Schwartz, and Jessell 2000; Kandel 2013). And the Kandel cartoons (and text explanations) use interlevel language (mixing organs, cells, receptors, second messengers, and ions, among other types of entities at different levels of aggregation)—not a language involving purely chemical entities interacting with other chemical signals. So this is no robust *unilevel* explanation of learning—even just in *Aplysia*—based solely on molecular mechanisms and chemical entities. The reasons for this have been suggested above: lack of any broad scope simple theories, plus the aggregated complexity of the parts of the mechanisms or models involved. Both of these reasons reflect the manner in which evolution has "designed" living organisms—by opportunistically utilizing whatever bits and pieces of mechanisms may be available and pulling them together in a Rube Goldberg assemblage—not pretty, but satisfactory if it wins in the fitness sweepstakes. (On this view, compare Jacob 1977.)

However, though we do not get sweeping reductions in the biological sciences, we do get extremely valuable potentially Nobel Prize-winning progress, albeit of a creeping sort. Thus, it is important to know at a general philosophical level what is occurring when we obtain these important results. The results are *like* reductions, but I think they are better described as *explanations,* using that term as an alternative to reduction because the e-word does not carry the conceptual freight of various reduction models and is a more appropriate general context within which to analyze what is actually occurring in the biomedical sciences. Such explanatory reductions are in a sense *complementary* to the sweeping theoretical reductions we can find in rare instances in the physical sciences.[3] Neither impugns the character of the other, and which type of reduction one finds will depend on the structure of the disciplines and empirical results. The present chapter focuses primarily on these explanatory reductions, but does so with the model of theoretical reduction as a backdrop.

A RETURN TO ROOTS AND A *VERY* BRIEF HISTORY OF SCIENTIFIC EXPLANATION

In point of fact, a revisiting of the wellspring of the major reduction model—that of Ernest Nagel—suggests it was a generalization or extension (but more accurately a specification) of an ancient Aristotelian model of deductive-nomological explanation, now what is often called the Popper-Hempel model (Hempel 1948; Popper 1959), which in the Hempel variants spawned 40–50 years of argument and criticism in the general explanation literature.[4] It is not possible to find textual evidence that Nagel was specifically generalizing Popper-Hempel, since the original publication of the Nagel model in his 1949 contains no bibliographic references. But the 1961 version places reduction within the context of explanation, and explanation itself has four patterns, according to Nagel, the first and oldest (actually Aristotelian) of which is the deductive model (Nagel 1961, 21). And Nagel did write in 1961 that "reduction, in the sense in which the word is here employed, is the *explanation* of a theory or a set of experimental laws established in one area of inquiry, by a theory usually though not invariably formulated for some other domain" (338; my emphasis).

On Nagel's views, then, reduction, in one important sense, is the *explanation* of a higher-level theory, or science, by a lower-level, more fundamental one (e.g., the reduction of biology by chemistry). In the Nagelian model of reduction, the *explanandum*—that which is to be explained—is a set of laws (theories) fully describing the higher-level or more primitive science to be reduced (e.g., biology or Newton's mechanics). The *explanans* or explainer is the set of laws (theories) fully describing the more fundamental or more recent science (e.g., molecular chemistry or Einsteinian relativity). And also needed in the Nagelian account are "connectability assumptions" (often called "bridge laws" and sometimes reduction functions) that define (or relate) the higher-level entities (e.g., genes) and properties (e.g., dominance) in terms of lower-level entities and properties (e.g., DNA and enzyme action).[5]

EXPLANATION AND EMERGENCE

Discussions of reduction are typically co-paired with an account of emergence—an almost Janus-related philosophical topic (see my 1993a, 411–22). In terms of explanation, one way to approach emergence is to define it as failure of any possible explanation of a whole *in terms of its parts and their relations* (and expressed only in the parts' language). To situate a discussion of emergence and its relations to reduction, I want to distinguish three types of emergence:

> *Innocuous.* The parts, without a specification of the interrelations, do not tell you what the whole will do. Example: the parts of an oscillator are a resistance, a capacitor, and a coil, plus a power source, but the system will not oscillate unless the connections are right, that is, the connections must be specified for the parts to be an oscillator. This is uncontroversial.
>
> *Strong.* All the information about the parts and the connections will *never* allow an explanation of the whole, and the behavior of the whole must be observed as the whole acts as a whole. This is *very* controversial; I suspect it is tantamount to substance pluralism. (For examples of such claims by Mayr and Weiss, see Schaffner 1993a, 415–17, though probably neither would have accepted substance pluralism as the natural implication of this position.)[6] Some of the DST proponents discussed in chapter 3 above seem to have held this view.
>
> *Pragmatic.* For the immediately foreseeable future, and maybe for many years, we will not have the analytical tools that allow us to infer the behaviors of the wholes (or sometimes even the next level up) from the parts and their connections. It is this pragmatic sense that runs through this chapter (and book). (For related views see Wimsatt 1976a; Simon 1981.)

FURTHER DATA-DRIVEN DEVELOPMENTS RELATED TO POST-NAGELIAN REDUCTION MODELS

Research into a wide range of examples in the biological and medical sciences was summarized in a perhaps overly long (115-page) reduction chapter in my book (1993a). The gist of this was that in biology most purported reductions are at best partial reductions in which corrected or slightly modified fragments or parts of the reduced science are reduced (explained) by parts of the reducing science, and that in partial reductions a causal/mechanical approach is better at describing the results than is a developed formal reduction model, which first appeared in my work (1967) as a general reduction model, and which evolved into an account that could tolerate some replacement, in what I called the "General Reduction/ Replacement Model" (GRR) (Schaffner 1977).

This GRR model (1993a), however, is still seen as a good executive summary and regulative ideal for unilevel clarified—and essentially static—science; and it also pinpoints where identities operate in reductions, and emphasizes the causal generalizations inherent in and sometimes explicitly found in mechanisms. (For recent support for or defenses of the GRR model see Butterfield 2011a, 2011b; Dizadji-Bahmani 2010, and for some recent criticism, but also an extension of the model, see Winther 2009; as well as my comments on these articles in Schaffner 2012.) As noted earlier, some such virtually complete reductions can, in point of fact, be found in the history of physics. The more common *partial reductions*, though usually simply termed "reductions," are, paradoxically, typically *multilevel* in both the reduced and reducing sciences, mixing relatively higher entities (and predicates) with relatively lower-level entities (and predicates); it is extremely rare that there are only two levels. What happens in these "reductions" is a kind of "integration"—to use Sterelny and Griffith's term—in the sense that there is a mixing and intermingling of entities and strategies from higher level and more micro domains in a consistent way (Sterelny and Griffiths 1999). In some ways this integration is reminiscent of what Kitcher and Culp (Culp 1989) termed an "explanatory extension," though I have disagreed with much of the unificatory and anticausal baggage that such a view seems to take (Schaffner 1993a, 499–500).

A table from my 1993 book is produced in table 5.1 to illustrate these conclusions. In the 1993 reduction chapter, I also elaborated—then using some of the

Table 5.1 Classical Mechanical (CM) and General Replacement-Reduction (GRR) Approaches in Different States of Completions of Reductions

State of completion/ approach	**CM**	**GRR**
Partial/patchy/ Fragmentary/interlevel	Box 1. CM approach is usually employed; interlevel causal language is more natural that GRR connections.	Box 2. Complex GRR model: The connections are bushy and complex when presented formally, but GRR does describe points of identity, as well as generalizations operative in mechanisms.
Clarified science/ Unilevel at both levels of aggregates	Box 3. Either approach could be used here, but where theories are collections of prototypes, the bias toward axiomatization or explicit generalization built into the GRR approach will make it less simple than a CM approach.	Box 4. Simple GRR model: this is best match between Nagelian reduction and scientific practice.

views of Wesley Salmon, though not accepting some key features of his causal approach[7]—on the strengths of a causal mechanical approach and what value the more formal GRR model might have as well.

In the sections that follow in this chapter, I develop and apply the explanation account found in chapter 6 of my 1993 book and also in outline form in my 2000 essay, and elaborated somewhat in later work (Schaffner 2006, 2008a) to offer a more detailed account of what occurs in partial reductions in the biological sciences.

A FURTHER RETURN TO ROOTS: A MINIMALIST EXPLANATION-REDUCTION MODEL EMPLOYING A CAUSAL MECHANICAL APPROACH

The Conditions for a Partial Reduction

In attempting to return to the explanatory roots of reductions, I will begin with what distinguishes a nonreductive explanation from one that is (at least partially) reductive. One way to work toward a minimalist set of distinguishing conditions is to look at strong candidates for reductive explanations in a science of interest, for which a general reduction account is desired. The following conditions were suggested by a review of the Kandel models for *Aplysia* learning that were discussed in my 1993 book (chapter 6) and are available in an updated and accessible form in many standard neuroscience texts (including Kandel, Schwartz, and Jessell 2000). Thus, the scientific details of those examples will not be re-presented in the current chapter. The general conclusion of my recent review is that successful (though partial) *reductions* are causal mechanical *explanations*, if, in addition to whatever we call adequate causal mechanical explanations (this will come later), the following three conditions hold. (I will state these in the material mode, though they can be rephrased so that they refer to sentences that describe the referents.) The first two of these are informal and the third is a formal condition that retains an important formal condition of the Nagel (and GRR model) regarding connectability assumptions:

1. The explainers (or *explanans*) (more on what these are later) are a part (or parts) of the organism/process, that is, they are a (partially) decomposable microstructure(s) in the organism/process of interest.[8]
2. The *explanandum* or event to be explained is a grosser (macro), typically aggregate property or end state.
3. Connectability assumptions (CAs, sometimes called bridge laws or reduction functions) need to be specified, which permit the relation of macrodescriptions to microdescriptions. Sometimes these CAs are causal consequences, but in critical cases they will be identities (such as gene ① = DNA sequence ①, or aversive stimulus = bacterial odor).[9]

So far I have said little about what these three conditions are conditions of, or to what account of causal mechanical explanation they need to be added, to reflect what we find in partial and patchy reductions. Before I sketch the explanation model, however, it is important to underscore a prima facie, somewhat paradoxical aspect of partial reductions. This is their *dual* interlevel character.

It's Interlevel All Over

Though the point is possibly underappreciated, I think it fair to say that it is reasonably broadly recognized that typical reducing/explaining models are interlevel (mixing together ions, molecules, cells, cell networks, and, not infrequently, even organs and organisms). Less appreciated, I think, is that the *reduced* theory/model is *also interlevel,* but not as fundamental or fine-structured as is the reducing model. In a long-running debate about reduction of Mendelian genetics by molecular genetics by myself, David Hull, and others, and offshoot debates among a number of others in the philosophy of biology throughout the 1970s and 1980s (see my 1993a, 432–87, for a summary), I do not think it was fully recognized, by me or others, that Mendel's theory of heredity was *itself* vigorously interlevel. Mendel had not only summarized his discoveries in genetics in terms of laws, but in the same article he also proposed an explanation in terms of underlying factors that segregated randomly, thus mixing in his theory phenotypes and what were later called genes. To underscore the intertwined and interlevel nature of Mendelian genetics, consider the following quotation from Mendel's 1865 paper:

> In our experience we find everywhere confirmation that constant *progeny* can be formed only when germinal *cells* and fertilizing *pollen* are alike, both endowed with the potential for creating identical individuals, as in normal fertilization of pure strains. Therefore we must consider it inevitable that in a hybrid *plant* also identical *factors* [= genes] are acting together in the production of constant *forms.* Since the different constant forms are produced in a single plant, even in just a single *flower,* it seems logical to conclude that in the *ovaries* of hybrids as many kinds of germinal cells (germinal *vesicles*), and in the *anthers* as many kinds of pollen cells are formed as there are possibilities for constant combination forms and that these germinal and pollen cells correspond in their internal make-up to the individual forms. (Mendel 1966, 20) (my italics).

A reduction of Mendel's laws and his process of factor segregation typically involve an appeal to entities intertwined from several levels of aggregation: cells, chromosomes, DNA strands and loci, and enzymes, so even this paradigm of reduction is also interlevel at the present time. The reduction is also partial because it is impossible (so far as I know) to account for *all* of the pea plant's phenotypes strictly in terms of molecular features and mechanisms, even today.

THE ELEMENTS OF A CAUSAL MECHANICAL EXPLANATION MODEL: FIELD AND PREFERRED MODEL SYSTEM

For this chapter, I am going to restrict an account of my model of explanation to what might be called "local" explanations. This notion of "local" is intended to indicate I am referring to explanations within a time slice of a field (sometimes called "synchronic") that use a currently accepted theory or class of mechanisms. I distinguish this type of explanation from "global" (or "diachronic") explanations that capture explanations across successive historical periods of scientific change—of the sort that Kuhn described as revolutions involving major paradigm change.[10] In an earlier summary of scientific explanation, I distinguished six components of an explanation (1993a, chap. 6), and though those are valuable background, for the purposes of partial explanation, the following account is more focused.

I want to argue that a satisfactory local explanation model, which I think can illuminate what occurs in partial reductions, has two main substantive components, with each substantive component having a closely related logical/epistemological aspect.[11] The first substantive component involves the scientific field, but more accurately *field elements*, or FE, and its epistemological aspect is a kind of inductive logic of comparative evaluation of plausible explanatory candidates, representing preliminary plausibility judgments. The second substantive component is what is focused on—the preferred (causal) model system or PCMS, which itself is an elaboration and extension of one of the plausible explanatory candidates of the first field element component. The epistemological aspect of the second component is a claim that the PCMS is a *causal* system representing a temporal process; the "logical" aspect of this second component is that such a system can be elaborated and tested using either *deductive logic* or *statistical* methodological *logic* or both. In a previous paper (Schaffner 2000) I have called this the *field and focus* model, which itself was a renaming of an account of explanation I developed in my book (1993a, chap. 6). In the present chapter (and in Schaffner 2006) I have also termed it a *field elements and preferred model system* or FE-PCMS account in an attempt to underscore the key constituent concepts involved in the explanation model. But an argument could be made that the idea should be kept simple, and that the terms "field" and "focus" or even, say, an *F2* model, would be more helpful. Each component, however, needs some additional discussion, and in the following section I also provide an illustration that relates the model to partial reductions; later in this chapter I also offer another illustration involving major depression.[12]

The Field and Field Element Component

I should begin by noting that the general sense of field used here is (probably) *not* the sense originally used in Darden and Maull's (1977) "interfield theory" approach to intertheoretic relations, including their different way of looking at reductions. I think a reading of their seminal 1977 paper suggests that *each* field is unilevel and that it is *interfield* explanations that are surrogates for (or alternatives to) reductions. That said, the approaches taken there and in my present chapter may well be fairly congruent, with any differences possibly more terminological than real. The field *elements* concept also has certain analogies with Shapere's (1977) notion of a "domain" as a set of "items of information" having "an association," but I think the field elements notion differs in being broader, and at the same time clearer, in the sense that *particular research articles define those field elements by specifying them.* (For additional comments on the pros and cons of the domain concept, see my 1993a, 52.)

The substantive *field* component in my approach contains most of the basic generalizations, mechanisms, experiments, and theories, typically introduced in a standard textbook for the field, and *field elements drawn from* the field. This makes the typical field (and usually the FE that selects portions from the field) vigorously interlevel, as well as (typically and also paradoxically) *inter*disciplinary, in virtually all instances with which I am familiar.[13] A textbook may, however, draw on several preexisting fields, for example, neuroscience and molecular biology, which can usually roughly be distinguished by referring to consensus *classic* texts in those fields. This *general* field component has possibly been captured implicitly in explanation models in the philosophical literature by Railton's (and also Salmon's) notion of an ideal explanatory text. Concrete examples of such texts in biology would be the Watson *Molecular Biology of the Gene* series of texts, or the Kandel and Schwartz *Principles of Neural Science* series of texts; in medicine this would be a standard medical textbook, such as *Harrison's Principles of Internal Medicine.* A more specific example is the field of *C. elegans* research, typified by what are known as the *Worm I* and *Worm II* collections of essays, and more recently, by the online textbook at www.wormbook.org. Though I do not believe anyone has ever done this, someone reasonably well acquainted with a field could make a list of major explanatory devices in a field by working through such a textbook. Some of these would go by the terms *model,* or *mechanism,* or *pathway,* or *law,* or *generalization,* or *theory,* or *hypothesis.* And they would not be independent, nor representable in a simple hierarchy, since some would be partial components of others and would reappear in slightly different forms multiple times. It is that richness and complexity that I believe we find and also have to deal with in real science.

In the kind of partial reductions I want to explicate in the present chapter, we should begin by considering a typical scientific journal research article (not a textbook nor collection of articles nor a review article, usually) in which an *explanation* is proffered. The typical article situates the phenomenon to be explained within a field (or sometimes in two and possibly more fields) and then presents a list of the classes of alternative possible explanations utilizing the field elements (possible explanations as seen by the authors as being viable in the field) for the phenomenon of interest. The alternatives are not exhaustive of other elements that can be found in the field as a whole, but are proposed, sometimes as a cluster, or sometimes seriatim, in the article.[14] The possible explanations in a scientific article often are evaluated and roughly ranked (though usually implicitly so) as best, better, good, worse, worst, which is, in an extended sense, a *logical aspect* of the first substantive (field) component.[15]

The Preferred (Causal) Model System (PCMS) Component

The second, and perhaps most salient, substantive component of my model of explanation—the *focus* as I also termed it—is the designation of a *preferred causal model system*, or PCMS, which implicitly or explicitly involves the laws or generalizations that are relevant to the particular problem or problems of interest to the investigator. Such a PCMS can be quite simple, as when one introduces a simple single-locus Mendelian model of a dominant/recessive gene pair, say as a Punnett-square representation, and then uses that model to explain the inheritance of Huntington's disease or cystic fibrosis. Alternatively, the PCMS can be more complex, as in an explanation of feeding behavior using specific mutants and neuron types in *C. elegans*, which will be discussed below, a Kandel cartoon depicting presynaptic sensitization in *Aplysia*, or a Hodgkin-Huxley classic sodium action potential model. The term "mechanism" is also sometimes used when referring to a PCMS—a term that evokes a set of issues (and recent) articles sometimes referred to as "the recently revived *mechanistic* philosophy of science" (Bickle 2006, 429–30). I discuss this at the end of this chapter.

The PCMS typically will utilize *a number* of subtly different types of causal sequences.[16] One type that is frequently cited in the biomedical sciences is a "pathway." A review of a comprehensive summary of research in *C. elegans* suggests that a pathway is a coordinated causal sequence that may contain entities at different levels of aggregation (and not necessarily molecular) with a defined endpoint, which may be a behavior or facet/component of behavior. Often these are called signaling or transduction pathways. They also may be termed regulatory or adaptive, depending on their function. Pathways can have genes as "entry points." Pathways are typically not fully detailed and (especially initially) may contain place markers and gaps. Interestingly, some proponents of the "new mechanism" tradition prefer to see pathways as *mechanism schemas* (Machamer, personal communication; Craver and Darden 2013), but that seems to me to expand the biologists' notion of "mechanism" beyond its typical usage—more on this below; also see the last section of this chapter.

The core meaning in the PCMS is that of a "model," which we can think of as representing or capturing one or more pathways, and which usually provides additional unification beyond a pathway, and also leading to a defined endpoint. A model can be at any level or typically interlevel. A model is more dynamic than a "circuit" (see below), which is more structural. Models are abstract—often in different degrees—and are typically idealized (see Schaffner 1993a, 98; Bogen, personal communication). The sense of "model" here is reasonably distinct from the sense of "model" in "model organisms," though there are some overlapping features.

The PCMS may also be termed a network, which can be characterized as an integrated set of pathways often involving adaptive or regulatory functions and oriented toward some general goal/behavior. Accordingly, one investigator may term a structure a network while another may call the same structure a model. In some instances, a network may be identified with a circuit, also seen as a model to account for a key behavior (see, e.g., figure 7 in Rogers et al. 2006 on *C. elegans* aggregation and O_2 response; and also Chang et al. 2006 for a "distributed network of oxygen-sensing neurons"). A circuit, however, is usually thought of as an anatomically existing *structure* built up from neurons and interneurons and excitatory/inhibitory connections (synapses) containing (implicitly) various pathways and mechanisms. The neurons may have "acting genes" as part of them, or even acting ion channels, such as K+ types or cGMP gated types.

Finally it should be noted that though "mechanism" language has become widely utilized in recent philosophy of science, it seems to me that the distinctions captured by the pathway, model, and network expressions are more faithful to the variety of ways that biologists think of their work. To me, and I think most biologists, a mechanism is a highly *specific* set of interactions (one could perhaps use the term "connecting activities" in place of interactions; Machamer, personal communication), interactions that are typically "molecular" but may be at a higher level of aggregation (e.g., cellular), and which may be found operating in various and quite diverse different types of pathways or models.

There is no formal limit on the degree of complexity of a PCMS, though it is always idealized to a greater or lesser extent. A critically important aspect of a PCMS is that there is a list of general assumptions embedded in the preferred model system that describe the system under study and that are believed to generalize to other like systems. (The extent to which generalizations, as opposed to "mechanisms," are needed in explanations is noted above but is also a somewhat contentious issue, which I shall return to later in this chapter.) This generalization, however, may have narrow or broad applicability: the generalization may be family or population limited, strain limited, or species limited, though possibly even broader, holding for all mammals, for example. The generalizations are typically *qualitative causal generalizations,* describing parts of mechanisms in a process, such as an inducer combining with repressor molecule in a *lac* operon model with the resultant loss of the repressor's affinity for the operator. In (fairly) rare cases, these generalizations will be mathematical formulas, such as a Nernst equation or a flux equation.[17] These generalizations that are instantiated in the models can be found in the text and especially in the figure legends of pictorial representations of models and mechanisms (and also referred to in the indexes) in standard biological textbooks, such as the Watson or Kandel series noted above. In my view, the explanatory elements in the biomedical sciences are a collection of (sometimes overlapping) model systems (PCMSs) (for details on this view see chapter 3 of my 1993a).

The epistemological aspect of this second substantive component of my explanation account is a claim about causality appealed to in the explanation.[18] Most explanations in basic science are causal mechanical, but they might involve a random probabilistic process, or even a human motivational account (in economics, or human psychology, for example). Example of a causal mechanical explanation strategy can be found in many of Wesley Salmon's examples.[19] A related logical aspect of this epistemological aspect is closely related to the type of the causality assumed in the proposed PCMS studied: deterministic systems can easily be elaborated using deductive logic; probabilistic causality suggests the need for an inductive logic.[20] The Popper-Hempel (or perhaps more accurately Aristotle-Mill-Popper-Hempel) model of explanation falls into the first type, involving deductive logic.

HOW THIS EXPLANATION AND PARTIAL REDUCTION MODEL IS ILLUSTRATED IN PRACTICE

A Recapitulation and Overview of the Explanation and Partial Reduction Process

A reasonably detailed illustration of how this two-component explanation and partial reduction model works, especially in partial reductions, may help clarify it. Building on chapter 3 and the *C. elegans* examples, I will use some further research published in 2002–2009 on the social-versus-solitary feeding behavior of the worm. This will serve two purposes simultaneously. First, it will show how this explanation and partial reduction model is illustrated in practice. (Later in chapter 7 I will again use this model and apply it to schizophrenia.) Second, the research done on social versus solitary feeding since the 1998 paper described in some detail in chapter 3 indicates how deeper research on genes and behaviors moves away from a "one mutation, one behavior" paradigm, and closer to the many-many paradigm captured in the eight rules of table 3.1 in chapter 3.

First we begin with a summary overview of how the explanation and partial reduction model works generally. To reiterate the general process: first, a typical scientific or medical research article provides explanations, for example of an organism's behaviors. But even in such focused research articles, the broader context of the problem(s) is sketched (however briefly) and assumptions are made that the reader is knowledgeable about the organism and familiar with the relevant parts of the field (the FEs) in neuroscience, or genetics, or molecular biology, and so on. Within this broad framework, such a research article quickly zeros in on several well-defined questions and then proceeds to present answers to the questions in terms of the advances that are the rationale for the publication of the paper. Within the context of these answers, it is possible to pick out a focus (or foci) and ask what specific PCMSs are used in the explanation. It is at this point, with a focus on a specific PCMS, that we can usefully begin to appeal to the nature of the law(s), mechanisms, component parts, and pathways, as well as to scrutinize the nature of the inference (deductive, statistical), and ask whether this explanation is causal (or perhaps unificatory), and if causal, what type(s) of causal conditions are operative. This general pattern of explanation is found in many of the papers in the study of the nematode and other model organisms. A useful preface to my specific example may be to first, and very briefly, summarize some basic facts about the worm for the readers of this chapter. In an important sense, the following section will introduce some of the field elements (FEs) needed to characterize an explanation (and a partial reduction) in the case below.

The Specific Example of a PCMS and a Partial Reduction, with complications

In this section we now turn to the specific example. Recall that in their 1998 *Cell* paper, de Bono and Bargmann contended that "a single gene mutation can give rise to all of the behavioral differences characteristic of wild and solitary strains" (1998, 680).

This wonderfully "simple" story of one gene that influences one type of behavior in the worm that was told in 1998 underwent further complications. Since then, additional work by de Bono and Bargmann, who did search for the cells in which *npr-1* acts and for the source of the NPR-1 ligands, has indicated that the story is more complex, and that complexity grew even further more recently. In follow-up work in 2002 to determine how such feeding behavior is regulated, de Bono and Bargmann proposed what were, at least initially, two separate pathways (de Bono et al. 2002; Coates and de Bono 2002). One pathway suggests that there are modifying genes that restore social feeding to solitary feeders under conditions of external environmental stress. The other pathway is internal to the organism and will be very briefly described at the conclusion of this section, as will a 2005 article that may synthesize the two pathways. (An accessible overview of the two initial pathways, and some possible very interesting connections with fly and honeybee foraging and feeding behaviors, can be found in Sokolowski's editorial accompanying the publication of the de Bono et al. 2002 and Coates and de Bono 2002 papers.)

The 2002 paper by de Bono et al. indicates how a partially reductive explanation works and also nicely illustrates the features of how general field and specific causal models systems work in tandem. The explanandum (again, the event to be explained) is the difference in social and solitary feeding patterns, as depicted in figure 3.3 in chapter 3. The explanation (at a very abstract level) is contained in the title of the 2002 paper. "Social Feeding in *Caenorhabditis elegans* Is Induced by Neurons That Detect Aversive Stimuli." The specifics of the explanation appeal to the 1998 study as background, and look at *npr-1* mutants, in a search to determine what *other* genes might prevent social feeding, thus restoring solitary feeding in *npr-1* mutants. A search among various *npr-1 mutants* (these would be social feeders) indicated that mutations in the *osm-9* and *ocr-2* genes resulted in significantly more *solitary* feeding in those mutant animals. (Both of these genes code for *components* of a sensory transduction ion channel known as TRPV [*t*ransient *r*eceptor *p*otential channel] that in vertebrates responds to the "*v*anilloid" [V] compound capsaicin found in hot peppers. Both the *osm-9* and *ocr-2* genes are required for chemoattraction as well as aversive stimuli avoidance.) Additionally, it was found that *odr-4* and *odr-8* gene mutations could disrupt social feeding in *npr-1* mutants. The *odr-4* and *odr-8* genes are required to localize a group of olfactory receptors to olfactory cilia. Interestingly, a mutation in the *osm-3* gene, which is required for the development of 26 ciliated sensory neurons, *restores* social feeding in the *odr-4* and *ocr-2* mutants. (Intrepid readers who have closely followed the account of the genetic influences on ion channels, other genes, and neurons thus far *are now entitled to a break!*)

De Bono et al. present extensive data supporting these findings in the article. Typically the reasoning with the data examines the effects of screening for single, double, and even triple mutations that affect the phenotype of interest (feeding behaviors), as well as looking at the results of gene insertion or gene deletion. This reasoning essentially follows Mill's methods of difference and concomitant variation (the latter because graded rather than all-or-none results are often obtained) and is prototypical causal reasoning. Also of interest are the results of the laser ablation of two neurons that were possibly involved in the feeding behaviors. These two neurons, known as ASH and ADL, are implicated in the avoidance of noxious stimuli and toxic chemicals. Identification of the genes noted above (*osm-9, ocr-2, odr-4,* and *odr-8*) allowed the investigators to look at where those genes were expressed (by using green florescent protein [GFP] tags). It turned out that ASH and ADL neurons were the expression sites. The investigators could then test the effects of laser beam ablation of those neurons and showed that ablation of both of them restored a solitary feeding phenotype, but that the presence of either neuron would support social feeding.

The net result of the analysis is summarized in a "model for social feeding in *C. elegans*" shown in figure 5.1.

The legend for the model (quoted from de Bono et al. 2002) reads as follows:

> The ASH and ADL nociceptive neurons are proposed to respond to aversive stimuli from food to promote social feeding. This function requires the putative OCR-2/OSM-9 ion channel. The ODR-4 protein may act in ADL to localize seven transmembrane domain chemoreceptors that respond to noxious stimuli. In the absence of ASH and ADL activity, an unidentified neuron (XXX) [involving *osm-3*] represses social feeding, perhaps in response to a different set of food stimuli. The photograph shows social feeding of a group of > 30 npr-1 mutant animals on a lawn of *E. coli.*

This model is what I termed a preferred causal model system (PCMS) for de Bono et al. *Nature* article. The model is what *does* the partial reduction—more on this

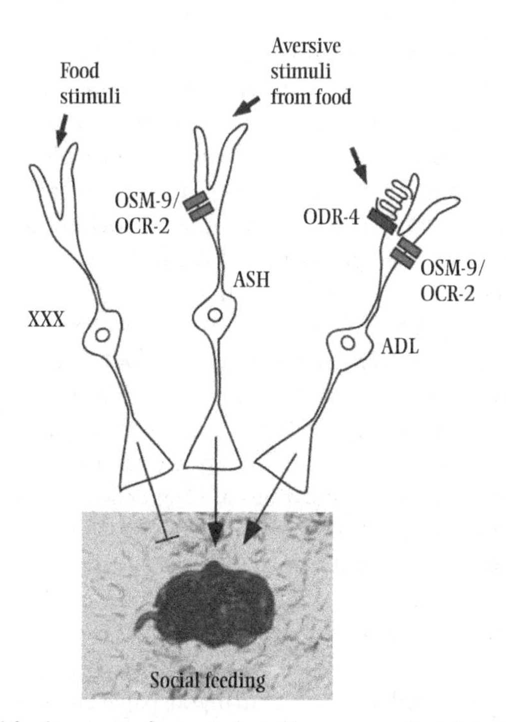

Figure 5.1 Social feeding in *C. elegans* induced by neurons that detect aversive stimuli.
SOURCE: de Bono et al. 2002.

in a few pages. The model is simplified and idealized and uses causal language such as "respond to" and "represses." (The causal verbs also include the word "act," about which much has been made in recent years in the philosophy of biology and neuroscience literature about "activities," as opposed to causation, which may be present in "mechanisms"—more on this later in this chapter.) The PCMS is clearly interlevel. I think it is best to approach such models keeping in mind the scientific field(s) on which they are based, and the specific "field elements" that a paper proposing the model needs to refer to in order to make the model intelligible to readers. Here the fields on which the model draws are molecular genetics and neuroscience, more specifically those sciences as especially applied to the nematode. Scattered throughout the article are occasional alternative but possible causal pathways (field elements) that are evaluated as not as good an explanation as those provided in the preferred model system presented. (One example is the dauer pheromone explanation, discussed on p. 899 of de Bono et al. 2002; another is the "reducing stimuli production" versus "reducing stimuli detection" hypotheses on p. 900 of that article.)

The preparation or *experimental system* investigated in the laboratory (this may include several data runs of the "same" experimental system) is conceptually identified in its relevant aspects with the preferred causal model system. At the abstract or "philosophical" level, the explanation proceeds by identifying the laboratory experimental system with the theoretical system—the preferred model system (PCMS)—and exhibiting the event or process to be explained (the explanandum) as the causal consequence of the system's behavior. The explanans (or set of explaining elements) here uses molecular biology and is mainly comparative rather than involving quantitative derivational reasoning, in the sense that in this paper two qualitatively different end states—the solitary and the social states of the worms—are compared and contrasted.[21] The theoretical system (the PCMS) utilizes generalizations of varying scope, often having to appeal to similarity analyses among like systems (e.g., the use of TRPV channel *family*) to achieve the scope, as well as make the investigation of interest and relevance to other biologists (e.g., via analogies of the NPR-1 receptor to Y receptors and the internal worm circuit to cyclic GMP signaling pathways found in flies and bees that control foraging and feeding behavior—see Sokolowski 2002). For those concerned with philosophical rigor, the preferred model system and its relations to model-theoretic explanation can be made more philosophically precise (and technical) (along the lines suggested in the "philosopher-speak" of note 22).[22]

The discussion sections of scientific papers are the usual place where larger issues are raised and where extrapolations are frequently found. This is also the case in this de Bono et al. (2002) paper, where the discussion section states that "food, food acquisition, and population density are important regulators of aggregation in a variety of species" (902). The paper concludes on an evolutionary note, tying the proximate cause model to a distal causal (i.e., evolutionary) advantage:

> The data in this paper and in the accompanying paper suggest that the regulation of social feeding behaviour in *C. elegans* is complex, involving several

> layers of positive and negative inputs. Such complexity may have evolved as a result of the tension between cooperation and competition that underlies social behaviour, and may be important to ensure that social behaviour is induced only when it offers a selective advantage.

Further work on the circuits that affect social and solitary feeding has been done in addition to what has just been described in detail above. Earlier I mentioned an essay that appeared simultaneously with the above paper in *Nature*. This second paper, by Coates and de Bono (2002), describes a regulatory circuit that senses the internal fluid in the worm and controls social versus solitary forms of behavior. It involves different neurons (AQR, PQR, and URX), and is affected by *tax-2* and *tax-4* gene mutations—genes that produce components of a cyclic GMP-gated ion channel.[23]

Also, in late 2003, de Bono's group was able to identify the ligands that stimulate the NPR-1 receptor (Rogers et al. 2003). These are a class of neuropeptides known as "FMRFamide and related peptides" (FaRPs) that stimulate foraging receptors in other species. In the worm, the relevant FaRPs are encoded by 22 different *flp* genes that can potentially produce 59 FaRP peptides by alternative splicings. It was also reported in this paper that comparative sequencing of the two NPR-1 variants (the F and V forms) as well as three other species of *Caenorhabditis* suggests that the *social* form of the receptor is ancestral, and that the behavior of solitary feeding arose later via a "gain of function" mutation. This is preliminary conclusion, and some insect researchers find it implausible, believing that social behaviors are likely to appear later than solitary activities (de Bono, personal communication). But that may depend on the different selection pressures experienced in different environments by different species.

More recently, additional complexity affecting social-versus-solitary feeding was discovered that initially further developed the role of oxygen as a significant environmental factor (noted in note 23 above). The internal circuit described by de Bono in 2002 involving AQR, PQR, and URX neurons, which were affected by *tax-2* and *tax-4* gene mutations, now seems to be a parallel pathway regulating feeding behavior. I will not detail these complications here but will present a summary figure (from de Bono and Maricq 2005) that combines several pathways of influence on aggregation behavior, and that emphasizes that O_2 is the main trigger switch for the solitary-social distinction (see figure 5.2).

Finally, though there is never a real "finally" in the worm world, Bargmann's group more recently identified the "central site of action" of the *npr-1* gene as the RMG inter/motor neuron, which functions in a "hub and spoke" circuit in the worm (Macosko et al. 2009). This helps clarify the nature of the social and solitary behaviors, but also embeds them in a larger context—one that also remarkably connects the circuit to the ASK neurons that this team found were implicated in male worm's attraction to hermaphrodite pheromones. Further work to additionally unravel these behaviors is ongoing.

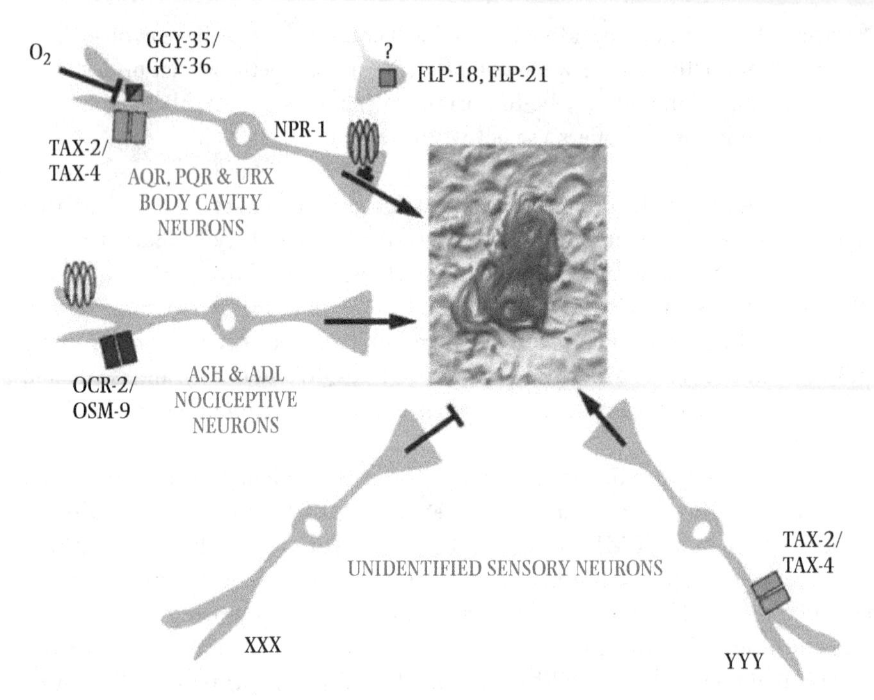

Figure 5.2 Neurons and signaling pathways that regulate aggregation behavior.
SOURCE: de Bono and Maricq 2005.

THIS EXPLANATION IS BOTH REDUCTIVE AND NONREDUCTIVE

The above example (both the simpler circuit and the 2005 augmented circuit) is typical of molecular biological explanations of behavior. Behavior is an organismic property, and in the example is actually a populational property (of aggregation), and the explanation appeals to entities that are *parts* of the organism, including molecularly characterized genes and molecular interactions such as ligand-receptor bindings and G-protein-coupled receptor mechanisms—thus this is generally characterized as a *reductive* explanation. But it represents *partial* reduction—what I termed reduction of the *creeping* sort—and it differs from *sweeping* or comprehensive reductive explanations because of several important features.

1. The first model does not explain *all* cases of social-versus-solitary feeding; a different though somewhat related model (that of Coates and de Bono 2002) is needed for the internal triggering of solitary behavior in *npr-1* mutants; the 2005 model does more but is still incomplete, as are the two speculative models outlined by Macosko et al. (2009).
2. Some of the key entities, such as the signal from bacteria that is noxious to the worms and the neuron represented by XXX in figure 5.2, have not yet been identified.
3. It utilizes what might be termed "middle level" entities, such as neuronal cells, in addition to molecular entities.

4. It is not a quantitative model that derives behavioral descriptions from rigorous general equations of state, but is causally qualitative and only roughly comparative.
5. Interventions to set up, manipulate, and test the model are at higher aggregative levels than the molecular, such as selection of the worms by their organismic properties (feeding behaviors), distributing the worms on an agar plate, and ablating the neurons with a laser.

The explanation *does* meet the three conditions that seem reasonable for a reductive explanation, namely:

1. The explainers (here the preferred model systems as shown in figures 5.1 and 5.2) are a partially decomposable microstructure in the organism/process of interest.
2. The explanandum (the social or solitary feeding behavior) is a grosser (macro) typically aggregate property or end state.
3. Connectability assumptions (CAs), sometimes called bridge laws or reduction functions, are involved, which permit the relation of macrodescriptions to microdescriptions. Sometimes these CAs are causal sequences as depicted in the model figure where the output of the neurons under one set of conditions causes clumping, but in critical cases the CAs are identities (such as social feeding = clumping, and aversive stimulus = [probably] bacterial odor).

Though etiological and reductive, the preferred model system explanation is not "ruthlessly reductive," to use Bickle's phrase, even though classical organismic biologists would most likely term it strongly reductionistic in contrast to their favored nonreductive or even antireductionist cellular or organismic points of view. It is a *partial* reduction.

This partial reduction for a very simple model organism does have some lessons for reductions related to far more complex human behavior, including psychiatric genetics. These additional lessons for reduction are best presented on the basis of some recent schizophrenia studies and will be deferred until we have had a chance to rediscuss psychiatric genetics, including the genetics of schizophrenia, in more depth (in chapter 7).

WILL "MECHANISM LANGUAGE" SUFFICE?

One recent philosophical alternative to classical models of theory reduction can be found in what Bickle (2006) calls "the recently revived *mechanistic* philosophy of science." This revival dates to the seminal article by Machamer, Darden, and Craver (2000) that stressed the importance of the "mechanism" concept as an alternative to law-based approaches to explanation and to reduction. In this approach, a mechanism is "a collection of entities and activities organized in the production of regular changes from start or set up conditions to finish or termination conditions" (Machamer, Darden, and Craver 2000, 3). The analysis has been applied to examples in the neurosciences and molecular biology, and even more broadly (see Darden 2008; Darden and Craver 2013) and recognizes that mechanisms need not be molecular, but can be multilevel (see Craver 2005, 2007). In some of its variants, the approach wishes to eschew causal language, causal generalizations, and any appeals to standard counterfactual analyses, which are typically developed as elucidations of causation (compare Schaffner 1993a, 296–312; Glennan 1996; and Woodward 2003 with Tabery 2004 and Bogen 2010).

An appeal to mechanisms, as a contrast with an emphasis on high-level general theories, is an eminently sensible approach. In biology there are few such general theories (with component laws) that are broadly accepted, though population genetics is a notable exception. An early commitment to theories such as population genetics as representing the best examples of biological theory (see Ruse 1973) is one, as I argued (1980; 1993a, chap. 3), that steered the appreciation of philosophers of biology away from better or more representative alternative approaches to theory structure and explanation. And in that 1980 article and in my 1993 book (chap. 3 as well as chaps. 6 and 9) I frequently utilized references to "mechanisms" as another way to describe the "models" that are so widely found in biology, and which function broadly as surrogates for theories in the biomedical sciences.

But the *strong* form of appeals to mechanisms, as in early arguments by Wimsatt (1976b) seemed to aim at avoiding any discussion of generalizations and laws of working of a mechanism, an avoidance that appeared both philosophically incomplete (see my 1993a, 494–95 for specifics), as well as contradicted by the way biologists present their own models. A paradigm case of how generalizations are articulated to form a model can be found in Jacob and Monod's classic paper on the operon model.[24] In their concluding section they write that "a convenient way of summarizing the conclusions derived in the preceding sections of this paper will be to organize them into a model designed to embody the main elements which we were led to recognize as playing a specific role in the control of protein synthesis; namely the structural, regulator and operator genes, the operon, and the cytoplasmic repressor" (Jacob and Monod 1961, 116). Jacob and Monod then state the generalizations that constituted the model.[25] Similar generalizations can be found in the figure legend from de Bono et al. (2002) quoted above.

This avoidance of generalizations by the revived mechanistic tradition is even more evident in the recent essays by Tabery (2004), by Darden (2005), and especially Bogen (2010), which also seem to me to try to replace the admittedly still problematic concept of causation with appeals to "activities"—a notion that I find much more opaque than causation. (In those places in scientific articles where terms like "acts" appear, I think a good case can be made that what is being referred to is plain old-fashioned causal action.)

More recently, the issue of generalizations versus mechanisms has resurfaced in the context of a discussion of biological complexity, for which the prime source is Mitchell (2009), who extends her earlier arguments about generalizations (Mitchell 2003; for a defense of generalizations in this context, also see Leuridan 2010). This still-developing debate suggests that a strong form of mechanism that abjures generalizations will not be sustained in the long run.

But in a weaker form, such as in Glennan's (1996) paper and in most of Machamer, Darden, and Craver's (2000), the revived mechanistic philosophy of science appears to me to be an important complement to the account of explanation developed in the present book, as well as to my 1993 book and 2000 essay. I had noted in my book that appeals to mechanisms that eschewed generalizations (such as Wimsatt 1976b) were problematic for a number of reasons, a chief one of which was that earlier writers in this tradition appeared to take "mechanism" as a largely unanalyzed term and place a very heavy burden on it. The new mechanistic philosophy of science remedies that problem by articulating a complex analysis of the terminology involved in appeals to mechanisms, but some of the stronger theses, such as those replacing causation by activities, seem to me to move in a less promising direction.

Finally, I should mention that several recent philosophical writers have suggested that general mathematical models capture important high-level equations of motion that are interestingly applicable in the neurosciences (see Chirimuuta 2015 and Ross 2015). However the neomechanists are critical of the explanatory abilities of these general mathematical models, as argued in the essay by Kaplan and Craver (2011).

SUMMARY AND CONCLUSION

In this chapter I began by proposing two theses and then examined what the consequences of those theses were for reduction and emergence. The first thesis was that what have traditionally been seen as robust reductions of one theory or one branch of science by another more fundamental one are a largely a myth in biology, though some rare, but extraordinarily important, instances of them can be found in physics. On closer inspection, and particularly in biology, these reductions seem to fade away, like the body of the famous Cheshire cat, leaving only a grin, or a smile . . . The second thesis was that these "smiles" are fragmentary patchy explanations, and often partial reductions, and though patchy and fragmentary, they are very important, potentially Nobel Prize–winning advances.

To get the best grasp of them, I argued that we needed to return to the roots of discussions and analyses of scientific explanation more generally, and not focus mainly on reduction models, though three conditions based on earlier reduction models are retained in the present analysis. This led us through a brief history of explanation and its relation to reduction models, such as Nagel's, and through a brief account of my own evolving views in this area. Though the account of scientific explanation I presented above is one I have discussed before, in this chapter I tried to simplify it and characterized it as involving a field and focus approach: more specifically field elements and a preferred causal model system abbreviated as FE-PCMS. This FE-PCMS account was then applied to a recent set of neurogenetic papers on two kinds of worm foraging behaviors: solitary and social feeding. One of the preferred model systems from a 2002 *Nature* paper was used to illustrate the FE-PCMS analysis in detail and was characterized as a partial reduction.

The chapter closed with a brief discussion of how this FE-PCMS approach partially differed from and partially was congruent with Bickle's "ruthless reductionism" (Bickle 2003) and the recently revived mechanistic philosophy of science of Machamer, Darden, Craver, et al. In that section I could only very briefly indicate some parallels of these approaches with the one developed in the present chapter. Clearly, discussion will continue on these topics for some time to come, and should deepen our appreciation of both the power and the limits of reductive explanations.

References

Bickle, J. 2003. *Philosophy and Neuroscience: A Ruthlessly Reductive Account.* Dordrecht: Kluwer.

Bickle, J. 2006. "Reducing mind to molecular pathways: Explicating the reductionism implicit in current mainstream neuroscience." *Synthese* 152: 411–34.

Bogen, J. 2010. "Analyzing causality: The opposite of counterfactual is factual." October 7. http://philsci-archive.pitt.edu/id/eprint/797. Accessed August 31, 2015.

Butterfield, J. 2011a "Less is different: Emergence and reduction reconciled." *Foundations of Physics* 41 (June): 1065–135.

Butterfield, J. 2011b. "Emergence, reduction and supervenience: A varied landscape." *Foundations of Physics* 41 (June): 920–59.

Chang, A. J., N. Chronis, D. S. Karow, M. A. Marletta, and C. I. Bargmann. 2006. "A distributed chemosensory circuit for oxygen preference in *C. elegans.*" *PLoS Biology* 4 (9): e274. doi:10.1371/journal.pbio.0040274.

Cheung, B. H., F. Arellano-Carbajal, I. Rybicki, and M. de Bono. 2004. "Soluble guanylate cyclases act in neurons exposed to the body fluid to promote *C. elegans* aggregation behavior." *Current Biology* 14 (12): 1105–11.

Chirimuuta, M. 2015. "Explanation in computational neuroscience: Causal and noncausal." Circulated draft.

Coates, J. C., and M. de Bono. 2002. "Antagonistic pathways in neurons exposed to body fluid regulate social feeding in *Caenorhabditis elegans.*" *Nature* 419 (6910): 925–29.

Craver, Carl. 2005. "Beyond reduction: Mechanisms, multifield integration and the unity of neuroscience." *Studies in History and Philosophy of Biological and Biomedical Sciences* 36 (2): 373–95.

Craver, Carl. 2007. *Explaining the Brain: Mechanisms and the Mosaic Unity of Neuroscience.* Oxford: Clarendon Press; New York: Oxford University Press.

Craver, Carl, and Lindley Darden. 2013. *In Search of Mechanisms: Discoveries across the Life Sciences.* Chicago: University of Chicago Press.

Culp, S., and P. Kitcher. 1989. "Theory structure and theory change in molecular biology." *British Journal for the Philosophy of Science* 40: 459–83.

Darden, L. 2005. "Relations among fields: Mendelian, cytological and molecular mechanisms." *Studies in History and Philosophy of Biological and Biomedical Sciences* 36: 349–71.

Darden, L. 2008. "Thinking again about mechanisms." *Philosophy of Science* 75 (5): 958–69.

Darden, L., and Carl Craver. 2013. *In Search of Mechanisms: Discoveries across the Life Sciences.* Chicago: University of Chicago Press.

Darden, L., and N. Maull. 1977. "Interfield Theories." *Philosophy of Science* 44: 43–64.
de Bono, M., and A. V. Maricq. 2005. "Neuronal substrates of complex behaviors in *C. elegans*." *Annual Review of Neuroscience* 28: 451–501.
de Bono, M., D. M. Tobin, M. W. Davis, L. Avery, and C. I. Bargmann. 2002. "Social feeding in *Caenorhabditis elegans* is induced by neurons that detect aversive stimuli." *Nature* 419 (6910): 899–903.
Dizadji-Bahmani, Foad, Roman Frigg, and Stephan Hartmann. 2010. "Who's afraid of Nagelian reduction?" *Erkenntis* 73: 393–412.
Einstein, Albert. 1949. "Autobiographical notes." In *Albert Einstein, Philosopher-Scientist*, edited by Paul Arthur Schilpp, 1–95. Evanston, IL: Library of Living Philosophers.
Glennan, S. 1996. "Mechanisms and the nature of causation." *Erkenntis* 44: 49–71.
Gray, J. M., D. S. Karow, H. Lu, A. J. Chang, J. S. Chang, R. E. Ellis, M. A. Marletta, and C. I. Bargmann. 2004. "Oxygen sensation and social feeding mediated by a *C. elegans* guanylate cyclase homologue." *Nature* 430 (6997): 317–22.
Hempel, C. G., and P. Oppenheim. 1948. "Studies in the logic of explanation." *Philosophy of Science* 15: 135–75.
Hertz, Heinrich. [1893] 1962. *Electric Waves, Being Researches on the Propagation of Electric Action with Finite Velocity through Space*. New York: Dover Publications.
Hodgkin, A. L. and Huxley, A. F. 1952. "A quantitative description of membrane current and its application to conduction and excitation in nerve." *Journal of Physiology* 117: 500–544.
Jacob, F. 1977. "Evolution and tinkering." *Science* 196 (4295): 1161–66.
Jacob, F., and J. Monod. 1961. "Genetic regulatory mechanisms in the synthesis of proteins." *Journal of Molecular Biology* 3: 318–56.
Kandel, Eric R. 2013. *Principles of Neural Science*. 5th ed. New York: McGraw-Hill.
Kandel, Eric R., James H. Schwartz, and Thomas M. Jessell. 2000. *Principles of Neural Science*. 4th ed. New York: McGraw-Hill, Health Professions Division.
Kaplan, M., and C. Craver. 2011. "The explanatory force of dynamical and mathematical models in neuroscience: A mechanistic perspective." *Philosophy of Science* 78 (4): 601–27.
Leuridan, B. 2010. "Can mechanisms really replace laws of nature?" *Philosophy of Science* 77: 317–40.
Machamer, P., L. Darden, and C. Craver. 2000. "Thinking about mechanisms." *Philosophy of Science* 67: 1–25.
Macosko, E. Z., N. Pokala, E. H. Feinberg, S. H. Chalasani, R. A. Butcher, J. Clardy, and C. I. Bargmann. 2009. "A hub-and-spoke circuit drives pheromone attraction and social behaviour in *C. elegans*." *Nature* 458 (7242): 1171–75. doi:nature07886 [pii] 10.1038/nature07886.
Mendel, Gregor. 1966. *The Origin of Genetics: A Mendel Source Book*. Edited by Curt Stern and Eva R. Sherwood. San Francisco: W.H. Freeman.
Mitchell, Sandra D. 2003. *Biological Complexity and Integrative Pluralism*. New York: Cambridge University Press.
Mitchell, Sandra D. 2009 *Unsimple Truths: Complexity, Science, and Policy*. Chicago: University of Chicago Press.
Nagel, Ernest. 1961. *The Structure of Science: Problems in the Logic of Scientific Explanation*. New York: Harcourt.
Popper, Karl Raimund. 1959. *The Logic of Scientific Discovery*. New York: Basic Books.

Rogers, C., A. Persson, B. Cheung, and M. de Bono. 2006. "Behavioral motifs and neural pathways coordinating O_2 responses and aggregation in *C. elegans*." *Current Biology* 16 (7): 649–59. doi:10.1016/j.cub.2006.03.023.

Rogers, C., V. Reale, K. Kim, H. Chatwin, C. Li, P. Evans, and M. de Bono. 2003. "Inhibition of *Caenorhabditis elegans* social feeding by FMRFamide-related peptide activation of NPR-1." *Nature Neuroscience* 6 (11): 1178–85.

Ross, L. N. 2015. "Dynamical models and explanation in neuroscience." *Philosophy of Science* 82 (1): 32–54.

Ruse, M. 1973. *Philosophy of Biology*. London: Hutchinson.

Sarkar, Sahotra. 1998. *Genetics and Reductionism*. New York: Cambridge University Press.

Schaffner, K. F. 1967. "Approaches to reduction." *Philosophy of Science* 34: 137–47.

Schaffner, K. F. 1968. "The Watson-Crick model and reductionism." *British Journal for the Philosophy of Science* 20: 325–48.

Schaffner, K. F. 1969. "Correspondence rules." *Philosophy of Science* 36: 280–90.

Schaffner, K. F. 1972. *Nineteenth Century Aether Theories*. Oxford: Pergamon Press.

Schaffner, K. F. 1977. "Reduction, reductionism, values, and progress in the biomedical sciences." In *Logic, Laws, and Life*, edited by R. Colodny, 143–71. Pittsburgh: University of Pittsburgh Press.

Schaffner, K. F. 1993a. *Discovery and Explanation in Biology and Medicine*. Chicago: University of Chicago Press.

Schaffner, K. F. 1993b. "Clinical trials and causation: Bayesian perspectives." *Statistics in Medicine* 12 (15–16): 1477–94; discussion 1495–99.

Schaffner, K. F. 2000. "Behavior at the organismal and molecular levels: The case of *C. elegans*." *Philosophy of Science* 67: s273–s278.

Schaffner, K. F. 2006. "Reduction: The Cheshire cat problem and a return to roots." *Synthese* 151 (3): 377–402.

Schaffner, K. F. 2008a. "Etiological models in psychiatry: Reductive and nonreductive." In *Philosophical Issues in Psychiatry*, edited by K. S. Kendler and J. Parnas, 48–90. Baltimore: Johns Hopkins University Press.

Schaffner, K. F. 2008b. "Theories, models, and equations in biology: The heuristic search for emergent simplifications in neurobiology." *Philosophy of Science* 75: 1008–21.

Schaffner, K. F. 2012. "Ernest Nagel and reduction." *Journal of Philosophy* 109 (8–9): 534–65.

Simon, H. 1981. *The Sciences of the Artificial*. Cambridge, MA: MIT Press.

Sokolowski, M. B. 2002. "Neurobiology: Social eating for stress." *Nature* 419 (6910): 893–94.

Sommerfeld, Arnold. 1950a. *Lectures on Theoretical Physics*. Vol. 3, *Electrodynamics*. New York: Academic Press.

Sommerfeld, Arnold. 1950b. *Lectures on Theoretical Physics*. Vol. 4, *Optics*. New York: Academic Press.

Sterelny, Kim, and Paul E. Griffiths. 1999. *Sex and Death: An Introduction to Philosophy of Biology*. Chicago: University of Chicago Press.

Suppe, F. 1977. *The Structure of Scientific Theories*. Urbana: University of Illinois Press.

Tabery, J. 2004. "Activities and interactions." *Philosophy of Science* 71: 1–15.

van Fraassen, Bas C. 1980. *The Scientific Image*. Oxford: Clarendon Press; New York: Oxford University Press.

Wimsatt, W. 1976a. "Reductionism, levels of organization, and the mind-body problem." In *Consciousness and the Brain: A Scientific and Philosophical Inquiry*, edited by G. G. Globus, G. Maxwell, and I. Savodnik, 205–67. New York: Plenum Press.

Wimsatt, W. 1976b. "Reductive explanation: A functional account." In *Proceedings of the 1974 Philosophy of Science Association*, edited by R. S. Cohen, C. A. Hooker, A. C. Michalos, and J. W. van Evra, 671–710. Dordrecht: Reidel.

Winther, R. B. 2009. "Schaffner's model of theory reduction: Critique and reconstruction." *Philosophy of Science* 76: 119–42.

Woodward, James. 2003. *Making Things Happen: A Theory of Causal Explanation*. New York: Oxford University Press.

6

Human Behavioral Genetics

Personality Studies, Depression, Gene-Environment Interplay, and the Revolutionary Results of GWAS

In this chapter I return to several of the themes raised in chapter 2, where we discussed human behavioral genetics. I revisit the novelty-seeking gene work mentioned earlier, reconsider the serotonin transporter gene, and discuss some further subtleties regarding replication problems in these two examples and more generally. In that context, I also consider difficulties with a simple *linear* approach to behavioral genetics, as well as some of the gene-environment interactions that initially appeared to be re-establishing a more optimistic view in current behavioral and psychiatric genetics, but have become more controversial in the past few years. These interaction approaches point behavioral and psychiatric genetics toward both environmental studies, in which genetics interacts significantly with the environment over humans' lifetimes, as well as toward more complex strategies that can still be characterized as partly reductionistic. Some of these reductionistic strategies have just recently begun to deal with *epistatic* (gene-gene) interactions that do not fit simpler linear models, and there is a high likelihood, which I will discuss in the following chapter, that *epigenetic* mechanisms are involved. Often these reductionistic approaches are using tools from the neurosciences, including brain imaging, conjointly with genetic research designs. I also briefly discuss some of the newer GWAS results in the area of personality genetics. These results in the personality genetics area are somewhat minimal at present, though the advent of GWAS in the past five years has significantly upended optimistic assessments of the ease with which personality results might be achieved. In the following chapter, I turn my attention to schizophrenia, in which there have been some very recent promising, though still perplexing, results on both the genetics and neuroscience fronts, also involving GWAS.

These complex reductionistic strategies described in this chapter are, I think, at the present time somewhat mixed in terms of their results and their methodological approach. Their strengths and shortcomings point the way toward deeper issues that are ultimately only partially reductionistic, multilevel, and multiperspectival. The importance of these latter approaches has been noted in several preceding chapters on *C. elegans*, and will also be revisited in chapter 8. The reductionistic methods of gene finding using candidate genes have in the past five years been critiqued as GWAS results have seriously called them into question. But GWAS approaches, as suggested in chapter 2, raise their own problems, such as "missing heritability" and the need for extremely large study numbers to identify thousands of genes with tiny effects. These themes are commented on in more detail below after some historical developments in personality genetics have been reviewed, and again in the following chapter, which focuses on schizophrenia. To foreshadow the views to come, skeptics of the ability of genetics (or genomics) to clarify personality will like the results and themes in the present chapter. Optimists will find support for the roles of genetics in the following chapter, which focuses on schizophrenia. Both chapters, however, are cautionary about the complexity and immense challenges that the brain, its normal workings, and its disorders pose for genetics and reductionistic strategies.

PERSONALITY GENETICS: A TALE OF MANY THEORIES

Several research programs in human behavioral genetics have proposed that fairly general personality features have a substantial genetic component. This idea has hoary roots. Darwin and Galton addressed these issues in the century before last. Darwin in particular explored the relation between emotions in humans and animals in a chapter in his *Descent of Man* (Darwin 1871) and more fully the following year in his *Expression of Emotions in Man and Animals* (Darwin 1979). And psychologist Jerome Kagan summarized his investigation into the temperament aspect of personality in his book *Galen's Prophecy* (1994), in which he traced all the way back to Galen of Pergamon the idea that much of the variation in human behavior is due to difference in temperament types. Early work in the area of behavioral genetics on "personality" analogues in animals, especially on temperament, focused on dogs. Two decades of their extensive work on the dog is summarized in Scott and Fuller's monograph (1965), which has played an important role in behavioral geneticists' thinking about the impact of genes on human behavior and personality (Kendler, personal communication, 2005; Plomin et al. 2008, 60–63; Plomin et al. 2013, 50–52). Plomin's route to his work in personality genetics was a consequence of his developing an allergy to mice, which resulted in his stopping his mouse behavioral research. Human personality genetics then seemed to be an attractive alternative area of study (Plomin, interview, March 11, 2004).

There are probably more than two dozen different theories of personality that attempt to capture and systematize in a general scheme those traits that we think of as associated with the lay meaning of the term "personality." Among these described in one introductory textbook are the psychodynamic theories of Sigmund Freud, Carl Jung, Alfred Adler, Karen Horney, Erich Fromm, Harry Stack Sullivan, and Erik Erikson. Also noted are the behavioral and cognitive theories of B. F. Skinner, Albert Bandura, and Julian Rotter. A third major approach includes the dispositional theories of Raymond B. Cattell, Hans Eysenck, and

Gordon Allport, and finally there are the humanistic/existential theories of George Kelly, Carl Rogers, Abraham H. Maslow, and Rollo May (see Feist and Feist 2001; a similar extensive range of personality theories can be found in John, Robins, and Pervin 2008). Further complicating the study of personality is a long-running controversy over whether normal populations exhibit the same or distinctive features of personality found in clinical subjects, some of whom will be diagnosed with "personality disorders."[1] This is an issue I return to below.

As Bouchard and Loehlin state in their review article on personality genetics, many of these personality theories can be described as "literary." Noting that though these are "fascinating . . . and often profound," they suggest that if one wishes examine empirically testable approaches, this will principally involve what are called "trait" theories of the sort that Eysenck developed. All of these trait theories propose key dimensions of both lower and higher orders that I shall describe shortly. The higher-order traits, which some have described as "superfactors," capture more general, higher in the hierarchy, features of personality (Plomin, DeFries, and McClearn 1990). Lower-order traits are often called "facets" (Plomin et al. 2008). Examples follow in a moment. To obtain empirical data, most trait theories use self-report questionnaires containing responses to such items as "I am easily angered," or "I usually act before thinking." It has often been emphasized that "people's responses to such questions are remarkably stable, even over several decades" (Plomin et al. 2008, 239; Plomin et al. 2013, 274). Bouchard has remarked that it is really the differences in personality that make most people interested in "individual differences" (interview, April 30, 2004). On the other hand, Turkheimer believes that though extensive work was done on "individual differences," including an interest in personality, from the 1930s on, that work was not sensitive to "negative traits," as found in many personality genetic studies and recent theories (interview, July 1, 2005).

Of all these trait theories, probably the most widely utilized one is known as the five-factor model (FFM), sometimes called the "Big Five." In agreement with many personality investigators, Avshalom Caspi, himself a major contributor to behavioral genetics in the past 15 years, calls the Big Five model "wonderful" (interview, April 21, 2005) (A similar five-factor model is described in Goldberg 1990, and some authors reserve the "Big Five" usage for Goldberg's approach and term the alternative, but more influential, account the FFM.) This FFM, originally proposed by McCrae and Costa (1989), identifies as its key dimensions or superfactors extraversion, neuroticism, agreeableness, conscientiousness, and openness to experience (culture). (A useful acronym for the five factors, with the name order rearranged, is OCEAN.) The first two in this list, extraversion (including friendliness, cheerfulness, and assertiveness) and neuroticism (including anxiousness, hostility, and depression) have moderate heritabilities of approximately .51 and .46, respectively (Loehlin 1992). The FFM has an associated instrument known as the Revised NEO Personality Inventory (NEO-PI-R)—the three letters corresponding to the three dimensions of neuroticism, extraversion, and openness. One important study that we will discuss later stated that the NEO-PI-R was the preferred instrument because of its "high retest reliability, item validity, longitudinal stability, consistent correlations between self and observer ratings, and a robust factor structure that has been validated in a variety of populations and cultures" (Lesch et al. 1996, 1528).[2] In further support of the FFM, a study of 656 personality disorders patients, 939 general population subjects, and 686 twin pairs that employed two sophisticated statistical tools—a *factor analysis* approach as well as multivariate genetic analyses—empirically confirmed the existence of close analogues to four of these Big Five superfactors (Livesley, Jang, and Vernon 1998).[3]

The FFM was largely the basis of a novel approach to redoing the traditional approach to personality disorders in the DSM-5 revision process. However, the American Psychiatric Association, reacting to extensive criticism of such a change, opted to place the new FFM-based approach in the DSM-5's Section III (761–81), and to retain the traditional account of personality disorders. For extensive commentary on these DSM-5 developments, see the article by Widiger and Krueger (2013) and the following additional articles in that issue of the journal *Personality Disorders*.

In addition to the FFM, other theories of normal personality in use include a revised form of Eysenck's approach called the "big three"—comprising neuroticism (N), extraversion (E), and psychoticism (P)—the Multidimensional Personality Questionnaire (MPQ) of Tellegen, and the Cattell 16 Personality Factor (16PF) personality inventory. In the field of personality disorders there are an even larger number of approaches and instruments currently pursued (see note 1 above for examples). In his very accessible text, Carey (2002) nicely illustrates both lower- and higher-order trait theories of personality, along with sample quantitative data, by using Tellegen's MPQ and Eysenck's theory. For readers wishing some how-to instruction in this area, the Carey volume is a good place to start.[4]

CLONINGER'S TRIDIMENSIONAL PERSONALITY THEORY, NOVELTY SEEKING, AND *DRD4*

In the text, I below will occasionally return to the FFM, which some influential contributors to the personality literature have recently characterized as representing a "Paradigm Shift to the Integrative Big Five Trait Taxonomy" (John, Naumann, and Soto 2008). However, it is Cloninger's well-known biologically based personality theory that initially measured three domains of temperament (and later seven) and was developed in the middle to late 1980s that is of particular interest to us in the first few sections of this chapter. Cloninger notes that his theory was based partly on animal studies and on "the evolution of learning ability across all animals," from unicellular animals on up to primates (interview, May 18, 2004). This interest in Cloninger arises in part because of the theory's close historical association with the *DRD4* gene, which constitutes a "teaching moment" in personality genetics. In addition, Cloninger has recently interpreted his theory within a more interactionist genetic framework, which I discuss later on in this chapter, as well as speculated on some intriguing philosophical implications of his developed account. In addition, my informal *PubMed* review of recent studies in personality genetics (done in August 2012 and essentially reconfirmed in June 2014) suggested that in spite of the "Paradigm Shift to the

Integrative Big Five Trait Taxonomy" noted above, about half of the genetic studies utilize the Cloninger approach, and the other half employ varieties of the FFM (approximately 150 studies each). However as I will discuss below, in even these recent studies we may be seeing an ongoing phase of research in personality genetics.

The original form of Cloninger's theory was operationalized in what is termed the "tridimensional personality questionnaire" (TPQ) (see (Cloninger 1987; Cloninger, Svrakic, and Przybeck 1993). Cloninger's three domains of temperament were novelty seeking (exploratory impulsiveness versus stoic frugality), harm avoidance (anxiety proneness versus outgoing vigor and risk taking), and reward dependence (social attachment versus aloofness). These three domains then had a fourth dimension explicitly added to them, persistence (industry versus underachievement), for a better fit with the data (Cloninger 2004, 40).[5] These dimensions of temperament were hypothesized to be biologically based on distinct chemical and genetic elements, with novelty seeking related to the neurotransmitter dopamine, reward dependence to norepinephrine, and harm avoidance to serotonin. Those scoring "higher on the TPQ Novelty Seeking scale were characterized as impulsive, exploratory, fickle, excitable, quick-tempered and extravagant, whereas those who score lower than average tend to be reflective, rigid, loyal, stoic, slow-tempered, and frugal," according to one team of behavioral geneticists (Ebstein et al. 1996, 78).

The January 1996 issue of *Nature Genetics* carried two important articles reporting a confirmation of Cloninger's prediction tying novelty seeking to the dopamine system in humans. The first article, by a group based in Israel and led by Ebstein, reported a statistically significant association between the personality trait of novelty seeking as measured by Cloninger's TPQ and a long allele (L) of the human D4 dopamine receptor gene known as *DRD4* (Ebstein et al. 1996). A second group, involving Benjamin at the National Institute of Mental Health and Dean Hamer from the National Cancer Institute, provided a replication of the Ebstein et al. study (Benjamin et al. 1996). The Benjamin et al. study used the NEO-PI-R personality inventory mentioned above as tied to the FFM, but developed a mapping between the NEO and TPQ, and also found a statistically significant relation of novelty seeking with the long allele of *DRD4*.

I discussed the *DRD4* locus briefly in chapter 2, but the gene, its allelic variants, and their functions require a bit more detail in order to better understand the checkered history of studies of this locus. This DNA locus, which is located on chromosome 11 and expressed mainly in the limbic (largely emotional) region of the brain, has 10 known alleles. As with other loci that have multiple forms, this sometimes is called a polymorphism. Since this area where the repeat polymorphism occurs is in the third exon or putative third loop of chromosome 11, it often is cited as the exon III polymorphism. These alleles are distinguished by the number of repetitions of a 48-base-pair sequence found in exon III: there may be from 2 to 11 repeats of that 48-base-pair sequence, with the 4-repeat variant being most common in most populations (Ding et al. 2002). The number of the repeats affects the structure of the receptor in which they are present, and test-tube experiments have shown that shorter alleles (2, 3, or 4 repeats) bind dopamine more efficiently than do larger alleles (6, 7, or 8 repeats). The dopamine receptor is coupled within cells to what are known as G-proteins, which are widely found mechanisms that can modulate downstream consequences of its signal, including attention and emotions (Cloninger 2004, 303).[6]

As already noted in chapter 2, *DRD4* studies use the method of allelic association to correlate individual differences in behavior such as novelty seeking with variations in the gene. Such an approach, as Ebstein et al. (1996) noted in the first report of the effect of this gene, works best when "they employ candidate genes that *a priori* make 'biological sense' and have functional significance in the determination of the trait" (1996, 79; also see Risch and Merikangas 1996; Lander and Schork 1994). Ebstein et al.'s source of their "a priori" hypothesis was, of course, Cloninger's personality theory and his TPQ.

It is important to understand that the long allele of *DRD4* (six to eight repeats), which was believed to be "causative" of novelty seeking, only accounted for a *small proportion* of the trait in the populations studied and reported on in the 1996 *Nature Genetics* articles. The Benjamin and Hamer group's article was more emphatic on this point than was the Ebstein et al. paper. Benjamin et al. wrote:

> Although the mean score for the L [long allele] subjects is greater than the S [short allele] subjects by 0.4 standard deviations (a moderate effect size), the distributions are highly overlapping and *DRD4* accounts for only 3 to 4% of the total variance. The broad heritability of Novelty Seeking has been estimated to be 41% from twin studies, and in our families there was a correlation of 0.23 for estimated TPQ-Novelty Seeking scores in siblings. *Thus DRD4 accounts for roughly 10% of the genetic variance, as might be expected if there are 10 or so genes for this complex, normally distributed trait. These results indicate that Novelty Seeking is partially but not completely mediated by genes, and that the DRD4 polymorphism accounts for some but not all of the genetic effects.* (83, 1996; my emphasis)

In figure 6.1 I provide a diagram from (Benjamin et al. 1996) that graphically depicts the subtle difference between individuals' scores related to the two groups.

The S group contained those individuals with only the "short" alleles, those, those with two to five exon III repeats. The L group comprised those with either one or two of the "long alleles," that is, those with six to eight exon III repeats.

Cloninger, Adolfsson, and Svrakic (1996) published a written a commentary on the Ebstein and Benjamin papers that appeared in the same January 1996 issue of *Nature Genetics*. This commentary makes several interesting points that are worth citing. First, Cloninger et al. argue that the TPQ approach was *designed* to be *genetically homogeneous*, in contrast to the NEO FFM-based personality questionnaire, and that this was confirmed by the two studies discussed. Second,

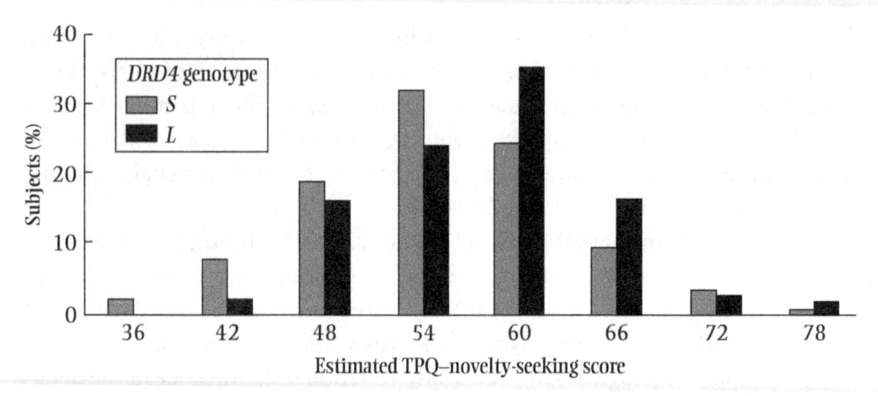

Figure 6.1 *DRD4* influence on novelty-seeking scores.
SOURCE: Benjamin et al. 1996.

Cloninger et al. suggest that "personality development is a complex dynamic process that has many influences on susceptibility to pathopsychology." More specifically, they write:

> A novelty seeker is likely to develop into an extravert with a mature creative character if he or she is also low in Harm Avoidance (optimistic), high in Reward Dependence (sociable) and high in Persistence. In contrast, a novelty seeker is more likely to become disorganized or schizotypal if they are also aloof (low in reward Dependence and average in other temperament dimensions).

They add:

> In contrast to the quick and clear replication of the *DRD4* association with Novelty Seeking by Ebstein et al. and Benjamin et al., replication of specific genetic contributions to genetically complex disorders like schizophrenia have been elusive. The exponential increase in risk of schizophrenia with increasing degree of genetic relationship indicates the importance of nonlinear interactions among multiple genetic factors. When a disease is caused by interactions among multiple susceptibility dimensions, each of which may be oligogenic, then replication of particular genes is unlikely in samples of practical size. (Cloninger, Adolfson, and Svrakic 1996, 4)

In a further commentary on the first two studies involving *DRD4* and novelty seeking Cloninger, Adolfsson, and Svrakic (1996) offered a proposal they thought might assist with the pervasive genetic replication problem of psychiatric disorders such as schizophrenia:

> It may be more fruitful to map genes contributing to temperament, which has a relatively simple genetic architecture, and can be quantified easily and reliably by questionnaire. Later susceptibility to complex disorders like schizophrenia and alcoholism can be evaluated in terms of the risk from

> heritable personality traits and possibly disease-specific factors. In this way, success in mapping genes for a normal personality may signal a fruitful way to map genes for pathopsychology as well. (Cloninger, Adolfsson, and Svrakic 1996, 4)

More recently, Fanous and Kendler suggested that looking for biological endophenotypes shared by schizotypy[7] (or schizophrenia personality disorder) and schizophrenia might clarify the genetic picture:

> It has been suggested that biological endophenotypes that are shared between schizophrenia and schizotypy are more specific indicators of underlying genetic factors, while those specific to schizophrenia may index environmental factors (Cannon *et al.*, 2002; Glatt *et al.*, 2003). If this is true, it could aid in identifying susceptibility genes for schizophrenia, as schizotypal traits may lay closer in the causal pathways leading from gene expression to psychotic illness without the complexity of the illness itself that is introduced by phenocopies, reduced penetrance, etc. (Fanous and Kendler 2004, 47)

These proposals *may* still be a fruitful path to follow, but they will have to do so in the context of an increasing complexity of personality genetics described below, a complexity that is reinforced by slightly later accounts of the nature of personality disorders (see the essays in Lenzenweger and Clarkin 2005 as examples). Furthermore, the true pathway may actually be *bi*directional, with schizophrenia studies directed at prodromal and first-episode psychosis patients providing an important perspective on the personality disorders, issues that I take up again in the following chapter. Further, recent studies in the molecular genetics of schizophrenia, also described in the next chapter, may provide a better purchase on schizophrenia directly than Cloninger envisioned in 1996 (for a later take on this issue see Cloninger 2002).

REPLICATION PROBLEMS AND ISSUES

As already indicated in my earlier discussion of human behavioral and psychiatric genetics, replication studies have been the Achilles heel of this subject, and continuing the *DRD4* story, Cloninger, Adolfsson, and Svrakic (1996) were right to emphasize these difficulties in their contrast between "the quick and clear replication of the *DRD4* association with Novelty Seeking" and schizophrenia. In point of fact, however, though Benjamin et. al. (1996) provided a putative "quick and clear replication" following on the Ebstein group's work, a number of other studies then *failed* to confirm this association. As noted in chapter 2, three meta-analyses published in 2002–2003 raised early questions about any widespread effect of the *DRD4* locus on novelty seeking (see Kluger, Siegfried, and Ebstein 2002; Schinka, Letsch, and Crawford 2002; Munafo et al. 2003). But the prima facie virtually null net effect presented in those meta-analyses needs to be contextualized within an understanding of both the difficulty of achieving replications and the possibility that there may be significant heterogeneity both of the alleles and of the populations studied, and that gene-gene interactions, referred to as epistasis, may further complicate this picture. The point about population heterogeneity is the more controversial one, however, and here I primarily summarize difficulties about replications about which there is a reasonably broad consensus.

This general consensus goes as follows. In general, replications or confirmations, particularly of complex traits, are difficult to obtain, partly because of weak gene effects, genetic heterogeneity, and environmental variation, and also for some subtle statistical reasons. Lander and Kruglyak made these points early and eloquently in an influential 1995 article:

> Failure to replicate does not necessarily disprove a hypothesis. Linkages will often involve weak effects, which may turn out to be weaker in a second study. Indeed there is a subtle but systematic reason for this: positive linkage results are somewhat biased because they include those weak effects that random fluctuations helped push above threshold [of statistical significance], but exclude slightly stronger effects that random fluctuations happened to push below threshold. Initial positive reports will thus tend to overestimate effects, while subsequent studies will regress to the true value. . . . Replication studies should always state their power to detect the proposed effect with the given sample size. Negative results are meaningful only if the [statistical] power is high. Regrettably, many reports neglect this issue entirely.
>
> When several replication studies are carried out, the results may conflict—with some studies replicating the original findings and others failing to do so. This may reflect population heterogeneity, diagnostic differences, or simply statistical fluctuation. Careful meta-analysis of *all* studies may be useful to assess whether the overall evidence is convincing. (Lander and Kruglyak 1995, 245)

To this set of caveats I might add that association studies, being in effect case-controlled retrospective designs employing many looks at the data as part of a data-mining approach, are likely to have high rates of false positive error. I will return to this problem of replications again later in this chapter after we have considered some of the recently acknowledged complexities of gene action involving *DRD4* and also another example to come, the *5-HTTLPR* gene.

FROM THE TPQ TO THE TCI AND BEYOND: CLONINGER'S EVOLUTION

Cloninger's TPQ underwent more than the already noted modification into a four-dimensional model, which explicitly broke out the trait of persistence. In several more years, by 1993, it evolved into a seven-factor model that also included three additional dimensions of "character" (Cloninger, Svrakic, and Przybeck 1993). Cloninger's recent account of this evolution is worth quoting. He writes of his experiences in applying the TPQ:

> I was shocked to observe that I could not distinguish my healthy friends from my patients based on temperament alone! . . . [S]ystematic clinical studies showed that there are other aspects of personality besides the temperament dimensions. Furthermore, these additional dimensions of personality are needed to determine whether someone has a personality disorder, or to specify the degree of maturity of particular individuals. (Cloninger 2004, 42)

These three additional aspects of personality arose from Cloninger's review of other personality measurement systems (Cloninger 2004, 44). He describes them as belonging to "character," which involves personal goals and values. The extension of the TPQ—now known as the Temperament Character Inventory (TCI), or the seven-factor model—added to the original four dimensions another three dimensions: self-directedness (a self concept), cooperativeness (a self in relation with other selves notion), and self-transcendence (our relations to the world as a whole). Representative descriptors of the self-directiveness dimension include responsible versus blaming, whereas cooperativeness includes such descriptors as helpful versus hostile. Self-transcendence includes descriptors such as idealistic versus practical, but also spiritual versus materialistic (Cloninger 2004, 45). Though the initial three traits were speculatively linked to neurotransmitters, these three additional aspects of personality do not appear to have any connection to such biochemical substrates.

Cloninger's TCI also serves as the basis for his exploration of personality theory to *consciousness* (see the text box for details). Relatedly, it should be noted here that the recent thrust of Cloninger's genetic research, and his interpretation of others' research, focuses on epistatic and gene-environment *interactions* rather than on the effects of individual alleles such as the *DRD4* locus discussed earlier in this chapter. The effects of such loci as *DRD4* are still viewed by Cloninger as significant ones, but as best interpreted within this larger interactive context. Cloninger's 2004 chapter that summarizes these recent studies is titled "The Epigenetic Revolution," and though it includes some discussion of true epigenetic mechanisms (which I review in the following chapter), most of the examples he was able to cite are probably more accurately termed gene-gene or gene-environment interactions and are worth discussing further. The difference between Cloninger's assessment of *DRD4*'s effects and some other behavioral geneticists' view of the role of *DRD4* may in part be due to their differing appreciation and expectations of the gene-gene interactions discussed in some detail in the section "Personality Genetics: On a Path to Additional Complexity."

Cloninger and Consciousness

It would take us beyond the scope of this chapter (and this book) to delineate the complex ways the seven-factor model is related to mental disorders as well as to a vision of personal growth and integration into society and the world or universe. Philosophers, especially, will be intrigued that much of Cloninger's account, developed systematically in his 2004 book, involves attention to consciousness.

This account includes a transcendental theory of consciousness development, which can be furthered by specific meditation exercises (described in some detail in chapter 3 of 2004). Individuals who become fully developed achieve a positive state of health and true happiness—and such an achievement results in their *Feeling Good,* the title of Cloninger's 2004 book. This development involves traversing "paths of the psyche," which reach their apogee in a Hegelian-like "third stage of consciousness." Cloninger writes that this "third stage" also corresponds to a parallel third stage of awareness that Krishnamurti (1991) proposed exists for solving human problems. Cloninger adds that "Hegel provides the philosophical foundation, whereas Krishnamurti describes a practical psychology of the same stages of consciousness" (Cloninger 2004, 80). This general analysis is intended to fall within a biopsychosocial approach to understanding human experience as well as employing that biopsychosocial approach in treatment contexts. My summary here is necessarily brief, and for elaborations of these themes the reader needs to be directed to Cloninger's 350-page book.

Cloninger's analysis is integrated with a variety of studies involving both genetic and neurophysiological approaches, though he is explicitly antireductionist. In addition, Cloninger is also antidualist but at the same time avowedly transcendentalist (he has a separate chapter in his 2004 book on American transcendentalist philosophers such as Emerson and Thoreau). I will not pursue an analysis of how the multiple facets of this metaphysical account might work together among themselves, and in concert with the genetics and neuroscience, since it would carry us far from the focus of this chapter.

PERSONALITY GENETICS: ON A PATH TO ADDITIONAL COMPLEXITY

As indicated, the general backdrop here for a preliminary account of gene interactions and gene-environment interaction is Cloninger's belief that "the personality dimensions relevant to the regulation of gene expressions in the brain are likely to be interactive with one another because of the nonlinear nature of complex adaptive systems" (2004, 290).[8] The seven dimensions of the TCI have been investigated genetically and provide results that indicate that both unique and interactive genetic effects are present (for data supporting this distinction see the different heritabilities in Cloninger's table 7.3 (2004, 292). Such interactive effects cannot easily be untangled by classical quantitative twin studies, since specific genes are not identified in classical studies,[9] but molecular methods do indicate some of the ways that such specific interactions may occur.

I first sketch the example of the *COMT* gene, since it has been reported that it interacts with *DRD4*. The *COMT* gene has also been found in some studies, though not others, to be linked to schizophrenia. As we have discussed, dopamine produces its effects via neuronal receptors, such as the one related to the *DRD4* locus. But dopamine is also itself regulated by enzymes that degrade it and *decrease* its availability. One of those enzymes, found in the prefrontal cortex of the brain, is known as *COMT,* an abbreviation for catechol-O-methyltransferase.

The *COMT* gene that produces that enzyme has two variants, one yielding an enzyme with the amino acid methionine (Met) at position 158, and the other with the amino acid valine at that locus. Most common is a heterozygous condition represented by a Val/Met combination, which metabolizes dopamine well, and the Val/Val homozygote works at about the same efficiency as the heterozygote. Met/Met homozygous individuals, however, turn out to have a fourfold *decrease* in the enzyme's activity. And these Met/Met individuals have high harm avoidance scores on the TCI, as well as higher risk for a form of alcoholism. This *COMT* variation, however, explains only a small proportion of the variance, and additional forms of regulation must be present.

COMT itself was reported to be involved in a *three-way* interaction with the *DRD4* locus and another gene, the human serotonin transporter gene (5-HTT), and more specifically with polymorphism in the regulatory promoter region of *5-HTTLPR* already introduced in chapter 2. Recall that serotonin (which more technically is chemically known as 5-hydroxytriptamine, or 5-HT) is the neurotransmitter most closely associated with anxiety and depression aspects of personality. The widely prescribed selective serotonin reuptake inhibitors (SSRIs) such as Prozac are employed to alleviate anxiety and depression, and are thought to act by increasing the amount of serotonin in treated individuals. The 5-HT transporter (5-HTT) is coded by a gene (*SLC6A4*) on chromosome 17q that functions to promote the reuptake of serotonin from the synaptic cleft. Transcriptional activity of the transporter gene is modulated by a regulatory region (a promoter) closely linked to the *5-HTT* gene. That regulatory region, known as *5-HTTLPR*, has two common alleles or polymorphisms: a "short" (S) one with 14 repeat elements and a "long" (L) variant with 16 repeat elements.

This three-way interaction emerged in part from a suggestion that "failure to replicate associations between personality factors and some genes may be due to the presence of additional modifying polymorphisms" (Benjamin et al. 2000, 98). These investigators supported this conjecture with a finding that in the absence of the short (S) *5-HTTLPR* alleles (i.e., *with* a 5HTTLPR L/L genotype) and in the presence of the high-activity *COMT* Val/Val genotype, novelty-seeking scores are higher in the presence of the *DRD4* seven-repeat allele than in its absence. This three-way interaction result has been replicated by independent investigators (Strobel et al. 2003), who wrote that their "study revealed a *DRD4* exon III main effect on Novelty Seeking when *5-HTTLPR* and *COMT* are considered as additional factors in the analysis of variance" (372). They also noted that "our findings underscore the notion that inclusion of additional genetic variations may help resolve some of the inconsistencies in human gene-personality/behavioral correlational studies." Though intriguing, no follow-up replication studies appear to have been done on this gene-gene set of interactions.

As already indicated in chapter 2, *DRD4* and its association with TCI novelty seeking is also regulated by interactions with the *environment*. This so-far nonreplicated result was found by a Finnish research group that has examined two large birth cohorts initially created as part of an ongoing cardiovascular risk study. As part of that research, 2,149 members of this group completed the TCI, and the researchers then selected 154 individuals comprising the top and bottom 10% on

TCI novelty-seeking scores. These selected individuals were then genotyped for the exon III polymorphism of *DRD4*. Recall that *DRD4* has many repeat variants (10 in all) in addition to the two alleles (the "long" 7-repeat and "short" 4-repeat forms that I have discussed to this point), and the Finnish research group also looked at the effects on novelty seeking of the 2- and 5-repeat alleles that are more common in the Finnish population. Of these, only 92 had mother-reported data available on childrearing practices, further restricting the relevant sample. Here I quote from the investigators' abstract from *Molecular Psychiatry* of March 2004, as it concisely and accessibly summarizes their findings:

> A sample of children (n = 92), derived from a representative population sample of healthy young Finns (n = 2149), was studied from childhood to adulthood over 14 years to determine whether the childhood environment moderated the effect of dopamine receptor gene (*DRD4*) polymorphism on novelty seeking (NS). A significant interaction between the *DRD4* alleles and environmental variables was observed. When the childhood-rearing environment was more hostile (emotionally distant, low tolerance of the child's normal activity, and strict discipline), the participants carrying any two- or five-repeat alleles of the *DRD4* gene had a significantly greater risk of exhibiting NS scores that were above the 10th percentile on a population distribution of 2149 adult Finnish women and men. The genotype had no effects on NS when the childhood environment was more favorable. Although the results are preliminary, pending replication, they nevertheless provide important information on the long-term effects of nurture and nature on NS temperament. (Keltikangas-Jarvinen et al. 2004)

This is suggestive of a gene by environment or G × E effect discussed in chapters 1 and 2, though the study still needs to be further replicated. This G × E result is quite similar in kind, however, to that found by the Caspi et al. studies of both MAOA on violence, and serotonin transporter gene (*5-HTTLPR*) and depression noted in chapter 2 (Caspi et al. 2002; Caspi et al. 2003), about which more below. Both studies follow on a working hypothesis that events early in life might show up in terms of antisocial or depressive problems on adolescence or adulthood (Moffitt, interview, March 11, 2004).

AN ANXIETY (AND MAJOR DEPRESSION) GENE: *5-HTT* (AND *5-HTTLPR*)

The dimensions of harm avoidance in the Cloninger theory, and neuroticism in the Big Five or FFM, also stimulated a search for genes that would influence these personality factors. One of the facets of harm avoidance includes anxiety, and neuroticism is assessed in part looking at scales for anxiety and depression. The most interesting of these research programs has focused on the serotonin transporter gene's *regulatory region*, the *5-HTTLPR* gene mentioned in the previous section. This gene continues to interest investigators and has led to some fascinating crossfire between groups of personality researchers, to be discussed further below.

In the same year that *Nature Genetics* published the first two articles on the *DRD4* gene and novelty seeking, an article with the title "Association of Anxiety-Related Traits with a Polymorphism in the Serotonin Transporter Gene Regulatory Region" (Lesch et al. 1996) appeared in *Science*. This article reported that the short variant of the polymorphism reduced the transcriptional efficiency of the *5-HTT* gene promoter, resulting in decreased *5-HTT* expression and decreased 5-HT uptake in the researchers' exploratory test system of human lymph system cells. And these authors added, "Association studies in two independent samples totaling 505 individuals revealed that the *5-HTT* polymorphism accounts for 3 to 4 percent of total variation and 7 to 9 percent of inherited variance in anxiety-related personality traits in individuals as well as sibships."

Lesch and his colleagues, including Dean Hamer, who had been a coauthor of the confirming *DRD4* novelty-seeking paper, assessed the role of *5-HTTLPR* in human personality using three instruments, with the FFM-based NEO personality inventory serving as the primary psychometric tool. They reported significant associations between the short polymorphism of *5-HTTLPR* and the neuroticism facets of anxiety, angry hostility, and impulsiveness. In addition, these authors also used Cattell's 16PF personality inventory and, employing a NEO → TPQ mapping similar to that used in the Benjamin et al. *DRD4* study discussed earlier, additionally assessed their subjects with Cloninger's TPQ instrument. Significant results linking Cattell's anxiety factor and Cloninger's harm avoidance score were also found.

The story of subsequent attempts to replicate this result has an eerie similarity to the *DRD4* story related earlier in this chapter. As in *DRD4*, most meta-analyses report that the cumulative assessment of studies of the effect of *5-HTTLPR* on anxiety-related traits dwindles to nonsignificance (see Munafo et al. 2003; Schinka, Busch, and Robichaux-Keene 2004; Sen, Burmeister, and Ghosh 2004; Munafo, Clark, and Flint 2004a). There are, however, two related advances, also akin to those positive developments found in the *DRD4* saga, which have emerged for the serotonin transporter regulator gene.

Readers might recall from chapter 2 that *5-HTTLPR* polymorphism has been found to have a substantial association with major depression disorder (MDD), albeit only in subjects who had experienced a number of stressful life events. An etiological relation between neuroticism and depression has been investigated quite extensively. Kendler and several of his coauthors have noted, based on twin studies and thus involving quantitative genetics and not molecular genetics, that most of the covariation between neuroticism and major depression is due to shared additive genetic factors (Fanous and Kendler 2004). Murphy and his coauthors, including Lesch, in a 2001 article speculated on the links among *5-HTT*, neuroticism, and depression (and other disorders) using figure 6.2 (from Murphy et al. 2001).

Though Fanous and Kendler (2004) wrote then that "at the present time, there are no established susceptibility genes for MD," the 2003 Caspi et al. study described briefly in chapter 2 suggested that when the environment is considered, *5-HTT* is such a gene.

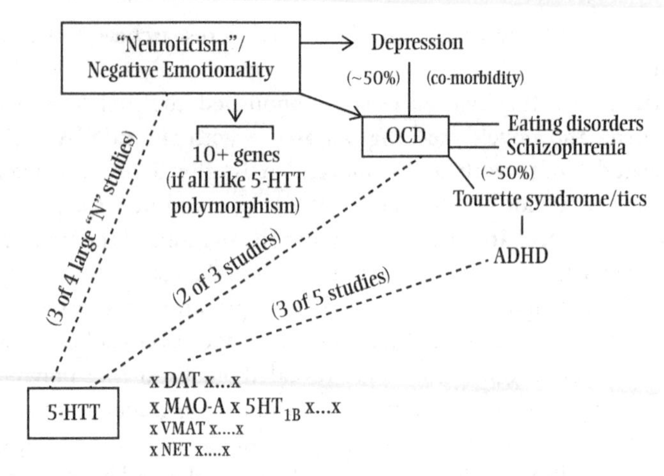

Figure 6.2 Speculative links among 5-HTT, neuroticism, and depression (and other disorders).
SOURCE: Murphy et al. 2001.

The key here was the G × E aspect of the effect of *5-HTTLPR*. In the absence of analyzing stressful life events in subjects carrying the short versus long alleles of *5-HTTLPR*, and sorting the subjects into subgroups based both on their putative genetic susceptibility *and* on their environmental experiences, the differential effect of *5-HTTLPR* remains hidden. Though it might seem plausible that studies of normal personality might *also* look at environmental differences as well as genetic polymorphisms to reveal stronger genetic effects, researchers in the field believe that this would not be a promising strategy. The reason is that the personality traits have been selected to be stable and invariant over long stretches of an individual's life, and thus would not likely show environmental influence of the type that G × E studies search for (personal communications, Kendler and Moffitt, 2005).

Also of considerable interest is a neuroendophenotype aspect of the *5-HTT* polymorphism. In 2002, Hariri et al. noted that there were good grounds, including animal model studies, for expecting an effect of the *5-HTTLPR* polymorphism on anxiety. But they also added, "Not surprisingly, however, the relation between *SLC6A4* [the 5-HTLPRR gene] genotype and subjective measures of emotion and personality has been weak and inconsistent . . ., likely reflecting the vagueness and subjectivity of the behavioral measurements, but also raising concern that the relation may be spurious" (Hariri et al. 2002, 401). Hariri et al., working in Weinberger's lab, looked instead for an effect on the amygdala, the brain organ typically engaged in fear and anxiety responses. Hariri et al. employed functional magnetic resonance imaging, or fMRI, a form of magnetic resonance imaging of the brain that registers blood flow to functioning areas of the brain, to examine the amygdala's response. Weinberger has noted that using imaging provides "huge statistical power," even though its resolution is low, because "you keep pinging the same environment" (interview, December 20, 2005). In the Hariri et al. 2002 study, two age-, gender-, and IQ-matched groups of healthy test subjects were formed, one with individuals with either one or two copies of the S allele (called the S group), with the other containing individuals homozygous for the L allele (the L group). The two groups were given an emotional task involving matching faces depicting the emotions of anger or fear. This test "has been shown to effectively and consistently engage the amygdala" (Hariri et al. 2002, 401). The groups showed significantly different responses to the task, with S allele individuals experiencing a heightened amygdala effect (see the fMRI data as summarized in figure 6.3).

In an early comment on this study, Hamer wrote: "The difference between the two genotype groups was nearly fivefold, accounting for 20% of total variance—an effect size nearly 10-fold higher than in typical experiments using subjective behavioral or personality measures as the outcome" (Hamer 2002, 71).

The Caspi-Moffitt 2003 study followed on, as noted in chapter 2, their MAOA-study published in the previous year, and foreshadowed their *COMT*-cannabis research published in 2005, as well as their study of breastfeeding and IQ (Caspi et al. 2007).[10] But it was the 2003 depression study that has had the most influence. In fact, Duncan and Keller wrote in a recent article that that "study has been extremely influential, having tallied over 3,000 citations and a large number of replication attempts" (Duncan and Keller 2011, 1041). The 2003 article and related G × E studies have also drawn sharp criticism in recent years, in part against the backdrop of GWAS results, and replications that seemed to flip back and forth between confirmations and falsifications. (For a 2011 meta-analysis and additional comments on this area see Blakely and Veenstra-VanderWeele 2011; Hardy and Low 2011; Karg et al. 2011). In this regard, the *5-HTTLPR* depression result is, as

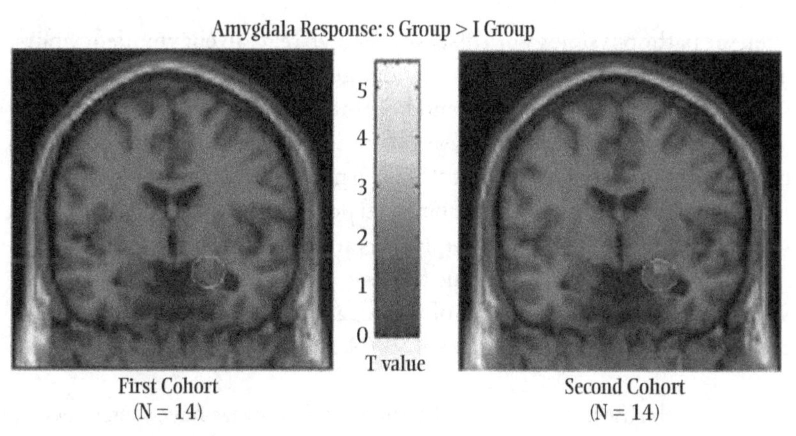

Figure 6.3 Genotype-based parametric comparisons illustrating significantly greater activity in the right amygdala of the s group versus the l group in both the first and second cohort. BOLD fMRI responses in the right amygdala (white circle) are shown overlaid onto an averaged structural MRI in the coronal plane through the center of the amygdala. (The differences here are very subtle in this grey scale rendering, and the reader is encouraged to consult the color original in the source.).

SOURCE: Adapted from Hariri et al. 2002. Reprinted with permission of the American Association for the Advancement of Science.

was noted above, eerily similar to what transpired regarding the *DRD4* finding and its subsequent history. The vagaries of such results had stimulated heightened interest in more rigorous standards in human genetics studies, and that topic is the theme of the following section. For an even more recent discussion of the controversies about G × E studies, see the article by Duncan, Pollastri, and Smoller (2014).

REPLICATION REDUX AND THE REVOLUTIONARY RESULTS FROM GWAS AND GCTA

As discussed in chapter 2, with the advent of GWAS in 2005 (and its burgeoning applications beginning in 2007), an alternative method to look for effects of genes that relate to traits and diseases became available.[11] Interestingly, though some previously discovered genes related to traits and diseases have been confirmed by GWAS results, for example in Crohn's disease, in the behavioral and psychiatric disorders this has not been the case. A companion method discussed briefly in chapter 2, genome-wide complex trait analysis (GCTA) has also begun to be used to confirm heritability estimates in personality and personality disorder research (Lubke et al. 2013). I discuss these methods in the area of schizophrenia research in the following chapter but here want to indicate some of the developments that affect areas more related to personality genetics and the issue of replication. (Below I will briefly describe the GCTA method and its significance.)

In important article published in 2009 in the *American Journal of Psychiatry* and authored by the Psychiatric GWAS Consortium Coordinating Committee made the following statement concerning candidate gene discoveries such as those discussed in this chapter, and frequently throughout this book:

> When the pathophysiology of a disease is known (e.g., an enzyme deficiency), it may be straightforward to define candidate genes and to determine which DNA sequence variants predict who becomes ill. For psychiatric disorders, pathophysiologies are unknown. Most candidate gene hypotheses are based on the effects of psychiatric medications on monoamine neurotransmission, focusing particularly on several functional polymorphisms in dopaminergic or serotonergic pathways (i.e., sequence variants that alter relevant receptor proteins or enzymes).... None has been shown to be associated with a psychiatric disorder with a level of significance that would lead to general acceptance of a finding. (Cichon et al. 2009, 541)

This comment clarifies, to an extent, the checkered story we have seen concerning not only a main effect of a candidate gene, such as *DRD4*, but leads into a similar negative or at least cautionary comment regarding the role of candidate genes in G × E studies (abbreviated as cG × E). In a recent review of such studies, also appearing in the *American Journal of Psychiatry* in late 2011, the authors write:

> Despite numerous positive reports of cG × Es in the psychiatric genetics literature, our findings underscore several concerns that have been raised about the cG × E field in psychiatry. Our results suggest the existence of a

> strong publication bias toward positive findings that makes cG × E findings appear more robust than they actually are. . . . The statistical power to detect cG × E effects is another important consideration. Unless cG × E effects are many times larger than typical genetic main effects, most cG × E studies conducted to date have been underpowered. . . . [which] increases the rate of false discoveries across a field. (Duncan and Keller 2011, 1047)

The authors do believe that G × E effects exist and are likely to be frequent, but do not think the standards for assessing and reporting these results have been sufficient to date.

In a long article published in 2010 in the *American Journal of Psychiatry* (Caspi et al. 2010), Caspi and Moffitt and their colleagues anticipated and responded to many of the problems that Duncan and Keller summarize. Caspi et al. argue that small samples can have the power to detect interesting hypothesis-driven G × E interactions, especially in experimental studies with balanced cell sizes (2010, 518). They also argue that cG × E studies may reveal significant and substantial gene effects that could be hidden from even much larger GWAS studies. Caspi et al. maintain that synthesizing evidence for such candidate genes by environment from across disciplines including neuroscience and animal model studies can provide sound results in the G × E area. Though Caspi et al. do not address the publication bias issue, they do argue that excess reliance on "a purely statistical (theory free) approach that relies wholly on meta-analysis," that is, GWAS studies, does not take into account an alternative "cultural" approach[12] involving "construct validation," which is "theory guided" (521). This view, and a related issue concerning the soundness of meta-analysis in general (on this see Stegenga 2011), will be a source of continuing debate.

Duncan and Keller argue that a key distinction between direct and indirect replications, particularly applied to the Caspi et al. HTTLPR example, is often either ignored or is not made sufficiently explicit by those conducting meta-analyses. The distinction is defined as follows: "Direct replications use the same statistical model on the same outcome variable, genetic polymorphism, and environmental moderator tested in the original report . . . [whereas] indirect cG × E replications, replicate some but not all aspects of an original report." They add that "indirect replications might sometimes be conducted to help understand the generalizability of an original report . . . and might in other cases be conducted out of necessity because available variables do not match those in the original report" (1046). With this distinction in mind, Duncan and Keller say they "believe it is important that only direct replications are considered when gauging the validity of the original cG × E finding," adding that after finding direct support, indirect replications can explore whether the original finding might be generalized. Duncan and Keller criticize a meta-analysis that found strong support for the HTTLPR finding (Karg et al. 2011) in contrast with two other meta-analyses (Munafo et al. 2009; Risch et al. 2009) that did not. The difference in assessments, Duncan and Keller argue, is the including of both direct *and indirect* replications in the Karg et al. analysis, and only direct replications in the negative reports. Duncan and Keller summarize their analysis: "Taken together, the pattern of results emerging from these three meta- and mega-analyses is surprisingly consistent: direct replication attempts of the original finding have generally not been supportive, whereas indirect replication attempts generally have" (1046).

This debate represented by the Psychiatric GWAS Consortium Coordinating Committee and the Duncan and Keller articles on one side and the Caspi et al. 2010 article on the other will almost certainly continue to generate extensive dialogue in the behavioral genetics area for several years to come. In point of very recent fact, see the putative replication of the Caspi et al. (2003) result in Rocha et al. (2015).

The concern with stricter standards of genetics investigations that emerged in the years 2005, 2007, and into the present did lead to a set of rigorous "points to consider in genotype-phenotype association reports," a set of "suggested criteria for establishing the soundness of an initial association report," and a list of "suggested criteria for establishing positive replication" in follow-ups (Chanock et al. 2007, 657–58). These proposals were generated by a prestigious NCI-NHGRI Working Group on Replication in Association Studies. The set of "points to consider in genotype-phenotype association reports" includes among its 50-plus specific suggestions the need for clear statements of the methods for ascertaining and validating affected or unaffected status and reproducibility of classification. In addition, the genotype quality control design for samples needs to be stated, including external control samples from standard accepted sets (such as HapMap) and internal control samples (duplicate samples). Important is an assessment of population heterogeneity and a clear indication of the genetic models tested such as dominant, additive, multiplicative, or trend.

Additional points include a discussion of choice of threshold for significance and the statistical basis for any adjustment for multiple testing and the relationship to overall study power. A sound study should also include a summary of replication and analysis attempts by the authors as well as summary of all known replication attempts by others, including nonreplications. Furthermore, genotyping data and their specifications for deposition in standard databases should be noted, including the data extraction and processing protocols. In terms of the reporting of results, there should be a statement of the strength of any observations, whether there is a suitably large sample size, and sufficiently stringent criteria for significance. Finally, it is desirable to have brief presentation of implications, especially as they relate to further follow-up of genetic markers and corroborative studies to investigate plausibility; and explanations of any notable findings. Appropriate alternative explanations should be proposed and briefly discussed. This list is a much abbreviated selection from the recommendations of the article, and reviewers and editors, as well as readers of articles describing genotype-phenotype association reports, would be well advised to utilize and attend to all of these proposals in assessing the credibility of such findings. Many of these points resonate with the recommendations made by Duncan and Keller in their article discussed above, but also with the views, albeit seen differently, of Caspi et al. as well.

The bottom line on the past 15 years of research into and "discoveries" of genetic effects on behaviors, whether these are "normal" personality traits or features of disorders such as anxiety or depression, needs to be a skeptical one. The field appears to have its optimistic periods (in the early days of the Caspi and Moffitt results, roughly 2002–2007) and its pessimistic periods (both 2000–2002—see Hamer 2002—and again from 2009 to the present, this due to GWAS results). In a very recent article, two researchers who have contributed extensively to personality studies and personality genetics write, "Even though molecular genetic approaches hold tremendous promise for unraveling the distal etiology of conscientiousness [one of the FFM traits], it is important to remember that molecular personality genetics is in its infancy as a field" (South and Krueger 2014, 1367). These investigators add that GWAS findings suggest that there are likely thousands of genes of small effect size that influence personality and that at present these results "tell us little about the biological pathways involved in personality and psychopathology" (1367). This assessment is not very different from the new state of research into causal pathways found in psychiatric genetics generally, as will be described in the next chapter.

At the end of a long review article on personality genetics, the president of the Behavioral Genetics Association, Eric Turkheimer (and his colleagues), recently wrote: "Null hypotheses cannot be confirmed, but the conclusion of this review is that in the genetics of personality, a paradoxical outcome that has been looming for a long time has finally come to pass: Personality is heritable, but it has no genetic mechanism" (Turkheimer, Pettersson, and Horn 2014, 535).

Thus this field of molecular behavioral genetics is still, at best, in its early stages, and the material it works with, human brains and behaviors, is extraordinarily complex. It may be that complexity that at present makes any "genetic mechanism" so occult. Newer methods, however, such as GWAS now appear to be achieving a few replicated results, but caution is advised for reasons mentioned in the present section as well as earlier. Larger studies, which will be expensive to fund and difficult to organize, appear to be one of the keys to achieving progress in this area. The small, if not vanishing, effects of *DRD4* and HTTLPR discussed above also raise the possibility that Cloninger, the FFM, and any other related approaches and instruments are not using constructs that are amenable to assessing genetic effects at the level of specific alleles or SNPs. There is the (theoretical) possibility that these trait categories, as plausible and well developed as they are, are still too close to folk psychology origins and do not parse the emotional and cognitive worlds of humans in quite the right way to disclose real effects at the genetic level. However, because there have been so many replicating studies at the classical (heritability and multivariate) levels, another possibility suggests itself. This is that the personality studies represent an area of "emergent simplifications," due to the robust nature of personality dimensions, though multiple pathways at the specific genetic level yield those robust simplifications (see chapter 5 for a discussion of emergent simplifications). Relatedly, a gloomy implication of this thesis for personality genetics research would be the possibility that there are so many genes of small effect working together in complex ways that despite the moderate heritability of the traits, gene-finding methods will fail to find individual gene effects on personality. Munafo and Flint

(2011) suggest that there may well be thousands of variants with small effects involved in influencing personality. Neuroendophenotypes may possibly provide stronger gene signals, as suggested in various areas of psychology and psychiatry (also see chapter 7 below), though integration of those signals with the mental life of the subjects will still be a sine qua non for useful explanations, and interventions in psychology and psychiatry.

The above paragraphs and this chapter have referred to the "heritability" notion that I analyzed in depth in chapter 1. The heritability concept has by no means disappeared from behavioral and psychiatric genetics; in point of fact, it continues to serve as the backdrop for current and still developing studies involving GWAS and GCTA methods (Plomin and Deary 2015; Visscher et al. 2012). The heritability concept was also raised in chapter 3 in connection with the developmentalist challenge. There it was noted that one of the major mistakes attributed to behavioral geneticists (BG) by the developmentalists was the BG claim that empirical studies can disentangle the contributions of heredity and environment into specific percentages. When we considered this claim from a molecular point of view, I noted that *C. elegans* researchers do distinguish the causal effects of DNA sequences, operating through protein synthesis and protein folding and assembly, from the effects of other molecules (e.g. pheromones) and conditions (e.g. heat/temperature). The causal schema is a complex web or network, but not an indivisible one from the point of view of analysis. At the level of quantitative genetics concepts, recall Kendler's levels 1 and 2 from chapter 1; no such clear separation claim was actually made. The concepts there involve simple heritabilities, which are acknowledged to incorporate effects of the environment in their very definitions. In more complex path analytical diagrams, the ACE models, various sophisticated distinctions are made to capture genetic and environmental interaction effects. Furthermore, path analytical methods *can* capture developmental effects, albeit at a high level of aggregation, as noted in the Kendler analysis of depression over five stages of a lifetime (see chapter 1). Individual developmental effects are represented nicely in the *C. elegans* example, particularly in the Chalfie material, where development and function are neatly parsed. Finally the claim by Lewontin and others that nonadditivity and environmental dependence of genetic effects vitiate the heritability concept (see chapter 1) is a claim not correctly directed. Evidence suggests that many genetic effects represented in classical quantitative genetics are in fact additive, and both the Kendler research at level 2 and the molecular designs of the Caspi team indicate how heritability can guide, in a cautionary way, molecular studies, which can capture such effects. Kendler has noted that the presence or absence of additivity can be empirically verified, but that doing so requires unusually large study sizes, for example 30,000 people (twins and relatives) (interview, December 14, 2003). Relatedly, Loehlin has suggested that the "equal environments hypothesis," discussed in chapter 1 and critiqued by Lewontin and his followers, is "in pretty good shape" despite the fact that "it really hasn't been investigated" (interview, June 7, 2011).

In the past five years, Visscher's team has developed an alternative method to determine narrow-sense heritability that was mentioned in chapters 1 and 2 as well as noted above, one that is arguably free from assumptions that might confound nature and nurture (Plomin and Deary 2015; Visscher et al. 2012). This method has been termed "the first new quantitative genetic method in a century" (Plomin and Deary 2015, 98). These authors also state that "the significance of the method is that it can estimate the net effect of genetic influence using DNA of unrelated individuals rather than relying on familial resemblance in groups of special family members such as monozygotic and dizygotic twins who differ in genetic relatedness" (100). As noted on the main source web page, "GCTA (Genome-wide Complex Trait Analysis) was originally designed to estimate the proportion of phenotypic variance explained by genome- or chromosome-wide SNPs for complex traits (the GREML method), and has subsequently extended for many other analyses to better understand the genetic architecture of complex traits."[13] Some of these additional analyses include the ability to "predict the genome-wide additive genetic effects for individual subjects and for individual SNPs" as well as to "partition the genetic variance onto individual chromosomes."

To the best of my knowledge, this approach to heritability and related issues has not yet been addressed by the developmental systems community, though a recent analysis does indicate that the general heritability concept might have some carefully defined uses in behavioral genetics (see chapter 6 of Griffiths and Stotz 2013). Heritability has *not* disappeared from contemporary behavioral and psychiatric genetics but is now understood in somewhat different ways, as just suggested. The core definition remains the same, but the caveats and the empirical source of heritability estimates have been modified.

References

Benjamin, J., L. Li, C. Patterson, B. D. Greenberg, D. L. Murphy, and D. H. Hamer. 1996. "Population and familial association between the D4 dopamine receptor gene and measures of novelty seeking." *Nature Genetics* 12 (1): 81–84.

Benjamin, J., Y. Osher, M. Kotler, I. Gritsenko, L. Nemanov, R. H. Belmaker, and R. P. Ebstein. 2000. "Association between tridimensional personality questionnaire (TPQ) traits and three functional polymorphisms: Dopamine receptor D4 (*DRD4*), serotonin transporter promoter region (*5-HTTLPR*) and catechol O-methyltransferase (COMT)." *Molecular Psychiatry* 5 (1): 96–100.

Blakely, R. D., and J. Veenstra-VanderWeele. 2011. "Genetic indeterminism, the *5-HTTLPR*, and the paths forward in neuropsychiatric genetics." *Archives of General Psychiatry* 68 (5): 457–58. doi:10.1001/archgenpsychiatry.2011.34.

Bouchard, Thomas J., and John Loehlin. 2001. "Genes, evolution, and personality." *Behavior Genetics* 31 (3) May: 243–73.

Byrd, A. L., and S. B. Manuck. 2014. "MAOA, childhood maltreatment, and antisocial behavior: Meta-analysis of a gene-environment interaction." *Biological Psychiatry* 75 (1): 9–17. doi:10.1016/j.biopsych.2013.05.004.

Carey, G. 2002. *Human Genetics for the Social Sciences*. Thousand Oaks, CA: Sage.

Caspi, A., A. R. Hariri, A. Holmes, R. Uher, and T. E. Moffitt. 2010. "Genetic sensitivity to the environment: The case of the serotonin transporter gene and its implications for studying complex diseases and traits." *American Journal of Psychiatry* 167 (5): 509–27. doi:10.1176/appi.ajp.2010.09101452.

Caspi, Avshalom, Joseph McClay, Terrie E. Moffitt, Jonathan Mill, Judy Martin, Ian W. Craig, Alan Taylor, and Richie Poulton. 2002. "Role of genotype in the cycle of violence in maltreated children." *Science* 297 (5582): 851–54.

Caspi, A., K. Sugden, T. E. Moffitt, A. Taylor, I. W. Craig, H. Harrington, J. McClay, et al. 2003. "Influence of life stress on depression: Moderation by a polymorphism in the 5-HTT gene." *Science* 301 (5631): 386–89.

Caspi, A., B. Williams, J. Kim-Cohen, I. W. Craig, B. J. Milne, R. Poulton, L. C. Schalkwyk, A. Taylor, H. Werts, and T. E. Moffitt. 2007. "Moderation of breastfeeding effects on the IQ by genetic variation in fatty acid metabolism." *Proceedings of the National Academy of Sciences USA* 104 (47): 18860–65. doi:10.1073/pnas.0704292104.

Chanock, S. J., T. Manolio, M. Boehnke, E. Boerwinkle, D. J. Hunter, G. Thomas, J. N. Hirschhorn, et al. 2007. "Replicating genotype-phenotype associations." *Nature* 447 (7145): 655–60. doi:10.1038/447655a.

Cichon, S., N. Craddock, M. Daly, S. V. Faraone, P. V. Gejman, J. Kelsoe, T. Lehner, et al. 2009. "Genome-wide association studies: History, rationale, and prospects for psychiatric disorders." *American Journal of Psychiatry* 166 (5): 540–56. doi:10.1176/appi.ajp.2008.08091354.

Cloninger, C. R. 1987. "A systematic method for clinical description and classification of personality variants: A proposal." *Archives of General Psychiatry* 44 (6): 573–88.

Cloninger, C. R. 2002. "The discovery of susceptibility genes for mental disorders." *Proceedings of the National Academy of Sciences USA* 99 (21): 13365–67.

Cloninger, C. R. 2004. *Feeling Good: The Science of Well-Being*. New York: Oxford University Press.

Cloninger, C. R., R. Adolfsson, and N. M. Svrakic. 1996. "Mapping genes for human personality." *Nature Genetics* 12 (1): 3–4.

Cloninger, C. R., D. M. Svrakic, and T. R. Przybeck. 1993. "A psychobiological model of temperament and character." *Archives of General Psychiatry* 50 (12): 975–90.

Darwin, Charles. 1871. *The Descent of Man, and Selection in Relation to Sex*. London: J. Murray.

Darwin, Charles. 1979. *The Expression of Emotions in Man and Animals*. London: Julian Friedmann; New York: St. Martin's Press.

Ding, Y. C., H. C. Chi, D. L. Grady, A. Morishima, J. R. Kidd, K. K. Kidd, P. Flodman, et al. 2002. "Evidence of positive selection acting at the human dopamine receptor D4 gene locus." *Proceedings of the National Academy of Sciences USA* 99 (1): 309–14.

Duncan, L. E., and M. C. Keller. 2011. "A critical review of the first 10 years of candidate gene-by-environment interaction research in psychiatry." *American Journal of Psychiatry* 168 (10): 1041–49. doi:10.1176/appi.ajp.2011.11020191.

Duncan, L. E., A. R. Pollastri, and J. W. Smoller. 2014. "Mind the gap: Why many geneticists and psychological scientists have discrepant views about gene-environment interaction (G × E) research." *American Psychologist* 69 (3): 249–68. doi:10.1037/a0036320.

Ebstein, R. P., and J. G. Auerbach. 2002. "Dopamine D4 receptor and serotonin transporter promoter polymorphisms and temperament in early childhood." In *Molecular Genetics and the Human Personality*, edited by J. Benjamin, R. B. Ebstein, and R. H. Belmaker, 137–49. Washington, DC: American Psychiatric Publishing.

Ebstein, R. P., O. Novick, R. Umansky, B. Priel, Y. Osher, D. Blaine, E. R. Bennett, L. Nemanov, M. Katz, and R. H. Belmaker. 1996. "Dopamine D4 receptor (D4DR)

exon III polymorphism associated with the human personality trait of novelty seeking." *Nature Genetics* 12 (1): 78–80.
Fanous, A. H., and K. S. Kendler. 2004. "The genetic relationship of personality to major depression and schizophrenia." *Neurotoxicity Research* 6 (1): 43–50.
Feist, Jess, and Gregory J. Feist. 2001. *Theories of Personality*. 5th ed. Boston: McGraw-Hill.
Goldberg, L. R. 1990. "An alternative 'description of personality': The big-five factor structure." *Journal of Personality and Social Psychology* 59 (6): 1216–29.
Griffiths, P. E., and K. Stotz. 2013. *Genetics and Philosophy*. New York: Cambridge University Press.
Hamer, D. 2002. "Genetics: Rethinking behavior genetics." *Science* 298 (5591): 71–72.
Hardy, J., and N. C. Low. 2011. "Genes and environment in psychiatry: Winner's curse or cure?" *Archives of General Psychiatry* 68 (5): 455–56. doi:10.1001/archgenpsychiatry.2011.35.
Hariri, A. R., V. S. Mattay, A. Tessitore, B. Kolachana, F. Fera, D. Goldman, M. F. Egan, and D. R. Weinberger. 2002. "Serotonin transporter genetic variation and the response of the human amygdala." *Science* 297 (5580): 400–403.
John, Oliver P., Laura P. Naumann, and Christopher J. Soto. 2008. "Paradigm shift to the integrative Big Five Trait Taxonomy: History, measurement, and conceptual issues." In *Handbook of Personality: Theory and Research*, edited by Oliver P. John, Richard W. Robins, and Lawrence Pervin, 114–58. 3rd ed. New York: Guilford Press.
John, Oliver P., Richard W. Robins, and Lawrence Pervin, eds. 2008. *Handbook of Personality: Theory and Research*. 3rd ed. New York: Guilford Press.
Kagan, Jerome. 1994. *Galen's Prophecy: Temperament in Human Nature*. New York: Basic Books.
Karg, K., M. Burmeister, K. Shedden, and S. Sen. 2011. "The serotonin transporter promoter variant (*5-HTTLPR*), stress, and depression meta-analysis revisited: Evidence of genetic moderation." *Archives of General Psychiatry* 68 (5): 444–54. doi:10.1001/archgenpsychiatry.2010.189.
Keltikangas-Jarvinen, L., K. Raikkonen, J. Ekelund, and L. Peltonen. 2004. "Nature and nurture in novelty seeking." *Molecular Psychiatry* 9 (3): 308–11.
Kluger, A. N., Z. Siegfried, and R. P. Ebstein. 2002. "A meta-analysis of the association between *DRD4* polymorphism and novelty seeking." *Molecular Psychiatry* 7 (7): 712–17.
Krishnamurti, J. 1991. *The Collected Works of J. Krishnamurti*. 17 vols. Dubuque, IA: Kendall/Hunt.
Krueger, R. F., and J. L. Tackett. 2003. "Personality and psychopathology: Working toward the bigger picture." *Journal of Personality Disorders* 17 (2): 109–28.
Lander, E. S., and L. Kruglyak. 1995. "Genetic dissection of complex traits: Guidelines for interpreting and reporting linkage results." *Nature Genetics* 11 (3): 241–47.
Lander, E. S., and N. J. Schork. 1994. "Genetic dissection of complex traits." *Science* 265 (5181): 2037–48.
Lenzenweger, Mark F. 2006. "Schizotaxia, schizotypy, and schizophrenia: Paul E. Meehl's blueprint for the experimental psychopathology and genetics of schizophrenia." *Journal of Abnormal Psychology* 115 (2): 195–200. doi:10.1037/0021-843X.115.2.195.
Lenzenweger, Mark F., and John F. Clarkin. 2005. *Major Theories of Personality Disorder*. 2nd ed. New York: Guilford Press.
Lesch, K. P., D. Bengel, A. Heils, S. Z. Sabol, B. D. Greenberg, S. Petri, J. Benjamin, C. R. Muller, D. H. Hamer, and D. L. Murphy. 1996. "Association of anxiety-related

traits with a polymorphism in the serotonin transporter gene regulatory region." *Science* 274 (5292): 1527–31.

Livesley, W. J., K. L. Jang, and P. A. Vernon. 1998. "Phenotypic and genetic structure of traits delineating personality disorder." *Archives of General Psychiatry* 55 (10): 941–48.

Loehlin, John C. 1992. *Genes and environment in personality development.* Newbury Park, CA: Sage Publications.

Lubke, G. H., C. Laurin, N. Amin, J. J. Hottenga, G. Willemsen, G. van Grootheest, A. Abdellaoui, et al. 2013. "Genome-wide analyses of borderline personality features." *Molecular Psychiatry.* doi:10.1038/mp.2013.109.

McCrae, R. R., and P. T. Costa Jr. 1989. "Reinterpreting the Myers-Briggs Type Indicator from the perspective of the five-factor model of personality." *Journal of Personality* 57 (1): 17–40.

Munafo, M. R., T. Clark, and J. Flint. 2004a. "Does measurement instrument moderate the association between the serotonin transporter gene and anxiety-related personality traits? A meta-analysis." *Molecular Psychiatry* 10: 415–19.

Munafo, M. R., T. G. Clark, L. R. Moore, E. Payne, R. Walton, and J. Flint. 2003. "Genetic polymorphisms and personality in healthy adults: A systematic review and meta-analysis." *Molecular Psychiatry* 8 (5): 471–84.

Munafo, M. R., C. Durrant, G. Lewis, and J. Flint. 2009. "Gene × environment interactions at the serotonin transporter locus." *Biological Psychiatry* 65 (3): 211–19. doi:S0006-3223(08)00731-2 [pii] 10.1016/j.biopsych.2008.06.009.

Munafo, M. R., and J. Flint. 2011. "Dissecting the genetic architecture of human personality." *Trends in Cognitive Sciences* 15 (9): 395–400. doi:10.1016/j.tics.2011.07.007.

Murphy, D. L., Q. Li, S. Engel, C. Wichems, A. Andrews, K. P. Lesch, and G. Uhl. 2001. "Genetic perspectives on the serotonin transporter." *Brain Research Bulletin* 56 (5): 487–94.

O'Connor, B. P. 2002. "The search for dimensional structure differences between normality and abnormality: A statistical review of published data on personality and psychopathology." *Journal of Personality and Social Psychology* 83 (4): 962–82.

Plomin, R., and I. J. Deary. 2015. "Genetics and intelligence differences: Five special findings." *Molecular Psychiatry* 20: 98–108.

Plomin, Robert, John C. DeFries, Valerie S. Knopik, and Jenae M. Neiderhiser. 2013. *Behavioral Genetics.* 6th ed. New York: Worth Publishers.

Plomin, Robert, J. C. DeFries, and G. E. McClearn. 1990. *Behavioral Genetics: A Primer.* 2nd ed. New York: W.H. Freeman.

Plomin, Robert, John C. DeFries, Gerald E. McClearn, and Peter McGuffin. 2008. *Behavioral Genetics.* 5th ed. New York: Worth Publishers.

Risch, N., R. Herrell, T. Lehner, K. Y. Liang, L. Eaves, J. Hoh, A. Griem, M. Kovacs, J. Ott, and K. R. Merikangas. 2009. "Interaction between the serotonin transporter gene (*5-HTTLPR*), stressful life events, and risk of depression: A meta-analysis." *JAMA* 301 (23): 2462–71. doi:301/23/2462 [pii] 10.1001/jama.2009.878.

Risch, N., and K. Merikangas. 1996. "The future of genetic studies of complex human diseases." *Science* 273 (5281): 1516–17.

Rocha, T. B., M. H. Hutz, A. Salatino-Oliveira, J. P. Genro, G. V. Polanczyk, J. R. Sato, F. C. Wehrmeister, F. C. Barros, A. M. Menezes, L. A. Rohde, L. Anselmi, and C. Kieling. 2015. "Gene-environment interaction in youth depression: replication of the 5-HTTLPR moderation in a diverse setting." *American Journal of Psychiatry* 172 (10): 978–85.

Schinka, J. A., R. M. Busch, and N. Robichaux-Keene. 2004. "A meta-analysis of the association between the serotonin transporter gene polymorphism (*5-HTTLPR*) and trait anxiety." *Molecular Psychiatry* 9 (2): 197–202.

Schinka, J. A., E. A. Letsch, and F. C. Crawford. 2002. "*DRD4* and novelty seeking: Results of meta-analyses." *American Journal of Medical Genetics* 114 (6): 643–48.

Scott, John Paul, and John L. Fuller. 1965. *Genetics and the Social Behavior of the Dog*. Chicago: University of Chicago Press.

Sen, S., M. Burmeister, and D. Ghosh. 2004. "Meta-analysis of the association between a serotonin transporter promoter polymorphism (*5-HTTLPR*) and anxiety-related personality traits." *American Journal of Medical Genetics B: Neuropsychiatriac Genetics* 127 (1): 85–89.

South, Susan C., and Robert F. Krueger. 2014. "Genetic strategies for probing conscientiousness and its relationship to aging." *Developmental Psychology* 50 (5): 1362–76. doi:10.1037/a0030725.

Stegenga, J. 2011. "Is meta-analysis the platinum standard of evidence?" *Studies in History and Philosophy of Science Part C: Studies in History and Philosophy of Biological and Biomedical Sciences* 42 (4): 497–507.

Strobel, A., K. P. Lesch, S. Jatzke, F. Paetzold, and B. Brocke. 2003. "Further evidence for a modulation of novelty seeking by *DRD4* exon III, *5-HTTLPR*, and *COMT* val/met variants." *Molecular Psychiatry* 8 (4): 371–72.

Turkheimer, E., E. Pettersson, and E. E. Horn. 2014. "A phenotypic null hypothesis for the genetics of personality." *Annual Review of Psychology* 65: 515–40. doi:10.1146/annurev-psych-113011-143752.

Visscher, P. M., M. A. Brown, M. I. McCarthy, and J. Yang. 2012. "Five years of GWAS discovery." *American Journal of Human Genetics* 90 (1): 7–24. doi:10.1016/j.ajhg.2011.11.029.

Widiger, T. A., and R. F. Krueger. 2013. "Personality disorders in the DSM-5: Current status, lessons learned, and future challenges. Introduction." *Personality Disorders* 4 (4): 341. doi:10.1037/per0000051.

7

Schizophrenia Genetics

Experimental and Theoretical Approaches

DISORDERS AS OVERLAPPING MODELS

This chapter looks in depth at the experimental basis of searches for schizophrenia susceptibility genes, at some new ways to think about different types of schizophrenia-related genes, and also at some preliminary results of the post-2005 genome-wide association studies (GWAS) and more recent genome-wide complex trait analysis (GCTA) results. In addition I discuss some epigenetic approaches to schizophrenia genetics and consider some general theories of schizophrenia's etiology. In connection not only with the latter, but also because of relations with experimental strategies, it may be useful at this point to consider *what kind of a thing* schizophrenia is, that is, as a mental disorder.

Before we can discuss the specifics of schizophrenia, we need to examine the concept of a (mental) disorder. The issue of how to define a "mental disorder" generally has been one of the most vexing for both psychiatrists and philosophers of psychiatry. There are very general issues and examples that cluster around this topic, including naturalistic approaches (e.g., by Wakefield 1992; Boorse 1977) as well as nonnaturalistic analyses (e.g., by Bolton 2008; Fulford 1990) to disease and disorder definitions. The present chapter is not able to address these more far-ranging issues, but confines itself to a brief discussion of some logical and epistemological features of psychiatric disorders related to schizophrenia genetics.

I will begin by briefly defending a view of a mental disorder as a family of overlapping prototypical multilevel models, an expansion of sorts of the PCMS introduced in chapter 5. This defense is based on some of my previous work on the role of theory in biology and medicine (Schaffner 1980, 1986, 1993), but a very similar view was then further and independently developed in the psychiatric literature in the late 1990s. Within psychiatry proper, this prototype approach is consistent with Bleuler's classic description of schizophrenia, as well as with emerging indications of an underlying genetic heterogeneity of schizophrenia(s). The prototype approach to mental disorders was argued for and underwent a test implementation in the work of Mezzich (Cantor et al. 1980), was subsequently commented on favorably by Frances et al. (1990), and then further developed by Lilienfeld and Marino (1995, 1999). This view of a psychiatric disorder also has implications for reductionistic (and molecular genetic and biological) methodology, in the sense that it allows for a more dimensional and less categorical approach to disorder reduction. More on this below. Finally, in the concluding section, I will consider briefly the ramifications that this analysis has for more precise disorder definition(s), etiology, diagnosis, and therapy, as well as some implications related to DSM-5 (and for later DSMs and ICD iterations), as well as for the emerging Research Diagnostic Criteria (RDoC) initiative of the National Institute of Mental Health.

What might be called the standard approach to contemporary psychiatry can be found in the DSM-IV-TR volume (American Psychiatric Association 2000) and the recent DSM-5 revision (American Psychiatric Association and DSM-5 Task Force 2013), in the ICD-10 "Mental Disorders" chapter (World Health Organization 1992), and in related treatment texts that describe a variety of ways to address the collection of "mental disorders" found in these diagnosis texts. This suggests that to the first approximation, progress in psychiatry would be directed at providing biological and even physical-chemical characterizations of those disorders as well as physical-chemical-biological explanations of them. There are some changes in disorder definition stated in DSM-5, and thus most likely also for ICD-11 (not due out until 2017), but these are not major departures, even though dimensional approaches figure somewhat more in these new developments.[1] Such disorders are in the main presented as categories, and later in this chapter we will look at the category of schizophrenia as a representative example, but the categories do not present necessary and sufficient conditions for the application (diagnosis) of the category in any given patient. Since there can be many specific cases that satisfy the conditions for a diagnosis of a disorder, they are best thought of as "polytypic" (or "polythetic") categories, or in my view, a collection of prototypes related by similarity.

There are other models that can capture such variation and relation by similarity, which will be remarked on further below, but for various reasons I believe that an overlapping prototype approach has the strongest case. From my perspective, what we encounter in biology, and I would argue eventually in psychiatry even as we might proceed reductively, are families of models or mechanisms with, in any given subdiscipline, a few prototypes. (See chapter 4 for a discussion of prototypes as developed in those more "basic science" contexts. Chapter 5's concept of a preferred causal model system (PCMS) can also function as a prototype.) These prototypes are typically interlevel: in biology they *intermingle* ions, molecules, cells, cell-cell circuits, and organs, in the same causal/temporal process. In psychiatry the currently intermingled levels are primarily behavioral data obtained from patients' acquaintances and patients' subjective reports, intercalated in research contexts with genetic and neurobiological/imaging information. In both biology and psychiatry, prototype models are related to each other by dimensional similarity, and there are many interforms. These partially overlapping interforms are often referred as a "spectrum" in psychiatry, as in "autistic spectrum" and "schizophrenia spectrum." The prototypes functioning as markers or signposts in spectra in this view are in effect *narrow classes*—ones that do allow for lawlike/probabilistically causal predictions, and explanations—including in some cases for individuals. Extensive variation among the models and mechanisms in biology is a natural consequence of the result of how evolution operates: by replicating entities with many small variations, and assembling odds and ends—thus what Jacob (1977) called "tinkering." And this kind of evolutionary variation is also currently evident in recent genetic studies of schizophrenia, about which more below.

A prototype approach has other independent arguments in its support in addition to consistency with the way that relationships among biomolecular models seem to behave. Psychologists, beginning with the pioneering work of Rosch in 1975, have argued that prototype representations are closer to the way that humans think naturally than are other more logically strict approaches to concepts (Rosch 1975; Rosch and Mervis 1975; for additional comments on Rosch's approach as well as two alternatives, see Machery 2009, chap. 4).

Let me conclude this section by drawing together some of the above themes and relating them back to the issue of reduction covered in chapter 5. Generally acknowledging a prototype approach in psychiatry allows a dimensional approach to be integrated naturally into the received view of psychiatric disorders. Since I have argued that a prototype approach is also encountered extensively in the biological sciences, including molecular biology and neuroscience, this may facilitate the cross-connections needed to accomplish the patchy reductions we find in the relation between psychiatry and molecular biological neuroscience, including the role of genetics.

THE CORE FEATURES IN THE DIAGNOSIS OF SCHIZOPHRENIA

Schizophrenia is a major mental disorder occurring in approximately one of every hundred individuals, a startlingly high prevalence for such a serious illness. In the United States alone, this translates into more than 2 million affected individuals, at enormous costs. (In 1995, the late Dr. Richard Wyatt, then at the NIMH, estimated the total costs at about $65 billion per year,[2] and similar recent assessments continue to support this amount.) It is found in virtually every culture and region of the world, with little variation among these regions and cultures (Jablensky 1999). Typically schizophrenia appears in early adulthood, with some early cases arising in the mid-teen years. Though the concept of schizophrenia has undergone changes over the past 100 years (see Hoenig 1983 and Gottesman and Wolfgram 1991), a current consensus view found in DSM-IV-TR as well as *DSM-5* lists five "characteristic" symptoms of this disorder:

(1) Delusions
(2) Hallucinations
(3) Disorganized speech (e.g., frequent derailment [ideas slip off track] or incoherence)
(4) Grossly disorganized or catatonic (motor problem) behavior
(5) Negative symptoms, that is, diminished emotional expression or avolition (inability to initiate or persist in goal-directed activities).

For a diagnosis, "Two (or more) of the above need to each be present for a significant portion of time during a 1-month period (or less if successfully treated)" (American Psychiatric Association and Task Force on DSM-IV 2000, 312). DSM-5 qualifies this list, stating that "at least one of these must be (1), (2), or (3)" (American Psychiatric Association and DSM-5 Task Force 2013, 99). In addition, social/occupational dysfunction must be present, and the symptoms need to be evident, though with some waxing and waning, for a *six-month* period. (There are also additional exclusion conditions, and time-course descriptions of different types of durational forms of the disorder found in those afflicted. For comparison the more narrative-like description from ICD-10 can be found in the appendix to this chapter.) Note that the first four of these symptom clusters are what have been called "positive" symptoms, that is, those kinds of symptoms that are not part of normal experience but appear for the first time during psychosis. The "negative" symptoms represent a loss or severe attenuation of what previously were normal characteristics of the subject.

There are some changes in the DSM-5 characterization of schizophrenia, though these seem to be relatively minor modifications. In the DSM-5 version summarized in the list above, there are five of what are termed "Criterion A" or active-phase symptoms of schizophrenia, but these are essentially identical with the list quoted above from DSM-IV-TR. And again, these five core symptoms require that two or more of them be present for a significant amount of time during a one-month period (excepting successful treatment), and, as noted, one of these symptoms must come from the first three listed, what are traditionally termed "positive symptoms." The disorder in some form still needs to be persistent for a six-month period, and there is a requirement that there be a significant decrease in various aspects of personal functioning. (The diagnosis also includes several exclusionary conditions, such as the symptoms not being caused by an abused drug or by a different medical condition.)

The picture of schizophrenia presented here both in IV-TR and the DSM-5 version is essentially that of a fully developed form of schizophrenia. Extensive research on the disorder, however, has indicated that it is part of a "schizophrenic spectrum," including a variety of less severe syndromes, such as schizotypal personality disorder, and this is noted explicitly in the DSM-5 section in which schizophrenia is characterized. That entire DSM section, beginning on page 87, is labeled "Schizophrenioa Spectrum and Other Psychotic Disorders." Also, related research has for some time suggested that schizophrenia is a *developmental* disorder, typically manifesting itself after puberty, but not after 40 years of age (Lewis and Lieberman 2000; Weinberger 1987). Schizophrenia also frequently begins with a somewhat ill-defined "prodrome," on which there has been considerable and continuing investigation as well as controversy. (For more discussion on this topic, including the notion of an "attenuated psychosis syndrome," see DSM-5, 783–86, as well as Schaffner and McGorry 2001, and additional commentary below.)

In the past decade, a number of investigators of schizophrenia have argued that a set of *cognitive* symptoms are more fundamental than these classical five and now "Criterion A" symptoms. In point of fact, the two reductionistic models summarized below focus in on these cognitive symptoms. The rationale for that focus is presented in that section, but toward the end of the discussion of the cognitive symptoms some concerns expressed by the DSM-5 schizophrenia task group concerning the current diagnostic *specificity* of the cognitive symptoms will be noted. These cognitive symptoms, which include deficits in working memory, can also be construed as endophenotypes or intermediate phenotypes that may provide a stronger signal of genetic variation on the clinical syndrome (see the discussion of endophenotypes below).

Schizophrenia was once generally characterized as a "thought disorder," though there are some emotional or affect disturbances found in the illness. Though this is largely outdated terminology, schizophrenia does exhibit significant thought disturbances (see Levy et al. 2010). Schizophrenia is, however, clearly a psychosis or has frequent periods of psychosis associated with it, insofar as there are serious disruptions in perceiving reality. And it needs to be distinguished from a lay notion of "schizophrenia" that implies conflict and the presence of "two minds." There is a mental illness known as "multiple personality disorder," or MPD, but this is not schizophrenia. (For a philosophical account of MPD, now called dissociative identity disorder in DSM-IV-TR as well as DSM-5, see Hacking 1995.)

So characterized, schizophrenia is a *syndrome*, and one without any definitive diagnostic laboratory or medical imaging tests that as yet might further objectify the construct of schizophrenia. Harrison and Weinberger recently summarized the currently available neuropathological findings in schizophrenia, which can be found in work by Jarskog, Miyamoto, and Lieberman (2007) and by Weinberger and Harrison (2011, esp. chap. 18). Ventricular enlargement has been a consistent, if nonspecific finding, and some small reductions in brain volume have been found. But generally the macroscopic, histological, and synaptic connectivity features of the disorder are primarily suggestive of some type of neurodevelopmental pathology, most likely involving synaptic connectivity and disruption at the level of the synapse (see Harrison and Weinberger 2005, 2–4).[3] The absence of definitive neuropathology and laboratory tests for the major psychiatric disorders is one reason that some writers have taken a strong social constructionist position on the "reality" (more accurately, "nonreality") of these disorders. This is a topic about which there is an extensive literature, but one that I will not address in this book. (For an introduction to these issues, see Zachar 2000, 45–48, and for a more extensive analysis, see Horwitz 2002; Schaffner and Tabb 2014).

In spite of the current syndromic characterization of schizophrenia, there is excellent inter-rater and test-retest reliability for diagnosing this disorder (Regier et al. 1994). Treatment options, though far from perfect, have improved significantly with the development of newer "atypical" antipsychotic drugs such as Risperdal (risperidone), Zyprexa (olanzapine), and several more recent atypicals.

Speculation that schizophrenia may mix together several *different* disorders has been considered in the literature. McGuffin, Farmer, and Gottesman (1987) reviewed a number of these proposals and the evidence available at that time, and argued that a spectrum notion of schizophrenia due to differences in familial-genetic loading accounted best for the observed heterogeneity. Some other studies have suggested that there are subtypes of schizophrenia, though those classical Kraepelin subtypes were dropped in DSM-5. It has also been suggested that individuals with severe negative symptoms (including anhedonia, apathy, blunted affect, poverty of speech and thought, loss of motivation, and lack of social interest) might have more relatives with schizophrenia, so this subtype might be more heritable (Malaspina et al. 2000). Familial-based heterogeneity has been reported using studies of the disorder as found in sib pairs (Kendler et al. 1997). These types of findings have led many psychiatrists to believe that the route to still better diagnostic precision and improved treatment options will come through an understanding of the genetic pathophysiology of schizophrenia, though the genetic studies at present also suggest that the most likely valid phenotype "extends beyond the core diagnosis of schizophrenia to include a *spectrum* of disorders including mainly schizophrenic schizoaffective disorder, and schizotypal personality disorder (Owen et al. 2002, citing McGuffin et al. 1987 and Kendler et al. 1995), an issue I pursue in more depth in the next section.

THE SCHIZOPHRENIC SPECTRUM

These types of considerations about heterogeneity, and the likely association of that heterogeneity with genetic liability, have generated broad interest in what can most generally be referred to as spectrum notions of schizophrenia and an even broader notion of a spectrum of psychosis. Similar to the issues I discussed in the previous chapter on dimensional approaches to personality disorders, several overlapping senses of the schizophrenic spectrum can be identified in the literature.

Continuous Variability of Schizophrenic Symptoms in the Clinic

A debate similar to the previous chapter's discussion about a dimensional as opposed to a categorical characterization of personality disorders has also arisen in the schizophrenia area. These considerations seem to be motivated by two major factors. The first, similar to what can be seen in the area of personality disorders, is the continuous variability of patient symptoms. A more precise way to think about this is in terms of the operationalized test instruments used to diagnose schizophrenia. There are many different, though somewhat overlapping, instruments used to assess such putatively schizophrenic patients (for examples and overviews see Rush, First, and Blacker 2008). Some of the more widely used ones are the Brief Psychiatric Rating Scale (BPRS), the Positive and Negative Syndrome Scale (PANSS), the Scale for the Assessment of Positive Symptoms (SAPS), and the Scale for the Assessment of Negative Symptoms (SANS).[4] The individual items in the instruments are scaled, typically with numerical judgments from 0 through 6, along such dimensions as anxiety, conceptual disorganization, and hallucinatory behavior. In the BPRS, the 0–6 seven-point scales represents judgments that a symptom is not present, very mild, mild, moderate, moderately severe, severe, and extremely severe features of that dimension. These scaled instruments underscore the need to clinically capture the extensive dimensional variation of patients' symptoms. Standard classifications such as DSM's and the ICD's, however, employ a categorical, present or absent, assessment of the patient.

Even DSM-IV-TR, however, did acknowledge the potential for a dimensional classification, and in the manual's appendix B the authors provided a three-factor dimensional model involving four degrees of severity, representing the extent to which the patient is psychotic, disorganized, or has negative symptoms. A more precise description is given in the text box.

Alternative Dimensional Descriptors for Schizophrenia

Specify: absent, mild, moderate, severe for each dimension. The prominence of these dimensions may be specified for either (or both) the current episode (i.e., previous 6 months) or the lifetime course of the disorder.

psychotic (hallucinations/delusions) dimension: describes the degree to which hallucinations or delusions have been present

disorganized dimension: describes the degree to which disorganized speech, disorganized behavior, or inappropriate affect have been present

negative (deficit) dimension: describes the degree to which negative symptoms (i.e., affective flattening, alogia, avolition) have been present.

Note: Do not include symptoms that appear to be secondary to depression, medication side effects, or hallucinations or delusions.

Source: American Psychiatric Association 2000, 766.

More recently, the DSM-5 "Psychotic Disorders" workgroup studying those disorders also used, as earlier noted, a proposed category named "Schizophrenia Spectrum and Other Psychotic Disorders" and developed a quasi-dimensional approach to eight features or domains of schizophrenia. These eight domains are hallucinations, delusions, disorganized speech, abnormal psychomotor behavior, negative symptoms (restricted emotional expression or avolition), impaired cognition, depression, and mania. Each of these eight is recommended to be assessed on a quasi-dimensional scale of 0 (not present) to 4 (present and severe. These domains are only advisory and are placed in Section III of DSM-5 (742–44).

Concern about Information (Including Etiological) Loss in Categorical Diagnoses

A second factor motivating interest in dimensional characterizations of schizophrenia is concern that forcing cases into categorical diagnoses may lose important information—information that might point better to the as yet unknown underlying etiology, especially including genetic factors, for the disorder. This aspect of the need to think more broadly about schizophrenia, and include various schizoid-like disorders, was foreshadowed in the early work of Kraepelin and Bleuler, then further developed in the investigations of Kety and his colleagues in the late 1960s and early 1970s (Kety et al. 1971); this approach can also be found in the now over 40-year-old analysis of Shields, Heston, and Gottesman (1975). And more recently this theme of the value of a spectrum or a continuum in pointing toward an genetic etiology of schizophrenia has been stressed by Fanous et al. (2001), who write,

> Establishing that normal and disease states represent end points of a single continuum of liability has important implications for understanding the genetic architecture of not only schizophrenia, but other complex disorders as well, such as hypertension and diabetes, where affection is defined quantitatively, not qualitatively. Establishing-such a continuum of liability in schizophrenia will inform the methodology of molecular genetic studies, where it has been difficult to define an optimal phenotype. Quantitative trait loci analysis may be statistically more powerful than traditional linkage methods in detecting susceptibility genes for complex disorders, but its use assumes the genetic continuity of normal and disease states. (660)

Below I will return to this suggestion, which Fanous et al. (2005) propose be implemented through gene searches for five or six different types of susceptibility genes related to schizophrenia, an implementation that does depend on conceiving of schizophrenia as a dimensional, and not only a categorical, disorder.

Another approach to characterizing a spectrum can be found in a study by Jablensky and his colleagues (Hallmayer et al. 2005). This study utilized a "grade of membership" (GoM) approach that identified a homogeneous familial subtype of schizophrenia that was characterized by a "pervasive neurocognitive deficit." A genome scan was conducted that linked this subtype to chromosome 6p24. They write that "our linkage findings coincide precisely with the strongest signal reported previously in the [1995] Irish high-density family study of schizophrenia" (473). Jablensky and his associates believe that the region that they have identified contains a novel susceptibility gene of relatively strong effect for schizophrenia. (For follow-ups also see Jablensky 2006; Chandler et al. 2010; Jablensky et al. 2012.)

The liability threshold model for schizophrenia, which I introduced in chapter 1, can support a dimensional model of the disorder, though the existence of a breakthrough threshold along a risk continuum is also consistent with a simple categorical model of schizophrenia (Kendler, personal communication, 2005), and that threshold model was recently re-endorsed for current GWAS analyses by Visscher's group (Lee et al. 2011). And a closely related idea can be seen in a renewed interest by the team of Tsuang, Faraone, and Stone in Meehl's notion of schizotaxia (Stone et al. 2012). This dimensional approach for a major psychiatric disorder also resonates with the vision of the structure of the biomedical sciences as involving continuities with prototype landmarks approximating current disorders reintroduced above. Regardless of the position that one takes regarding spectrums in schizophrenia, however, there is ample evidence from both classical quantitative and more recent molecular studies that the disorder has significant genetic components, and it is to these genetic features of the disorder that we turn next, after a brief discussion of yet another aspect of the spectrum of schizophrenia—its prodrome.

THE PRODROME

The "prodrome" concept in psychiatry, namely a period preceding the active phase of psychosis (schizophrenia) in which there is "clear deterioration from a previous level of functioning" (DSM-III-R, 190) is controversial. The prodrome generally refers to that period before an individual experiences his or her *first* psychotic break. The concept is controversial because it is vaguely defined—a vagueness that probably led to the concept being downgraded from its brief appearance in DSM-III-R to its absence in the subsequent version of the American Psychiatric Association's *Diagnostic and Statistical Manual,* DSM-IV. In connection with the development of DSM-5, there was a strong controversy about a close cousin of the prodrome, "attenuated psychosis syndrome," or APS. The result of the controversy was placement of that disorder in Section III of DSM-5 (783–86). Researchers have done much over the past 20 years to make the concept more precise, though research continues.

The prodrome concept is, however, of vital importance because it may identify individuals *at risk* of psychosis *before* the full-blown disorder takes hold. There is suggestive evidence, first reviewed by Wyatt in 1991, that psychosis in and of itself may be biologically toxic to the brain, and thus that intervening early in psychosis may be beneficial in both the short and long term (Wyatt 1991). Even if this *biological* toxicity hypothesis turns out not to be confirmed (see Lieberman and Fenton 2000), it is the case that there is *psychosocial* toxicity of the impact of the subthreshold prodromal symptoms, and those that mimic them but turn not be precursors of psychosis. These symptoms, such as reduced concentration, depressed mood, and anxiety, are typically nonspecific but are disturbing to those experiencing them (Yung and McGorry 1996). (This *psychosocial* toxicity is thought to be more than sufficient to justify considering intervening in those who will accept help.) One aspect of the prodrome that is of particular philosophical interest relates to early and subtle changes in the concept of the "self," and its property of "ipseity," or selfhood. (For accounts of how changes in ipseity can be studied as well as references to brain circuits that may be responsible for these early changes in mental processing in developing schizophrenia see Nelson et al. 2009; Parnas et al. 2005; Nelson, Parnas, and Sass 2014). Research on the prodrome continues and is summarized by Addington and Heinssen (2012); for more discussion on this research and APS see Schaffner (2014).

QUANTITATIVE SCHIZOPHRENIA GENETICS

It has been known for many years that schizophrenia runs in families, and a number of family, twin, and adoption studies have indicated that genetic factors account for an important component. One of the major early pioneers in this area was Irving Gottesman.[5] I discussed the nature of these studies, their methodology, and their findings about schizophrenia in chapter 1, which also cited the importance of Heston's landmark 1966 adoption study of schizophrenia. (Further discussion of the assumptions involved in adoption studies can be found in Carey 2002, 301–3, and additional details and references are available in Owen et al. 2002). Twin studies of the disorder conducted over the last century have strongly tended to support a genetic liability for schizophrenia (reviewed in Cardno and Gottesman 2000).[6] Here it will be helpful to present more explicitly the variation of the concordances or familial risks noted in chapter 1, by providing a diagram from Gottesman's work that summarizes these many studies by many different authors (figure 7.1).

This type of familial risk distribution implies that the underlying genetics is multifactorial and thus that the presence of a single gene of major effect is very unlikely. Gottesman, writing with McGue, also provides a figure (figure 7.2) depicting the difference in fit between a multifactorial threshold model (MFT) such as outlined in chapter 1 and a generalized single locus (GSL) hypothesis. A comparison of the graph of the empirical data with those of the GSL and MFT predictions indicates strong support for the MFT model. Relatedly, Risch has calculated that the recurrence risk picture of schizophrenia is strongly incompatible with a single-locus model, though the specifics involve some nuance (see Risch 1990 for details). In contrast to the strong genetic liability of schizophrenia as evidenced above,

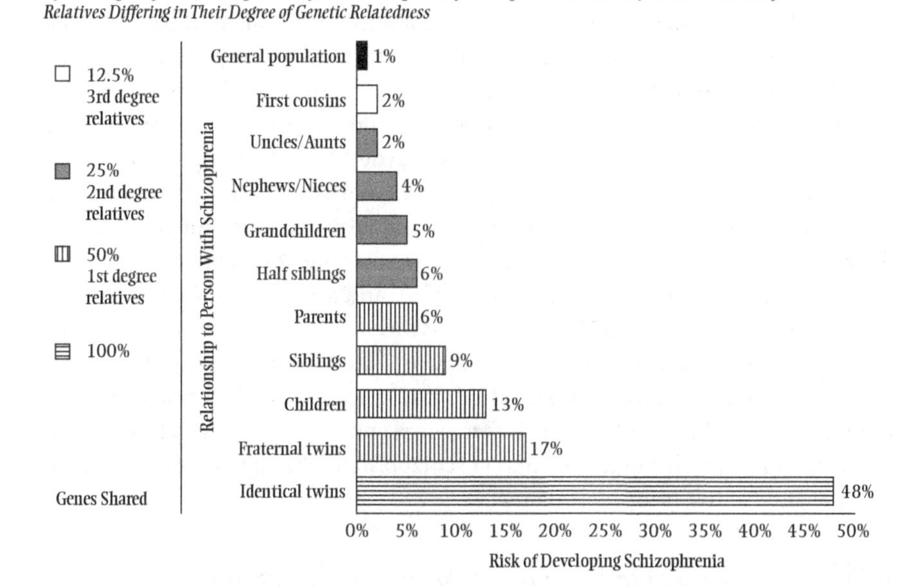

Figure 7.1 Lifetime age-adjusted averaged risks for the development of schizophrenia-related psychoses in classes of relatives differing in their degree of genetic relatedness. SOURCE: Gottesman 1991, © Irving I. Gottesman, with the copyright holder's permission.

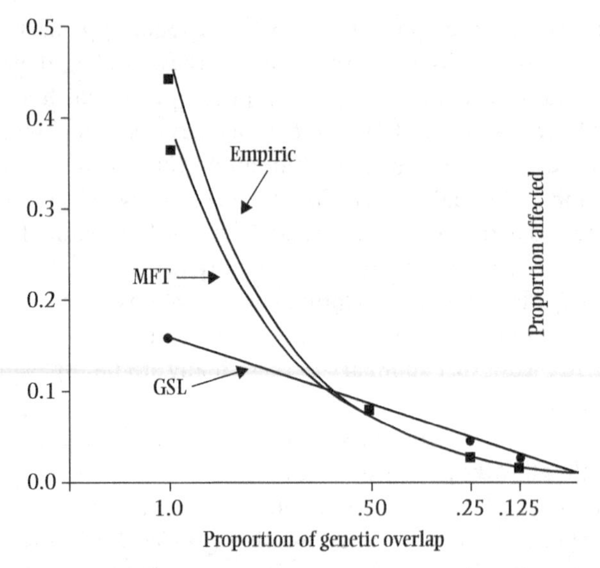

Figure 7.2 Recurrence risk in various classes of relatives for developing schizophrenia under a generalized single locus (GSL) model and under a multifactorial threshold (MFT) model, both contrasted with the combined empirical observations.
SOURCE: Adapted from McGue and Gottesman 1989.

environmental factors have been difficult to pin down, though obstetric complications, winter birth, and low social-economic status seem to be minor risk factors (see Heinrichs 2001), as well as cannabis use (Arseneault et al. 2004; Dragt et al. 2012) and in utero malnourishment found in famines (St Clair et al. 2005).[7] That environmental factors *are* involved in a major way, however, is pointed to by the "only" 48% concordance ratio for genetically identical twins, indicating that some significant environmental differences (possibly in utero and postnatal) are at work.

In recent years more attention has been turned both to the general role of environment in schizophrenia and to gene-environment interactions and related gene-environment correlations. Both kinds of relations were introduced and defined in chapter 1. A new term has been proposed for this topic, namely "ecogenetics" (van Os, Rutten, and Poulton 2008), and a recent *PubMed* search for this term produced over 2,000 citations. This interest has been fueled by both the findings and extrapolations of the work of Caspi and Moffitt discussed in earlier chapters, as well as by recent investigations into epigenetics, which will be treated more extensively below. One recent, somewhat controversial focus of the effects of maladaptive family environments on schizophrenia has been research on the effect of child abuse (see Sideli et al. 2012).

MOLECULAR GENETICS OF SCHIZOPHRENIA: HISTORY AND META-ANALYSES

Sherrington et al. (1988) published evidence of a genetic linkage of two DNA polymorphisms on the long arm of human chromosome 5 to schizophrenia, based on seven British and Icelandic families with multiple affected members. Those results were not able to be confirmed, however, and for years these (most probably) false positives also fueled criticisms of the entire endeavor seeking a precise genetic contribution to mental disorders. These false positive reports, however, have also engendered a much more sophisticated and critical approach to the methodology of complex traits, of which mental disorders are paradigm examples.

Chapter 2 provided an overview of the molecular methods of linkage and association that have proven mathematically sound and empirically effective in uncovering genes strongly influencing such diseases and disorders as cystic fibrosis, Huntington's chorea, and Alzheimer's disease(s). Replicating contributory chromosomal regions and identifying specific genes related to other mental disorders, including schizophrenia, has been much more difficult. The themes of replication problems and interaction that we encountered in chapter 6 related to personality genetics in general, and *DRD4* and *5-HTT* in particular, reprise themselves in gene-finding studies in bipolar disorder (manic depressive disorder) and in schizophrenia. After a period in the years roughly from 2002 through 2007 when the picture regarding schizophrenia genes seemed to be clearing, the situation has again become considerably more complex.[8]

SCHIZOPHRENIA MOLECULAR GENETICS: GWAS SAYS, "NOT SO FAST!"

GWAS Developments

During the period roughly from 2002 through 2007, the discovery of two strong candidates as risk genes, dysbindin and neuregulin, and of several others as quite probable genetic liability factors, fueled optimism that we had "turned the corner" as regards molecular psychiatric genetics (Kendler 2004).[9] These and a small number of additional genes related to schizophrenia replicated to some extent, but their support now seems somewhat weaker, though these assessments could still change as active research in schizophrenia genetics continues. In several cases *chromosomal* abnormalities that point to specific genes, such as *DISC 1*, have also been associated with schizophrenia.

Beginning initially in 2005, but developing rapidly in 2007, SNP results from genome-wide analyses began to be available and become increasingly influential, as described in chapter 2. As previously noted, these GWAS results identify loci, but not necessarily genes—though it is suspected relevant genes are close by (or may be identical with) the loci. Psychiatric geneticists have seen "a flood of such data" from GWAS, including a large consortium study of schizophrenia published in September 2011 in *Nature Genetics* (Ripke et al. 2011). Many of these newer studies also involve CNVs (copy number variations) as well as SNP studies. But the results have only recently become more productive of consistent replications, but not of odds ratios (ORs) > 1.2. (A gene or SNP with an OR of 1.2 roughly translates into an increase in risk of

20%, so from 1 in 100 for population-wide schizophrenia risk to 1 in 80.) This may be changing as new results from GWAS analyses continue to be reported (such as Ripke et al. 2013 and even more recently Consortium 2014), but these studies have not yet really illuminated the molecular pathways involved in schizophrenia (Kendler 2013).

Including those with relatively weak support, the current literature that reports GWAS findings, including a very recent large study (Consortium 2014), suggests that the schizophrenia susceptibility genes number over 100. Other analyses that initially cast their net even more widely (and are more tolerant of very weak evidence) have suggested that the genes may number in the thousands (see Ayalew et al. 2012). Discoveries continue at a rapid pace, and reporting on them can overwhelm even the databases tracking gene discoveries for psychiatric researchers. (In fact, a list of schizophrenia susceptibility genes that was provided until 2011 at http://www.szgene.org/ has been "frozen," presumably awaiting coding advances as well as new reports of loci and SNPs, if not genes, such as the very recent *Nature* study (Consortium 2014).)

But GWAS results in the past eight years, as well as CNV identification, have begun to produce some tantalizing results. (A summary of these can be found in a recent book chapter, Saunders, Duan, and Gejman 2012, and also in Consortium 2014.) One of the strongest results identifies the mixed histocompatibility complex (MHC) of genes on chromosome 6. These results suggest that one aspect of schizophrenia may be related to disorders of the immune system. These could include failure to ward off infections that dispose to schizophrenia, or they could involve autoimmune aspects. Though these results are promising, it is too early to determine what the MHC signal means. Further investigation, probably involving large studies, as well as advances in the basic biology of genetic regulation, will be needed before these puzzles can be adequately addressed, a view developed in some detail by Saunders, Duan, and Gejman (2012).

In this chapter, I will only briefly discuss one of the frequently favored susceptibility genes related to schizophrenia, namely the neuregulin-1 gene. I picked this gene in part because of its role in the second reductionistic model covered below. Neuregulin, however, has an effect size that is quite small, only ~1.2 in terms of odds ratios. In contrast, Mendelian disorder genes, such as the Huntington's chorea gene, have effect sizes about 100 times as strong (Kendler 2005).

Neuregulin-1 (*NRG1*) as an Example of a Susceptibility Gene

This gene was found on chromosome 8p21-22 by Stefansson et al. (2002) using the deCODE Icelandic population database (and also data from a mouse model). A whole genome scan provided suggestive evidence for the 8p12-21 region, and a follow-up study using higher-resolution techniques focused in on two large-risk haplotypes. Application of an early application of the SNP strategy then disclosed a core haplotype with a highly significant association with schizophrenia and which contained the neuregulin-1 (*NRG1*) gene (Stefansson et al. 2003). This gene produces a molecule affecting neuronal growth and development, as well as glutamate and other neurotransmitters, and also glial cells (see Moises, Zoega, and Gottesman 2002). *NRG1* may additionally affect myelin in the brain. Neuregulin is believed to be regulated by a postsynaptic cell receptor known as ErbB3/4. The *NRG1* effect is small, accounting for approximately 10% of the schizophrenia risk in the Icelandic population, but the association has also been confirmed in a Welsh study and one Chinese population, though not in another Chinese population (Hong et al. 2004) nor in the Irish high-density study (Thiselton et al. 2004). Again, the effect size is tiny in comparison with Mendelian genes, being less than 1.2 in terms of odds ratios, and there is extensive population heterogeneity (Gong et al. 2009). The *NRG1* gene, the *ERB4* gene, and its pathway, however, figure prominently in our second model of schizophrenia, as recently summarized by Law et al. (2012), and other studies have also supported a role for neuregulin more generally (Morar et al. 2011). More will be said about *NRG1* below in connection with our reduction examples.

GENES AS FUNDAMENTAL? OR MULTILEVEL MODELS IN A REDUCTIONIST GUISE?

The Fundamental Unit of Genetic Influence and Replication?

Most accounts of genetics in schizophrenia are actually simplifications of the more complex study results. Often one gets what Harrison and Weinberger (2005, 14) refer to as "replication in a broad sense (i.e., to variation in the gene) but *non*replication in detail (i.e., to a particular allele or haplotypes or to risk v. 'protective' haplotypes)." This, they noted, is "perplexing" and will require additional studies to resolve. Possibly we are seeing the results of allelic heterogeneity, environmental effects, or complex regulatory interactions, possibly also including epigenesis, in these disparate details (more about the latter issue later). It is a real possibility that some of these genes, for example dysbindin, may work in the regulatory capacity (Funke et al. 2004), but the mechanism or mechanisms by which dysbindin may influence schizophrenia remain speculations, including converging pathways and influences on dendritic spines (Williams, O'Donovan, and Owen 2005; Walker et al. 2011). But even assuming that specific alleles or haplotypes (a small collection of alleles) are replicated in detail, the effect size is highly likely to remain quite small, on the level of an odds ratio or relative risk size of less than 1.5 and frequently lower than that.

The fundamental genetic unit of influence is typically thought of as the allele, in comparison to a marker (which is just another allele or polymorphism) or a haplotype (which is a collection of closely linked alleles); recall those haplotypes cited above in connection with the schizophrenia-related genes. Relatedly, search strategies, whether they be linkage based or association based, have traditionally seen as their postsequencing endpoint the identifications of key allelic variants in a *gene*. However, in their influential article, Risch and Merikangas (1996) urged that the SNP or single nucleotide polymorphism be the focus of future association studies, which has turned out to be the case in GWAS analyses. This tack encounters difficulties of replication in part, however, because of large numbers of SNPs that likely require extraordinarily low levels of significance (1×10^{-8}) in order to compensate for low prior probabilities and multiple tests. Meeting such requirements would be possible if very large numbers of subjects were available, but they have initially not easily been found (see Neale and Sham 2004, but also a more optimistic assessment also citing the recent Consortium study with 37,000 individuals, Tansey, Owen, and O'Donovan 2015). In their still intriguing article, Neale and Sham (2004) argued that *gene-based* strategies, in contrast to both SNP and haplotype designs, are likely to be optimal in discovering functional variants related to disease etiology. They further state:

> We propose a shift toward a gene-based approach in which all common variation within a candidate gene is considered jointly. Inconsistencies arising from population differences are more readily resolved by use of a gene-based approach rather than either a SNP-based or a haplotype-based approach. A gene-based approach captures all of the potential risk-conferring variations; thus, negative findings are subject only to the issue of power. In addition, chance findings due to multiple testing can be readily accounted for by use of a gene-wide-significance level. (353)

The distinction among units of genetic influence (and optimal search design) also has implications for replications. Kendler has suggested (personal communication, 2005) we have four potential empirical levels: that of the specific allele, the individual marker, the haplotype, and the gene. Depending on what one thinks is the evolutionary history of the mutation that an investigator is trying to study, one would predict different "units of replication." If you are examining a unique mutation occurring once on a specific haplotypic background, then you would expect replication at all four levels. However, if this situation is quite different, for example, with multiple mutational events on different haplotypic backgrounds, then you would only expect replication of the last, or gene level. Though GWAS has resulted in a number of changes in the genetic picture, Neale still held to this general position as of 2012 (personal communication, October 2012).

Multilevel Models in a Reductionist Guise?

Some sense of the interacting complexity of these various schizophrenia-influencing genes might be appreciated by looking at several of the speculative models that have been published. A relatively early model by Owen, Williams, and Harrison was typical and is reproduced in figure 7.3. The types of gene action (and interaction) are noted in the figure.

Two alternative models with somewhat different (though partially overlapping) foci are discussed in the following sections. These models resonate with

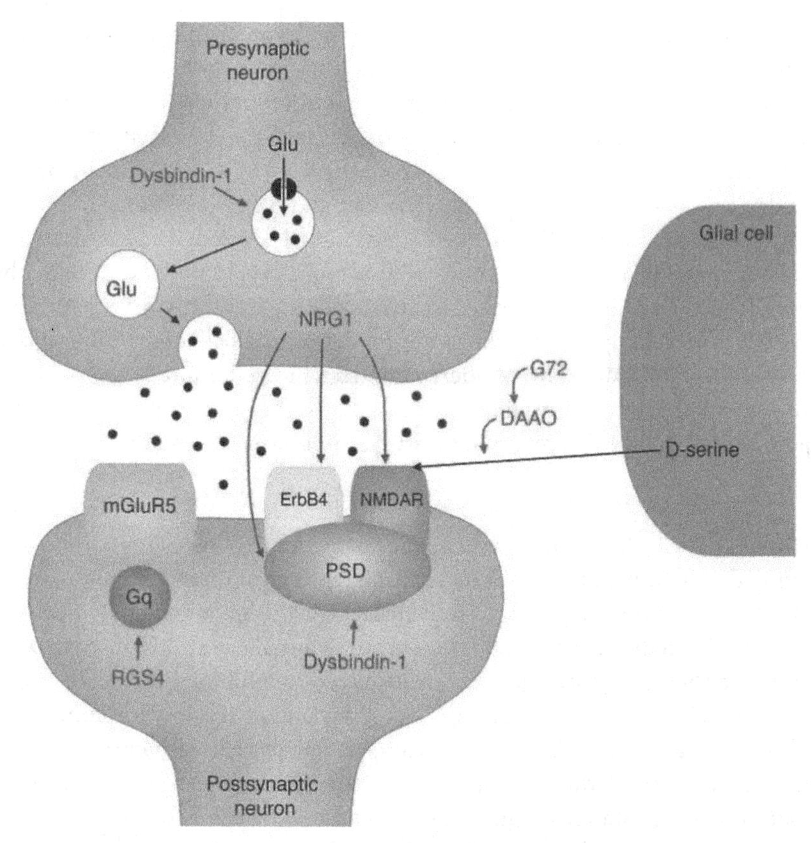

Figure 7.3 A speculative schematic representation of a glutamergic synapse involving several interacting schizophrenia susceptibility genes.
SOURCE: Owen et al. 2004, with permission.

issues raised in our reduction discussions, especially chapter 5, but are more directly related to psychiatry and to human studies than the model organism orientation in chapter 5.

The value of such models is that they organize our knowledge and may, as additional data are generated on both genetic and neuropathological fronts, including structural and functional neuroimaging, begin to produce falsifiable models that can account for the development of, and point to interventions for, this devastating disorder.

The "Lewis Model of Schizophrenia"

This subsection begins by providing an overview of what we might call the "Lewis Model of Schizophrenia." The model is situated within a broader context of evolving concepts of schizophrenia, including the additional attention since the late 1990s to cognitive dimensions of the disorder, for example, disturbances in memory. As noted above, a number of investigators see these deficits as a "core domain" of schizophrenia, perhaps underlying a number of aspects of the disorder, though they are not as dramatic as the traditional hallucinations and delusions. An important aspect of the Lewis model is that it provides mechanisms and pathways that terminate in effects on human brain gamma frequency oscillations, now believed to be one of the key underlying factors in brain function and dysfunction, such as schizophrenia. The Lewis model fits into the increased attention to the neurotransmitters glutamate and GABA, which are as relevant to schizophrenia today as was dopamine originally (Kendler and Schaffner 2011). The Lewis approach also intercalates with an overlapping model developed by Harrison and Weinberger, our second model, which is also more explicitly backgrounded by the role of the *NRG1* gene mentioned earlier, including its downstream pathways. Though this section focuses on the Lewis model, at its close some of the significant features of the Harrison-Weinberger model will also be noted.

The attention to gamma oscillations, as perhaps a key factor in schizophrenia, is quite recent. A useful recent review article summarizes some of this work as follows:

> There has been a fascinating convergence of evidence in recent years implicating the disturbances of neural synchrony in the gamma frequency band (30–100 Hz) as a major pathophysiologic feature of schizophrenia. Evidence suggests that reduced glutamatergic neurotransmission via the N-methyl-D-aspartate (NMDA) receptors that are localized to inhibitory interneurons, perhaps especially the fast-spiking cells that contain the calcium-binding protein parvalbumin (PV), may contribute to gamma band synchrony deficits, which may underlie the failure of the brain to integrate information and hence the manifestations of many symptoms and deficits of schizophrenia. (Woo, Spencer, and McCarley 2010, 173)

This review article adds further that gamma oscillation disturbances might affect important developmental aspects of the schizophrenia pathophysiology, as well as introduce "excitotoxic or oxidative injury to downstream pyramidal neurons, leading to further loss of synapses and dendritic branchings" (173), in both prodromal and early phases of the disorder. It should also be noted that gamma oscillations have been considered related to the "binding problem" of consciousness, whereby multiple sense streams are bound into one common perception (Merker 2013).

The extent to which gamma oscillation pathologies can be identified with standard schizophrenia symptoms is still under discussion and requires further research in psychiatry. The review article just cited, however, does provide a number of references that support these connections, stating that "gamma abnormalities have been found to be associated with various symptom domains of schizophrenia, such as hallucinations, thought disorder, disorganization, and psychomotor poverty" (Woo, Spencer, and McCarley 2010, 176, which provides references.) As noted above, the Lewis model is one of the more detailed models that account for gamma oscillation disturbances in schizophrenia.

Lewis, who along with most of his research collaborators is at the University of Pittsburgh, was profiled in a recent issue of *Nature*. There, the current director of the National Institutes of Health, Thomas Insel, wrote of the Lewis investigations into schizophrenia that the model "provides something this field really needed: a framework for linking observations at the molecular, cellular and systems levels. We haven't had a story that crossed those levels of explanation before. And his story, whether it pans out in all its details or not, is invaluable for doing that" (Dobbs 2010, 155). In addition to levels integration, the Lewis model also has had some preliminary practical results and has led to some phase II clinical trials conducted with novel compounds predicted to improve cognitive dysfunction in schizophrenia.

The Lewis model argues for the importance of cognitive dimensions of schizophrenia, noting that "cognitive deficits are present and progressive years before the onset of psychosis, and the degree of cognitive impairment is the best predictor of long-term functional outcome" (Lewis et al. 2012, 57).

In one of the Lewis group's most recent reviews, they write:

> Deficits in cognitive control, a core disturbance of schizophrenia, appear to emerge from impaired prefrontal gamma oscillations. Cortical gamma oscillations require strong inhibitory inputs to pyramidal neurons from the parvalbumin basket cell (PVBC) class of GABAergic neurons. (Lewis et al. 2012, 57)

The central element in the Lewis model, then, is the circuit that involves these inputs from the parvalbumin (PV) basket cells to the brain's pyramidal neurons using the GABA neurotransmitter, though pyramidal cells themselves use the neurotransmitter glutamate. (There are actually two types of basket cells, which we need not discuss in this simplified account, and there is also another type of cell, chandelier cells, which assist in the regulation of the oscillatory activity of the pyramidal neurons as noted briefly below.) In addition, there are two important input-regulating receptor types for glutamate; one is termed the NMDAR (N-Methyl-D-aspartate receptor), introduced briefly above in the discussion of gamma oscillations, these being under the influence of several type(s) of schizophrenia susceptibility genes, including *NRG1*. There is also an additional glutamate receptor type, abbreviated as AMPAR (a-amino-3-hydroxy-5-methyl-4-isoxazolepropionic acid receptor), which mediates fast synaptic neurotransmission. (The NMDA receptor is the main molecular mechanism controlling memory function, but a balance between both types of receptors may explain some features of pyramidal cell oscillation regulation. For details on this see Lewis et al. 2012; Gonzalez-Burgos and Lewis 2012; also see figure 7.4.)

The Lewis model has evolved in the past five years from the form originally presented (Lewis and Sweet 2009) to one that now focuses more on the basket cell interactive role with pyramidal cells (see Lewis et al. 2012; Lewis 2014). This is an evolution that has largely been driven by additional empirical findings over these past five years. The current form of the model is most easily presented in the form of a published figure that contrasts the process leading to gamma oscillations in normal cells versus schizophrenia-affected cells (see figure 7.4). The figure shows several alterations in disordered cells.

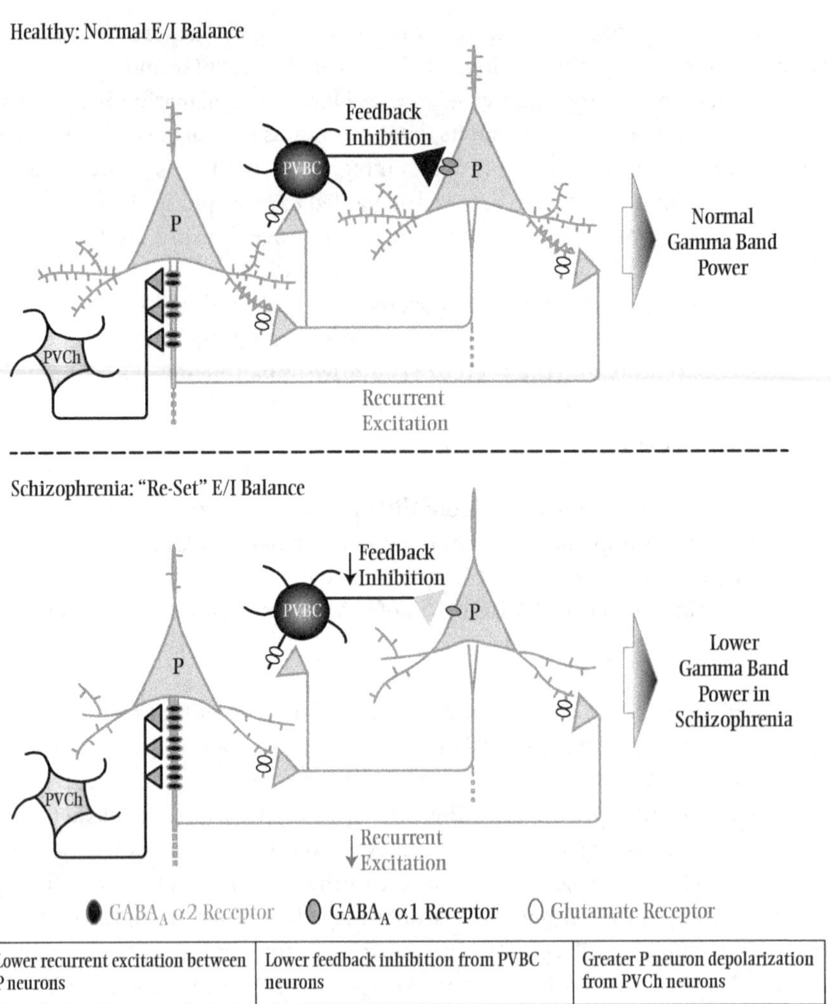

Lower recurrent excitation between P neurons	Lower feedback inhibition from PVBC neurons	Greater P neuron depolarization from PVCh neurons
• Smaller dendritic arbor • Fewer dendritic spines	• Reduced GABA synthesis • Increased suppression of GABA release • Fewer GABA$_A$ α1 receptors • Reduced chloride influx	• Reduced GABA re-uptake • More GABA$_A$ α2 receptors

Figure 7.4 Alterations in disordered cells in schizophrenia leading to gamma band power changes. See text for explanation.
SOURCE: Adapted by David Lewis from Lewis et al. 2012, with permission from Elsevier.

First, the key pyramidal cells making up 75% of the DLPFC cortex engage in recurrent excitations regulated by feedback inhibition by the basket cells to produce stable gamma oscillations. However, the pyramidal cells are smaller in schizophrenia and also have fewer dendritic spines because of the disorder. (Dendritic spines—tiny buttons found on the dendrites—act both to facilitate synaptic strength and to assist in the transmission of electrical signals to the neuron's cell body.) The affected pyramidal cells thus cannot output as strong a signal to the regulating PV basket cells, which also may be down-regulated by less effective NMDA receptors on those cells. This decreased signal induces a compensatory response in the basket cells, in the form of decreased inhibition via less (or less active) glutamate enzyme, itself controlled by the *GAD67* gene (thus there is an equivalent increased excitation by the basket cells). Recent evidence that the chandelier cells are excitatory suggests that they too are part of a (still insufficient) compensatory response. This net undercompensation results in a gamma oscillation output that is still suboptimal in comparison with healthy pyramidal cortex cells, yielding symptoms of disordered cognitive control.

A considerably oversimplified summary of the Lewis model might be the following:

> Pyramidal cell signal strength ↓ (via smaller cells and decreased dendritic spines in schizophrenia) plus NMDAR ↓ → PV basket cell activity ↓ (as feedback compensation) → pyramidal cell oscillation activity still insufficiently increased = schizophrenic symptoms (the types of symptoms varying depending on the circuits involved, thus memory in some areas and decreased tone sensitivity in the auditory region).

This model as described thus far does not examine in any detail what may be causing the decrease in NMDAR activity, which may be an important factor in schizophrenia, beyond an effect on the basket cells. An examination of NMDAR activity related to the Lewis model remains somewhat equivocal, with a recent article suggesting that "additional research is required to determine the particular cell type(s) that mediate dysfunctional NMDAR signaling in the illness" (Gonzalez-Burgos and Lewis 2012). Various additional hypotheses have been developed, one that identifies the neuregulin gene (*NRG1*) malfunction and another more recent proposal by the Lewis group that cites dysbindin gene effects (Gonzalez-Burgos and Lewis 2012). The former (*NRG1*) hypothesis is the most developed in the literature, having been extensively researched by Harrison's group at Oxford, also working with Weinberger and his colleagues at the NIH (Weinberger is now director at the Lieber Institute for Brain Development in Baltimore, Maryland). An overview of the multiple pathways through which *NRG1* may affect schizophrenia can be found in my article on models (Schaffner 2008), and more recent hypotheses related to NMDA and LTP (including effects on pyramidal cells in the hippocampus) are developed by Pitcher et al. (2011), Nicodemus et al. (2010), and Deakin et al. (2011). A recent communication from the Harrison and Weinberger groups, based in part on animal model studies (mice and rats) as well as human lymphocytes and brain tissue, proposes "a genetically regulated, pharmacologically targetable, risk pathway associated with schizophrenia and with [*NRG1* and] ErbB4 genetic variation involving increased expression of a PI3K-linked ErbB4 receptor (CYT-1) and the phosphoinositide 3-kinase subunit, p110δ (PIK3CD)" (Law et al. 2012, 1).

Close inspection of the Lewis model publications, as well as alternative approaches such as the Harrison-Weinberger pathway just cited, indicate that a number of questions remain to be answered about the details of the cells and circuits affecting schizophrenia. Furthermore, a focus on the cognitive deficits of schizophrenia emphasized by the Lewis model must not be interpreted as arguing for a major *clinical* diagnostic focus on these features; cognitive deficits will continue to be but one aspect of schizophrenia symptoms.

In point of fact, the DSM-5 category "Schizophrenia Spectrum and Other Psychotic Disorders" has included cognition in the eight features or domains of schizophrenia, but only as advisory in the DSM-5 "Assessment Measures" section. As noted earlier, these eight domains are hallucinations, delusions, disorganized speech, abnormal psychomotor behavior, negative symptoms (restricted emotional expression or avolition), impaired cognition, depression, and mania. Each of these eight is recommended to be assessed on a quasi-dimensional scale of 0 (not present) to 4 (present and severe). Impaired cognition is *not*, however, included in the five (Criterion A) major diagnostic features of the disorder because the workgroup believed that cognitive impairment lacked the necessary *specificity*, being present in a number of other psychiatric disorders as well. They originally wrote that a "wealth of data suggest that this separation [from other disorders] is not sufficient to justify inclusion of cognition as a Criterion A symptom of schizophrenia."[10]

The Lewis model and the overlapping Harrison-Weinberger model appear largely directed at capturing fully developed schizophrenia. There are other nascent models that aim at representing brain circuits that may be responsible for prodromal aspects of the disorder, but studies of these alternative circuits are in the early stages of research. One aspect of the prodrome that is of particular philosophical interest relates to early and subtle changes in the concept of the "self" and its property of "ipseity," or selfhood. For accounts of how changes in ipseity can be studied as well as references to brain circuits that may be responsible for these early changes in mental processing in developing schizophrenia see articles by Nelson et al. (2009), Parnas et al. (2005), and Nelson, Parnas, and Sass (2014).

In spite of these diagnostic limitations, including, as noted earlier, incomplete knowledge of the details of the cells and (probably multiple) circuits affecting schizophrenia, we can abstract from these reductionistic models (with a focus on the Lewis model) what a reductionistic account of a representative psychiatric disorder involves.

A Partial Reduction Model from Philosophy of Science Applied to Schizophrenia

As noted in chapter 5, there have been two contrasting approaches to reduction that I described as *sweeping* and *creeping*. A more formal variant of the sweeping type of reduction found in the philosophy of science has been the Nagel account of theory reduction, wherein large domains of a scientific subject are reduced by a theory developed in a quite distinct scientific subject area. Examples in the reduction literature include thermodynamics' reduction by statistical mechanics and optics' reduction by electrodynamics. (For references to this literature and an extended optics example see Schaffner 2012.) In an earlier essay on reduction in psychiatry with applications to schizophrenia (Schaffner 1994), I sketched a slightly modified version of that sweeping approach, which some refer to as the

"Nagel-Schaffner model" (Dizadji-Bahmani 2010). But in subsequent years, I have come to believe that a "creeping" or partial reduction approach is more applicable both in the biological sciences, as described in chapter 5, and in psychiatry. I continue to hold, however, that in physics something like a Nagelian type of theory reduction is largely applicable (again see Schaffner 2012 for details). Let me briefly sketch in summary form some of the reasons for this shift of position; also see chapter 5 for additional details.

It turns out, after a general, careful review of putative reduction examples in the history of science, that ultimately all attempts at theory reductions in science are incomplete, partial, or patchwork in character. The nature and degree of the incompleteness varies with the type of science, however, and in physics, because of its "Euclidean" form of theories, virtually "systematic" or "sweeping" Nagelian-type reductions seem possible. It was the prospect of these sweeping reductions that motivated the original Nagel model of reduction and its application to the now canonical example of thermodynamics and statistical mechanics. However, for classical Nagelian-type reductions to work well, the branches of science that enter into a reduction relation need to be axiomatizable in a fairly straightforward manner, that is, by using an integrated, small number (three to six) of core scientific laws. Also, connections between the reduced and reducing fields need to be straightforward and relatively simple, though the connections may well be far from obvious. (These connections are frequently referred to as "bridge laws," though I think other expressions such as "reduction functions" or even better "connectability assumptions" carry less philosophical baggage. This issue of "connectability" has generated a very large and often contentious literature in the philosophy of science; see Schaffner 2012 for references.)

Even in physics, though reductive expositions can give the *appearance* of a complete systematic reduction, closer inspection reveals this systematicity fails at the margins (Schaffner 2012). Further, in more complex sciences such as molecular genetics and neuroscience, both of these stringent conditions, simple axiomatizability and simple connectability, fail in significant ways, though that they do fail, or would fail, was not necessarily obvious at the beginning of the Watson-Crick era of comparatively uncomplicated molecular biology. Furthermore, in psychology and psychiatry, where *conscious experience* figures prominently, connectability will ultimately need to involve the thorny and contentious issue of what are termed "neural correlates of consciousness" (NCC) (Koch 2004). Some philosophers and scientists believe these "correlates" will ultimately need to be *identities* (of mind entities and properties with brain entities and properties), and not just correlates of them. And this issue will also likely generate variants of what is termed in the philosophy of mind "the hard problem" (Chalmers 1996).

In summary, more than 60 years of explorations and refinements of alternative reduction possibilities strongly support a more "creeping" form of reductions, particularly in the biological sciences and the neurosciences. But this view, however, does not impugn the importance of such partial reductions. Such partial reductions amount to potentially Nobel Prize–winning accomplishments in unifying and deepening significant, though "local," areas of scientific investigation.

The Background and Locus of Partial Reductions in Biology and the Neurosciences

The models depicted in figure 7.4 from the Lewis et al. (2012) paper discussed in detail above *are* what I referred to in chapter 5 as *preferred causal model systems* (PCMSs) for the specific scientific article in which they appear. The model (or in this case two contrasting models representing healthy and schizophrenically disordered brains) is what accomplishes the partial reduction—more on this in the next few pages. The model is simplified and idealized and uses causal language, such as "provide recurrent excitation," "increased depolarization," and "resulting in." The PCMS is clearly interlevel, showing cells, parts of cells (dendrites), receptors, and connections among them (circuits), ultimately resulting in differing prefrontal cortex gamma band oscillations. I find that it is best to approach such models keeping in mind the scientific fields on which they are based, and the specific alternative explanations (model sketches in a sense) that a scientific article proposing the preferred model provides as contrasts within the fields. Here, the fields on which the model draws are generally molecular genetics and especially neuroscience. Scattered throughout the article are occasional alternative but possible causal pathways or contrasting alternatives that are evaluated as not as good an explanation as those provided in the preferred model system. (One example in the Lewis article is the previous role of the chandelier cells, now modified in this article. Another alternative is noted in the second to last paragraph at the end of the article, where what is explicitly termed an "alternative model" involving NMDA receptor hypofunction resulting in a GAD67 deficit in the PV neurons is noted, but mainly rejected.)

I will not say more about how fields and alternative explanations are characterized here, but interested readers can see chapter 5 as well as other work (Schaffner 2006) for additional details, albeit with somewhat different terminology. I should note here that the notion of a "pathway" appears to be especially prominent in models in the biomedical sciences, and here including psychiatry. Pathways appear to be more general than any specific mechanism that may be part of a pathway, but like "mechanisms," they are also typically interlevel in character. On this point, compare my previously expressed views (Schaffner 2011, 153) and Machamer's (2000) on whether "mechanism language"—a prominent development since the later 1990s in philosophy of science—will suffice; also see comments in chapter 5 on pathways in PCMSs as well as the concluding section of chapter 5.

The preparations or specific *experimental systems* investigated in the clinic and laboratory (this may include several data runs of the "same" experimental system) are identified in their relevant aspects with the (more abstract) preferred causal model system. In this Lewis article, which is partly a review article, several such data sources or experimental systems are noted, ranging from human patients with cognitive deficits due to schizophrenia, to stem cells derived from patients, to results based on mice and rat studies, some of which involve using hippocampal or brain slices. At the abstract or "philosophical" level, the reductive explanation proceeds by identifying the laboratory or clinical experimental system with the theoretical system—the preferred model system (PCMS)—and exhibiting the event or process to be explained (in philosophy this is called the explanandum) as the causal consequence of the system's behavior. The explanans (or set of explaining elements) here uses molecular neuroscience and is mainly comparative rather than involving strict quantitative derivational reasoning, in the sense that in this paper two qualitatively different end states—the prefrontal gamma band power—are compared and contrasted. The theoretical system (the PCMS) utilizes generalizations of varying scope, some from basic biochemistry involving the balance of chloride ions. Often these generalizations appeal to similarity analyses among like systems, especially different animal models and humans, to achieve the desired scope, as well as make the investigation of interest and relevance to other neuroscientists and psychiatrists. For those concerned with philosophical rigor, the preferred model system and its relations to model-theoretic explanation can be made more philosophically precise (and technical), along the lines suggested in other publications (including Schaffner 2006) and chapter 5.

The discussion sections of scientific papers are the usual place where larger issues are raised, and where extrapolations are frequently found and future research proposed. In this Lewis paper, which is partly a review article and thus is not structured in the usual way that more specific scientific and medical papers are, with materials and methods, results, and a discussion section, the extrapolations and more general and future-looking comments are found at the end of the paper. These comments are partly cautionary, noting that the current model is based on a "limited portion of cortical circuitry," and propose that additional studies involving connectivity patterns in the DLPFC, better methods to assess compensation, and further animal models are needed.

This Explanation Is Both Reductive and Nonreductive

The Lewis example is typical of molecular neuroscience psychiatric explanations of behavior, including cognitive aspects. Behavior is an organismic property, and in the example the explanation appeals to entities that are *parts* of the organism, including molecularly characterized genes and molecular interactions such as receptor bindings and G-protein coupled receptor mechanisms—thus this is generally characterized as a *reductive* explanation. But it represents *partial* reduction—what I termed reduction of the *creeping* sort—and, similarly to what we saw in chapter 5 in connection with reductionistic explanation of worm feeding behaviors, it differs from *sweeping* or comprehensive reductive explanations because of several important features.

1. The model does not explain *all* cases of schizophrenia; and different though somewhat related models can address other features of the disorder, for example, prodromal features of schizophrenia.
2. Some of the key entities, such as the signal producing the gamma oscillations, have not yet been fully identified.

3. The causal account utilizes what might be termed "middle level" entities, such as neuronal cells, in addition to molecular entities. Further work that is even more reductionistic can address ion channels in the neurons, but this research is yet to come. (I discuss some of this ion channel work in connection with worm touch sensitivity in chapter 3 above.)
4. It is not a quantitative model that derives behavioral and cognitive properties from a small set of axioms or even from rigorous general equations of state, but is causally qualitative and only roughly comparative.
5. Interventions to set up, manipulate, and test the model are at higher aggregative levels than the molecular, such as the traditional diagnosis of patients afflicted with schizophrenia and the use of brain slices.

The explanation does meet three conditions that seem reasonable for a reductive explanation, namely:

1. The explainers (here the preferred model system shown in figure 7.4) are a partially decomposable microstructure in the organism/process of interest.
2. The explanandum (the cognitive deficits) is a grosser (macro) typically aggregate property or end state.
3. Connectability assumptions (CAs), sometimes called bridge laws or reduction functions, are involved, which permit the relation of macrodescriptions to microdescriptions. Sometimes these CAs are causal sequences as depicted in the model figure where the output of the neurons under one set of conditions causes strong gamma band power, but in critical cases the CAs are provisional identities (such as weak gamma oscillations = cognitive deficits), which later may be expanded, perhaps to identities underlying neural correlates of consciousness (Koch 2004).

Again, though etiological and reductive, the preferred model system explanation is not "ruthlessly reductive," to use Bickle's phrase (2003), even though classical organismic biologists would most likely term it strongly reductionistic in contrast to their favored nonreductive or even antireductionist cellular or organismic points of view. It is a *partial* reduction.

ENDOPHENOTYPES REDUX

As noted in both chapters 2 and 6, the search for endophenotypes that might provide a stronger signal of gene action on a trait or disorder has been of major recent interest in behavioral and psychiatric genetics. The endophenotype approach has also been pursued by a number of groups working on schizophrenia.[11] One additional thought on the use of endophenotypes is that they might provide a more scientific approach to the *reclassification* of mental disorders, which have been inherited by the psychiatric community from the early years of clinical observation (Gottesman and Gould 2003).

There have been a number of candidate endophenotypes proposed based on biochemical, neurophysiological, cognitive, and neuropsychological measures. Gottesman and Gould (2003) propose five criteria for endophenotypes:

1. The endophenotype is associated with illness in the population.
2. The endophenotype is heritable.
3. The endophenotype is primarily state-independent (manifests in an individual whether or not illness is active).
4. Within families, endophenotype and illness cosegregate. Subsequently, an additional criterion [criterion 5] that may be useful for identifying endophenotypes of diseases that display complex inheritance patterns was suggested. . . .
5. The endophenotype found in affected family members is found in nonaffected family members at a higher rate than in the general population. (639)

Candidate endophenotypes that have been well researched include sensory-motor gating and eye tracking abnormalities in schizophrenia. Cognitive endophenotypes have also come in for special interest, including deficits in working memory. Interestingly, Heinrichs notes that "average effect sizes derived from common clinical tests of attention, memory, language, and reasoning are twice as large as those obtained in structural magnetic resonance imaging and positron emission tomography studies" (Heinrichs 2005, 229). Interestingly, gamma band oscillations that figure importantly in the Lewis model above have been suggested as a useful endophenotype by Hall et al. (2011) and more recently by Buzsaki et al. (2013). However, in spite of high promise for endophenotypes to elucidate the nature of disorders such as schizophrenia, and well-funded research programs in the subject,[12] no dramatic breakthroughs in the use of this important tool have yet occurred.

Extensive research into plausible endophenotypes, however, continues. A very recent summary of 94 candidate genes is suggestive of extensive interactions among these endophenotypes (Greenwood et al. 2012; Greenwood et al. 2011; Greenwood et al. 2013). Multiple testing is an issue in endophenotypes research as in other GWA studies, though a creative bootstrapping method was developed for use with a custom microarray in this set of analyses. There are various caveats about replicability in this area, as we found in other genomic investigations, so though extremely promising as well as neuroscientifically informed and plausible, this is no magic bullet as yet. The importance of pathway analyses, discussed more extensively below, is also stressed in these recent articles, and intriguing depictions of the various pathways, especially the importance of glutamate signaling, are underscored (see Greenwood et al. 2011, figure 4; Greenwood et al. 2012, figure 3).

The role that endophenotypes can play in a search for etiological dimensions of schizophrenia is graphically depicted in a diagram from Gottesman (figure 7.5), which parallels a similar picture of the role of genetics in IQ, as shown in chapter 2

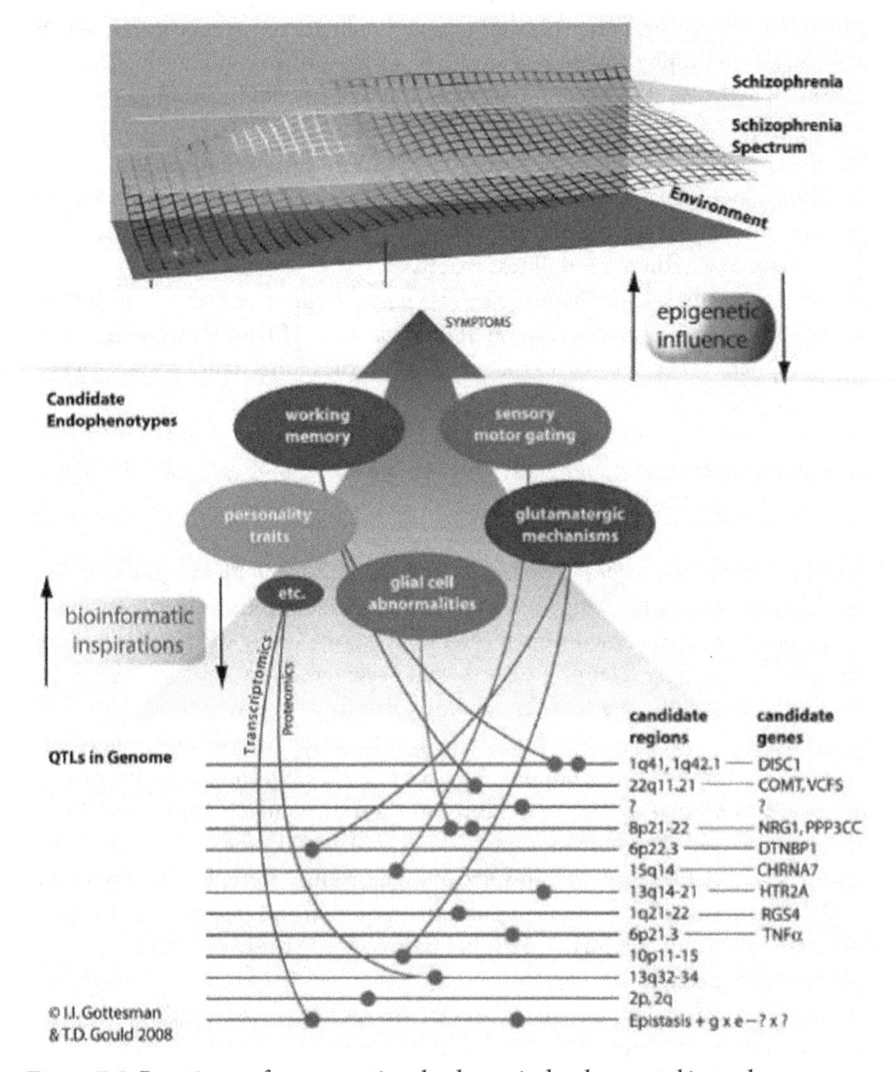

Figure 7.5 Reaction surface suggesting the dynamic developmental interplay among genetic, environmental, and epigenetic factors that produce cumulative liability to developing schizophrenia. Gene regions where linkage findings are more consistent are in bold, while gene regions corresponding to candidate genes or endophenotypes are shown in normal lettering. Many of these endophenotypes are discussed in detailed reviews addressing overall strategies for schizophrenia discriminators, sensory motor gating, oculomotor function, working memory (sometimes synonymous with informational processing, executive function, attention), and glial cell abnormalities. None of the sections of this figure can be definitive; many more gene loci, genes, and candidate endophenotypes exist and remain to be discovered (represented by question marks). Linkage and candidate gene studies have been the topic of recent reviews. The figure is not to scale. For supporting references see Gottesman and Gould 2003. SOURCE: Gottesman and Gould 2003, © Irving I. Gottesman and Todd Gould, with the copyright holder's permission.

above. Though this figure is now about 10 years old, the approach represents a vitally important tool in the search for clues related to schizophrenia as noted in the discussion above.

SCHIZOPHRENIA EPIGENESIS? EPIGENETICS AND BEYOND

In the past dozen years there has been increasing attention paid to "epigenetics," and the epigenetic viewpoint is beginning to have an influence in behavioral and psychiatric genetics, including in the *C. elegans* research discussed in chapters 3 and 4. I noted Cloninger's views about the general importance of this approach in chapter 6. In 2000, an article by 11 behavioral and psychiatric geneticists appeared entitled "Psychiatric Epigenetics: A New Focus for the New Century" (Petronis et al. 2000). Here I will describe the meaning of the term "epigenetics" more precisely, indicate what some of the epigenetic mechanisms are, and briefly discuss ways that it is impacting the study of schizophrenia genetics. I close this section by speculating on additional directions that research in this area might go.

The major reason for looking at epigenetics has to do with the fact that identical twins are only about 50% concordant for schizophrenia, a difference that has traditionally been attributed to unknown environmental factors. But another rationale for the heightened interest in epigenetics is that DNA sequence per se increasingly seems impoverished as a biological explainer. The fact that the human genome seems to be only about slightly larger than the worm's (25,000 genes compared to 20,000 for *C. elegans*) suggests that the answer to human complexity may lie in a variant of the old realtors maxim, "Location, location, location," namely "Regulation, regulation, regulation." Epigenetics provides additional regulatory mechanisms by which complex diversity may be realized and, if awry, may explain variations, including pathological changes, in the behavioral area. Another rationale for examining the role of epigenetic mechanisms is that it may actually turn out to account for some of the putative "environmental" variation found in behavioral studies, including schizophrenia as just noted, and which has been difficult to identify specifically (see chapter 1 and above, as well as Oh and Petronis 2008). Reference to epigenetics may also have some attraction to those who have concerns about genetic determinism and heredity as explainers of individual differences, since, in a way, epigenetic modifications are partially environmental influences, though possibly random ones, that are implemented via DNA.

The term "epigenetics" is itself not one that has a univocal meaning, some understanding it very broadly to mean "all the mechanisms that lead to the phenotypic expression of genetic information in an individual" (Jablonka and Lamb 1995, 80), and others proposing that we understand the term more narrowly, as "modifications in genetic expressions that are controlled by heritable but potentially reversible changes in DNA methylation and/or chromatin structure" (Henikoff and Matzke 1997, 293). Cloninger in his writings on the subject appeared to take the broader sense, though he also reviewed methylation mechanisms in some detail (Cloninger 2004, 278–85). Petronis prefers a slight modification of the Henikoff definition, proposing that "epigenetics refers to regulation of various genomic functions that is controlled by partially stable modifications of DNA and histones" (personal communication, 2005). (For a more general overviews, see Petronis 2001; Labrie, Pai, and Petronis 2012.) In this book I will generally adhere to the narrower interpretation of epigenetics and characterize the broader class of non-sequence-based mechanisms as "regulatory" in a more general sense, with the proviso that most of these regulatory modifications are not typical rapidly reversible forms of regulation, but are of a more stable and heritable kind.

The first class of epigenetic mechanisms, DNA methylation, involves adding a methyl group ($-CH_3$) to one of the carbon atoms in the cytosine (or guanine) nucleotide molecule in a DNA sequence. This adding or tagging process called methylation can change the transcription of the tagged (or untagged) DNA, and thus result in an absent or different gene product, at the RNA or the protein level. Methylation is widely found as a regulating mechanism in the mammalian genome. The second class of epigenetic mechanisms involves the histone proteins that are part of the chromosomes. DNA sequences are wrapped around histone complexes forming the nucleosomal structure of the chromosomes. At one point in time these histone complexes were thought to be inert structures, but subsequent research indicated they play an active role in remodeling chromatin and in gene activity (Jenuwein and Allis 2001). The lengthy article in *Science* by Jenuwein and Allis described the role of the histones as "a 'histone code' that considerably extends the information potential of the genetic code" (1074). Histone acetylation change appears to be a significant mechanism involved in histone-based epigenetics (Fraga, Ballestar, Paz, et al. 2005; Fraga, Ballestar, Villar-Garea, et al. 2005). More recently, there has been the recognition that these two mechanisms can engage in "cross-talk" (Cedar and Bergman 2009).

A study specifically applying the epigenetic perspective to schizophrenia was published in the *Schizophrenia Bulletin* in 2003 with the title of "Monozygotic Twins Exhibit Numerous Epigenetic Differences: Clues to Twin Discordance?" (Petronis et al. 2003). This investigation noted that though DNA sequences remain basically stable throughout a lifetime, both DNA methylation and chromatin remodeling are dynamic processes and might account for identical twin differences within a neurodevelopmental framework. The researchers selected two pairs of identical twins at random from a population of 66, in which one twin pair was discordant for schizophrenia (for a six-year period), but in the other pair both had developed the disorder. The authors write: "The main goal of this study was to demonstrate proof of principle that there are numerous epigenetic differences in MZ twins, and such epigenetic variation can serve as a new source of putative molecular substrates for discordance of MZ twins" (170). The genetic region examined in both pairs of twins was a fragment of the regulatory region of the dopamine D2 receptor gene, obtained from twins' lymphocyte cells. There is some evidence that this gene may be involved in schizophrenia, though the authors also admit that extrapolation from lymphocytes to brain cells and schizophrenia is speculative. After presenting their data, the authors summarize by saying that "this study demonstrates that fine mapping of DNA modification detects numerous molecular differences in genetically identical organisms. This is in contrast to previous DNA sequence-based studies of MZ twins that detected either no . . . or rare . . . differences" (172).

Petronis et al. (2003) conclude this study noting the difficulty of pursuing epigenetic studies, in part because the experimental method used is quite labor intensive, and also because of more general concerns, such as "a significant confounding factor given that our understanding of the critical epigenetic regulatory regions is very superficial, and in some cases such regions can be localized up to hundreds of kilobases away from the coding sequence." They also note that

> at present there are no good tools to differentiate epigenetic "noise" from phenotype relevant epigenetic signals. Furthermore, the degree of somatic mosaicism of epigenetic signals may vary across individuals, tissues, and genes, and the representative number of clones of bisulfite modified DNA is not straightforward. Access to large numbers of samples of *post mortem* brain tissues of MZ twins is unrealistic. (174)

These results and the just-cited sobering caveats appeared to represent where the epigenetic approach was in its first few years of investigation. Petronis and his colleagues and others have pursued additional studies; , see the review article by Wong, Gottesman, and Petronis (2005). Two more recent studies are by Dempster et al. (2011) and by Mill et al. (2008). A general overview on the epigenetics of schizophrenia and published recently is by Cheung, Jakovcevski, and Akbarian (2012).

It may be that the epigenetic perspective will in the next several years begin to bear more specific fruit in the form of more direct evidence of a link of these regulatory variations to schizophrenia and other mental disorders, though as of 2012, Petronis admits there is still no "direct evidence" of these types of effects on psychosis (Labrie, Pai, and Petronis 2012). It is also possible that studies in other areas of the *regulatory* systems of genetics, such as the machinery of multiple DNA splicing, will disclose novel approaches to schizophrenia. Multiple splicing has already been implicated in apoptosis or cell death, and several diseases, including a neurological disorder, have been found to be associated with gene-splicing defects. A tantalizing finding in this area is that a group of primate specific alternatively spliced exons known as Alus may be involved in rapid gene expansion (see Ast 2005). Whether this mechanism might account in part for the development of species-specific structures in the primate and human brains, as well as pathological variants in these elements that could produce mental disorders, is quite speculative at present, but may be where general genetic *regulation*, rather than exon sequence mutations, plays a key role. Though epigenetic mechanisms are unlikely to be the indicators of splicing points, such epigenetic factors can control the activity of different splicing factors, that is, regulate the regulators (Petronis, personal communication, 2005). In the past few years, epigenetics has also begun to assimilate the results of GWA studies and that technology, leading to what has been termed an "EWAS" approach (Rakyan et al. 2011).

Epigenetics is a major growth area, both in cancer biology (see Fraga, Ballestar, Paz, et al. 2005; Fraga, Ballestar, Villar-Garea, et al. 2005; Boumber and Issa 2011) and in the behavioral realm (Pidsley and Mill 2011; Dempster et al. 2013). Because there seems to be extraordinarily flexibility regarding the kinds of mechanisms and interactions that can be classified as epigenetic (compare Petronis 2001; Labrie, Pai, and Petronis 2012), the field will have to be especially rigorous in specifying its mechanisms and its falsifiable hypotheses in order to make solid empirical progress. A recent philosophical analysis of epigenetics, with special reference to the DST themes discussed in chapter 3, can be found in chapter 5 of Griffiths and Stotz (2013).

FIVE (OR SIX) TYPES OF SCHIZOPHRENIA GENES AND RESEARCH DESIGN IMPLICATIONS

In several of the previous sections, I described both the need for a dimensional model of schizophrenia and notion of the schizophrenic spectrum. This variation in the presentation of schizophrenia is the backdrop for an intriguing proposal distinguishing five or six types of schizophrenia-influencing genes. Fanous and Kendler, the authors of the paper outlining these different senses of genes, begin by reminding readers of this variation:

> It has long been noted that there is substantial variation in the presentation of schizophrenia. Individual patients may differ greatly with respect to age of onset, course of illness, premorbid psychopathology, interepisode recovery, and the prominence of a range of clinical features such as hallucinations, delusions, thought disorder, and negative symptoms. Indeed, experienced clinicians often remark that it is quite possible to encounter patients with no features in common. (Fanous and Kendler 2005, 6)

These authors also note that various studies, some of which I have cited above, "suggest that the familial factors influencing the clinical features of schizophrenia have effects in the same direction on related personality traits in nonpsychotic individuals" (8).

These considerations lead to the suggestion that in addition to the now standard notion of susceptibility genes related to schizophrenia—these are the five or six genes covered above—there are probably genes that do *not* affect the susceptibility to the disorder, but *do* affect the form and features of the disorder when it is manifest. These Fanous and Kendler call "modifier genes." The possible existence of susceptibility genes and modifier genes raises the prospect that such genes might be classifiable as involving five different types of gene action. These five types are labeled I through V, and are defined as follows (the italics are mine):

> Type I, or "pure" susceptibility genes (herein referred to as S) are defined as genes that increase the risk of schizophrenia but do not *preferentially* impact any of its associated clinical features. . . .

Type II, or mixed susceptibility/modifier genes (herein referred to as MS). This class of genes would increase the susceptibility to schizophrenia, but to certain presentations more than others. Examples might include dysbindin and *COMT*. . . .

Type III, or "pure" modifier genes (herein referred to as M). These genes would affect the clinical features of the illness, but would *not* increase susceptibility to illness by themselves. . . .

Type IV is defined as MS genes that influence not only the liability to subtypes of schizophrenia in which specific features are prominent but, *in addition* also influence attenuated versions of the same features in *nonpsychotic* relatives. We henceforth call these MS_B for *broad* genes. . . .

Finally, Type V is denoted as M genes that act *broadly*, influencing clinical features in psychotic, as well as related traits in nonpsychotic individuals, herein referred to as M_B genes [but do not affect susceptibility, i.e., move the individual past the threshold to diagnosable schizophrenia]. (Fanous and Kendler 2005, 9)

The actions of these different types of genes would manifest themselves in different types of outcomes in empirical studies. For example, the pure susceptibility gene, Type I, or the S type, would, in ill relatives, not produce any *correlation* of the clinical features of the disorder. On the other hand, modifier genes, the M type, would not predict the *risk* of the disease if found in relatives, but "would cause the resemblance of relative pairs for clinical features, as well as greater MZ than DZ correlations for the same features" (9). Fanous and Kendler also speculate on additional types of empirical findings that would likely arise from these five types of genes. It remains to be seen how more recent GWAS analyses might intersect with these subtly different forms of susceptibility genes.

This is a preliminary set of proposals, and Fanous and Kendler caution readers that "the actual biological functions of putative susceptibility genes, and therefore, the consequences for brain structure and function of risk alleles in them are only vaguely understood. There are few published studies of the neurobiology of such genes in schizophrenia" (9). There may also be additional types of genes, and mechanisms of gene action that go beyond the five proposed classes. For example, one such possibility might be an SB type of gene that would increase risk not for diagnosable schizophrenia but rather for the schizophrenia spectrum, for example, for the prodrome, or for schizotaxia, to use two potential prototypes from the schizophrenia spectrum. I suggested this to Fanous and Kendler in correspondence, and they are open to such a possibility and are considering a sixth type of gene further (personal communications with Fanous and Kendler, January 2005). Still more speculative is that one subtype of a susceptibility gene that might influence the development of prodromal or schizotaxic features could in principle be a "protective" allele. This intuition here is that this allelic form might nudge the individual into prodromal or schizotaxic symptoms, but *not* further into full-blown schizophrenia. But such intuitions are only speculative at present and will require empirical research to confirm or disconfirm them.

CONCLUSION AND PRELIMINARY IMPLICATIONS FOR VALID BEHAVIORAL CONSTRUCTS AND THE ROLE OF GENETICS

Several themes have been developed in this chapter that both echo and go beyond some similar themes discussed in chapter 6. Replication problems resurfaced in the present chapter, as did problems of defining the phenotype, the use of endophenotypes, and spectrum perplexities. Especially perplexing are the replications of schizophrenia susceptibility genes that seemed to replicate in the narrow regional sense, but not in the specific allelic version sense; some ways to consider different units of replication along lines suggested by Kendler were sketched. The replication issue is also affecting GWAS results, with the general view that ever larger studies will be needed to sort out those problems (Visscher et al. 2012), and as suggested above, more results continue to appear (Tansey, Owen, and O'Donovan 2015). Additionally, I speculated on the epigenetic aspects of schizophrenia and the difficulties of identifying empirically and assessing those forms of regulation. Finally we considered newer forms of modifier and joint modifier-susceptibility genes, which are only beginning to be pursued further. Implementing these newer ideas along with the research designs they suggest may represent an important dimension of future research in this area.

Many of these issues triangulate on the problem of establishing the validity of psychological constructs, such as the personality categories discussed in chapter 6, and the validity of psychiatric disorders, foreshadowed in the personality disorder references of chapter 6 and considered in connection with schizophrenia in the present one. I would argue that these issues fit naturally into and support the framework reintroduced above, including a grade of membership (GoM) approach to disorder definition (see Hallmayer et al. 2005; Green et al. 2013). Recall that this framework suggested that the typical theory in the biomedical sciences is a structure of prototypical, overlapping, interlevel, temporal models. The models of such a structure usually constitute a series of idealized prototypical multilevel mechanisms and variations (mutants) that bear family or similarity resemblances to each other, and characteristically each has a (relatively) narrow scope of straightforward application to (a few) pure types. These prototypes are thus salient points along a (multidimensional) continuum, and relations of individuals to the prototypes can be characterized by a GoM function.

In the following chapter, I try to bring the various messages of these seven chapters together, looking at the issues of why we frequently are concerned about genetic results, especially in the behavioral area, whether we should be concerned, and what the prospects are for future developments in the areas of behavioral and psychiatric genetics.

APPENDIX: ICD-10 DEFINITIONS

(Source: ICD-10-MBD Clinical Descriptions and Diagnostic Guidelines. Mental and Behavioural Disorders, 1992, Section F20 Schizophrenia, pages 78–79, with permission.)

F20 Schizophrenia

The schizophrenic disorders are characterized in general by fundamental and characteristic distortions of thinking and perception, and by inappropriate or blunted affect. Clear consciousness and intellectual capacity are usually maintained, although certain cognitive deficits may evolve in the course of time. The disturbance involves the most basic functions that give the normal person a feeling of individuality, uniqueness, and self-direction. The most intimate thoughts, feelings, and acts are often felt to be known to or shared by others, and explanatory delusions may develop, to the effect that natural or supernatural forces are at work to influence the afflicted individual's thoughts and actions in ways that are often bizarre. The individual may see himself or herself as the pivot of all that happens. Hallucinations, especially auditory, are common and may comment on the individual's behavior or thoughts. Perception is frequently disturbed in other ways: colors or sounds may seem unduly vivid or altered in quality, and irrelevant features of ordinary things may appear more important than the whole object or situation. Perplexity is also common early on and frequently leads to a belief that everyday situations possess a special, usually sinister, meaning intended uniquely for the individual. In the characteristic schizophrenic disturbance of thinking, peripheral and irrelevant features of a total concept, which are inhibited in normal directed mental activity, are brought to the fore and utilized in place of those that are relevant and appropriate to the situation. Thus thinking becomes vague, elliptical, and obscure, and its expression in speech sometimes incomprehensible. Breaks and interpolations in the train of thought are frequent, and thoughts may seem to be withdrawn by some outside agency. Mood is characteristically shallow, capricious, or incongruous. Ambivalence and disturbance of volition may appear as inertia, negativism, or stupor. Catatonia may be present. The onset may be acute, with seriously disturbed behavior, or insidious, with a gradual development of odd ideas and conduct. The course of the disorder shows equally great variation and is by no means inevitably chronic or deteriorating (the course is specified by five-character categories). In a proportion of cases, which may vary in different cultures and populations, the outcome is complete, or nearly complete, recovery. The sexes are approximately equally affected by the onset tends to be later in women.

Although no strictly pathognomonic symptoms can be identified, for practical purposes it is useful to divide the above symptoms into groups that have special importance for the diagnosis and often occur together, such as:

(a) thought echo, thought insertion or withdrawal, and thought broadcasting;
(b) delusions of control, influence, or passivity, clearly referred to body or limb movements or specific thoughts, actions, or sensations; delusional perception;
(c) hallucinatory voices giving a running commentary on the patient's behavior, or discussing the patient among themselves, or other types of hallucinatory voices coming from some part of the body;
(d) persistent delusions of other kinds that are culturally inappropriate and completely impossible, such as religious or political identity, or superhuman powers and abilities (e.g. being able to control the weather, or being in communication with aliens from another world);

(e) persistent hallucinations in any modality, when accompanied either by fleeting or half-formed delusions without clear affective content, or by persistent over-valued ideas, or when occurring every day for weeks or months on end;
(f) breaks or interpolations in the train of thought, resulting in incoherence or irrelevant speech, or neologisms;
(g) catatonic behavior, such as excitement, posturing, or waxy flexibility, negativism, mutism, and stupor;
(h) "negative" symptoms such as marked apathy, paucity of speech, and blunting or incongruity of emotional responses, usually resulting in social withdrawal and lowering of social performance; it must be clear that these are not due to depression or to neuroleptic medication;
(i) a significant and consistent change in the overall quality of some aspects of personal behavior, manifest as loss of interest, aimlessness, idleness, a self-absorbed attitude, and social withdrawal.

Diagnostic Guidelines

The normal requirement for a diagnosis of schizophrenia is that a minimum of one very clear symptom (and usually two or more if less clear-cut) belonging to any one of the groups listed as (a) to (d) above, or symptoms from at least two of the groups referred to as (e) to (h), should have been clearly present for most of the time during a period of 1 month or more. Conditions meeting such symptomatic requirements but of duration less than 1 month (whether treated or not) should be diagnosed in the first instance as acute schizophrenia-like psychotic disorder and are classified as schizophrenia if the symptoms persist for longer periods.

Viewed retrospectively, it may be clear that a prodromal phase in which symptoms and behavior, such as loss of interest in work, social activities, and personal appearance and hygiene, together with generalized anxiety and mild degrees of depression and preoccupation, preceded the onset of psychotic symptoms by weeks or even months. Because of the difficulty in timing onset, the 1-month duration criterion applies only to the specific symptoms listed above and not to any prodromal nonpsychotic phase.

The diagnosis of schizophrenia should not be made in the presence of extensive depressive or manic symptoms unless it is clear that schizophrenic symptoms antedated the affective disturbance. If both schizophrenic and affective symptoms develop together and are evenly balanced, the diagnosis of schizoaffective disorder should be made, even if the schizophrenic symptoms by themselves would have justified the diagnosis of schizophrenia. Schizophrenia should not be diagnosed in the presence of overt brain disease or during states of drug intoxication or withdrawal.

References

Addington, J., and R. Heinssen. 2012. "Prediction and prevention of psychosis in youth at clinical high risk." *Annual Review of Clinical Psychology* 8: 269–89. doi:10.1146/annurev-clinpsy-032511-143146.

American Psychiatric Association. 2000. *Diagnostic and Statistical Manual of Mental Disorders: DSM-IV-TR*. 4th ed. Arlington, VA: American Psychiatric Association.

American Psychiatric Association and DSM-5 Task Force. 2013. *Diagnostic and Statistical Manual of Mental Disorders: DSM-5*. 5th ed. Arlington, VA: American Psychiatric Association.

American Psychiatric Association and Task Force on DSM-IV. 2000. *Diagnostic and Statistical Manual of Mental Disorders: DSM-IV-TR*. 4th ed. Washington, DC: American Psychiatric Association.

Anastasi, Anne. 1958. *Differential Psychology: Individual and Group Differences in Behavior.* 3rd ed. New York: Macmillan.

Arseneault, L., M. Cannon, J. Witton, and R. M. Murray. 2004. "Causal association between cannabis and psychosis: Examination of the evidence." *British Journal of Psychiatry* 184: 110–17.

Ast, G. 2005. "The alternative genome." *Scientific American* 292 (4): 40–47.

Ayalew, M., H. Le-Niculescu, D. F. Levey, N. Jain, B. Changala, S. D. Patel, E. Winiger, et al. 2012. "Convergent functional genomics of schizophrenia: From comprehensive understanding to genetic risk prediction." *Molecular Psychiatry.* doi:10.1038/mp.2012.37.

Bickle, J. 2003. *Philosophy and Neuroscience: A Ruthlessly Reductive Account.* Dordrecht: Kluwer.

Bolton, Derek. 2008. *What Is Mental Disorder? An Essay in Philosophy, Science, and Values*. Oxford: Oxford University Press.

Boorse, Christopher. 1977. "Health as a theoretical concept." *Philosophy of Science* 44: 572–73.

Boumber, Y., and J. P. Issa. 2011. "Epigenetics in cancer: What's the future?" *Oncology (Williston Park)* 25 (3): 220–26, 228.

Buzsaki, G., N. Logothetis, and W. Singer. 2013. "Scaling brain size, keeping timing: Evolutionary preservation of brain rhythms." *Neuron* 80 (3):751–64. doi: 10.1016/j.neuron.2013.10.002.

Cantor, N., E. E. Smith, R. S. French, and J. Mezzich. 1980. "Psychiatric diagnosis as prototype categorization." *Journal of Abnormal Psychology* 89 (2): 181–93.

Carey, G. 2002. *Human Genetics for the Social Science*. Thousand Oaks, CA: Sage.

Cedar, H., and Y. Bergman. 2009. "Linking DNA methylation and histone modification: Patterns and paradigms." *Nature Reviews Genetics* 10 (5): 295–304. doi:10.1038/nrg2540.

Chalmers, David John. 1996. *The Conscious Mind: In Search of a Fundamental Theory.* New York: Oxford University Press.

Chandler, D., M. Dragovic, M. Cooper, J. C. Badcock, B. H. Mullin, D. Faulkner, S. G. Wilson, et al. 2010. "Impact of Neuritin 1 (NRN1) polymorphisms on fluid intelligence in schizophrenia." *American Journal of Medical Genetics B: Neuropsychiatric Genetics* 153B (2): 428–37. doi:10.1002/ajmg.b.30996.

Cheung, Iris, Mira Jakovcevski, and Schahram Akbarian. 2012. "The epigenetics of schizophrenia." In *The Origins of Schizophrenia,* edited by Alan S. Brown and Paul H. Patterson, 227–52. New York: Columbia University Press.

Cloninger, C. Robert. 2004. *Feeling Good: The Science of Well-Being.* New York: Oxford University Press.

Consortium, Schizophrenia Working Group of the Psychiatric Genomics. 2014. "Biological insights from 108 schizophrenia-associated genetic loci." *Nature* 511 (7510): 421–27. doi:10.1038/nature13595.

Deakin, I. H., W. Nissen, A. J. Law, T. Lane, R. Kanso, M. H. Schwab, K. A. Nave, et al. 2011. "Transgenic overexpression of the Type I isoform of neuregulin 1 affects working memory and hippocampal oscillations but not long-term potentiation." *Cerebral Cortex*. doi:10.1093/cercor/bhr223.

Dempster, E. L., R. Pidsley, L. C. Schalkwyk, S. Owens, A. Georgiades, F. Kane, S. Kalidindi, et al. 2011. "Disease-associated epigenetic changes in monozygotic twins discordant for schizophrenia and bipolar disorder." *Human Molecular Genetics* 20 (24): 4786–96. doi:10.1093/hmg/ddr416.

Dempster, E. L., J. Viana, R. Pidsley, and J. Mill. 2013. "Epigenetic studies of schizophrenia: Progress, predicaments, and promises for the future." *Schizophrenia Bulletin* 39 (1): 11–16. doi:10.1093/schbul/sbs139.

Dizadji-Bahmani, F., R. Frigg, and S. Hartmann. 2010. "Who's Afraid of Nagelian Reduction?" *Erkenntis* 73: 393–412.

Dobbs, D. 2010. "Schizophrenia: The making of a troubled mind." *Nature* 468 (7321): 154–56. doi:10.1038/468154a.

Dragt, S., D. H. Nieman, F. Schultze-Lutter, F. van der Meer, H. Becker, L. de Haan, P. M. Dingemans, et al. 2012. "Cannabis use and age at onset of symptoms in subjects at clinical high risk for psychosis." *Acta Psychiatrica Scandinavica* 125 (1): 45–53. doi:10.1111/j.1600-0447.2011.01763.x.

Fanous, A., C. Gardner, D. Walsh, and K. S. Kendler. 2001. "Relationship between positive and negative symptoms of schizophrenia and schizotypal symptoms in nonpsychotic relatives." *Archives of General Psychiatry* 58 (7): 669–73.

Fanous, A. H., and K. S. Kendler. 2005. "Genetic heterogeneity, modifier genes, and quantitative phenotypes in psychiatric illness: Searching for a framework." *Molecular Psychiatry* 10 (1): 6–13.

Farmer, A. E., P. McGuffin, and I. I. Gottesman. 1987. "Twin concordance for DSM-III schizophrenia: Scrutinizing the validity of the definition." *Archives of General Psychiatry* 44 (7): 634–41.

Fraga, M. F., E. Ballestar, M. F. Paz, S. Ropero, F. Setien, M. L. Ballestar, D. Heine-Suner, et al. 2005. "Epigenetic differences arise during the lifetime of monozygotic twins." *Proceedings of the National Academy of Sciences USA* 102 (30): 10604–9.

Fraga, M. F., E. Ballestar, A. Villar-Garea, M. Boix-Chornet, J. Espada, G. Schotta, T. Bonaldi, et al. 2005. "Loss of acetylation at Lys16 and trimethylation at Lys20 of histone H4 is a common hallmark of human cancer." *Nature Genetics* 37 (4): 391–400.

Frances, A., H. A. Pincus, T. A. Widiger, W. W. Davis, and M. B. First. 1990. "DSM-IV: Work in progress." *American Journal of Psychiatry* 147 (11): 1439–48.

Fulford, K. W. M. 1990. *Moral Theory and Medical Practice*. New York: Cambridge University Press.

Funke, B., C. T. Finn, A. M. Plocik, S. Lake, P. DeRosse, J. M. Kane, R. Kucherlapati, and A. K. Malhotra. 2004. "Association of the DTNBP1 locus with schizophrenia in a U.S. population." *American Journal of Human Genetics* 75 (5): 891–98.

Gong, Y. G., C. N. Wu, Q. H. Xing, X. Z. Zhao, J. Zhu, and L. He. 2009. "A two-method meta-analysis of Neuregulin 1(*NRG1*) association and heterogeneity in schizophrenia." *Schizophrenia Research* 111 (1–3): 109–14. doi:10.1016/j.schres.2009.03.017.

Gonzalez-Burgos, G., and D. A. Lewis. 2012. "NMDA receptor hypofunction, parvalbumin-positive neurons and cortical gamma oscillations in schizophrenia." *Schizophrenia Bulletin* 38 (5) (September): 950–57. doi:10.1093/schbul/sbs010.

Gottesman, I. I. 1991. *Schizophrenia Genesis: The Origins of Madness*. New York: W.H. Freeman.

Gottesman, I. I., and T. D. Gould. 2003. "The endophenotype concept in psychiatry: Etymology and strategic intentions." *American Journal of Psychiatry* 160 (4): 636–45.

Gottesman, Irving I., James Shields, and Paul E. Meehl. 1972. *Schizophrenia and Genetics: A Twin Study Vantage Point.* New York: Academic Press.

Gottesman, Irving I., and Dorothea L. Wolfgram. 1991. *Schizophrenia Genesis: The Origins of Madness.* New York: W.H. Freeman.

Green, M. J., M. J. Cairns, J. Wu, M. Dragovic, A. Jablensky, P. A. Tooney, R. J. Scott, and V. J. Carr. 2013. "Genome-wide supported variant MIR137 and severe negative symptoms predict membership of an impaired cognitive subtype of schizophrenia." *Molecular Psychiatry* 18 (7) (July): 774–80. doi:10.1038/mp.2012.84.

Greenwood, T. A., L. C. Lazzeroni, S. S. Murray, K. S. Cadenhead, M. E. Calkins, D. J. Dobie, M. F. Green, et al. 2011. "Analysis of 94 candidate genes and 12 endophenotypes for schizophrenia from the Consortium on the Genetics of Schizophrenia." *American Journal of Psychiatry* 168 (9): 930–46. doi:10.1176/appi.ajp.2011.10050723.

Greenwood, T. A., G. A. Light, N. R. Swerdlow, A. D. Radant, and D. L. Braff. 2012. "Association analysis of 94 candidate genes and schizophrenia-related endophenotypes." *PLoS One* 7 (1): e29630. doi:10.1371/journal.pone.0029630.

Greenwood, T. A., N. R. Swerdlow, R. E. Gur, K. S. Cadenhead, M. E. Calkins, D. J. Dobie, R. Freedman, et al. 2013. "Genome-wide linkage analyses of 12 endophenotypes for schizophrenia from the Consortium on the Genetics of Schizophrenia." *American Journal of Psychiatry* 170 (5): 521–32. doi:10.1176/appi.ajp.2012.12020186.

Griffiths, P. E., and K. Stotz. 2013. *Genetics and Philosophy.* New York: Cambridge University Press.

Hacking, Ian. 1995. *Rewriting the Soul: Multiple Personality and the Sciences of Memory.* Princeton, NJ: Princeton University Press.

Hall, M. H., G. Taylor, P. Sham, K. Schulze, F. Rijsdijk, M. Picchioni, T. Toulopoulou, et al. 2011. "The early auditory gamma-band response is heritable and a putative endophenotype of schizophrenia." *Schizophrenia Bulletin* 37 (4): 778–87. doi:10.1093/schbul/sbp134.

Hallmayer, J. F., L. Kalaydjieva, J. Badcock, M. Dragovic, S. Howell, P. T. Michie, D. Rock, et al. 2005. "Genetic evidence for a distinct subtype of schizophrenia characterized by pervasive cognitive deficit." *American Journal of Human Genetics* 77 (3): 468–76.

Harrison, P. J., and D. R. Weinberger. 2005. "Schizophrenia genes, gene expression, and neuropathology: On the matter of their convergence." *Molecular Psychiatry* 10 (1): 40–68.

Heinrichs, R. W. 2005. "The primacy of cognition in schizophrenia." *American Psychologist* 60 (3): 229–42.

Henikoff, S., and M. A. Matzke. 1997. "Exploring and explaining epigenetic effects." *Trends in Genetics* 13 (8): 293–95.

Hoenig, J. 1983. "The concept of Schizophrenia: Kraepelin-Bleuler-Schneider." *British Journal of Psychiatry* 142: 547–56.

Hong, C. J., S. J. Huo, D. L. Liao, K. Lee, J. Y. Wu, and S. J. Tsai. 2004. "Case-control and family-based association studies between the neuregulin 1 (Arg38Gln) polymorphism and schizophrenia." *Neuroscience Letters* 366 (2): 158–61.

Horwitz, Allan V. 2002. *Creating Mental Illness.* Chicago: University of Chicago Press.

Jablensky, A. 1999. "The concept of schizophrenia: Pro et contra." *Epidemiology and Psychiatric Sciences* 8 (4): 242–47.

Jablensky, A. 2006. "Subtyping schizophrenia: Implications for genetic research." *Molecular Psychiatry* 11 (9): 815–36. doi:10.1038/sj.mp.4001857.

Jablensky, A., D. Angelicheva, G. J. Donohoe, M. Cruickshank, D. N. Azmanov, D. W. Morris, A. McRae, et al. 2012. "Promoter polymorphisms in two overlapping 6p25 genes implicate mitochondrial proteins in cognitive deficit in schizophrenia." *Molecular Psychiatry* 17 (12): 1328–39. doi:10.1038/mp.2011.129.

Jablonka, Eva, and Marion J. Lamb. 1995. *Epigenetic Inheritance and Evolution: The Lamarckian Dimension*. New York: Oxford University Press.

Jacob, F. 1977. "Evolution and tinkering." *Science* 196 (4295): 1161–66.

Jarskog, L. F., S. Miyamoto, and J. A. Lieberman. 2007. "Schizophrenia: New pathological insights and therapies." *Annual Review of Medicine* 58: 49–61. doi:10.1146/annurev.med.58.060904.084114.

Jenuwein, T., and C. D. Allis. 2001. "Translating the histone code." *Science* 293 (5532): 1074–80.

Kendler, K. S. 2004. "Schizophrenia genetics and dysbindin: A corner turned?" *American Journal of Psychiatry* 161 (9): 1533–36. doi:10.1176/appi.ajp.161.9.1533.

Kendler, K. S. 2005. "'A gene for . . .': The nature of gene action in psychiatric disorders." *American Journal of Psychiatry* 162 (7): 1243–52.

Kendler, K. S. 2013. "What psychiatric genetics has taught us about the nature of psychiatric illness and what is left to learn." *Molecular Psychiatry* 18 (10): 1058–66. doi:10.1038/mp.2013.50.

Kendler, K. S., L. Karkowski-Shuman, F. A. O'Neill, R. E. Straub, C. J. MacLean, and D. Walsh. 1997. "Resemblance of psychotic symptoms and syndromes in affected sibling pairs from the Irish Study of High-Density Schizophrenia Families: Evidence for possible etiologic heterogeneity." *American Journal of Psychiatry* 154 (2): 191–98.

Kendler, K. S., and K. F. Schaffner. 2011. "The dopamine hypothesis of schizophrenia: An historical and philosophical analysis." *Philosophy, Psychiatry, and Psychology (PPP)* 18 (1): 41–63.

Kety, S. S., D. Rosenthal, P. H. Wender, and F. Schulsinger. 1971. "Mental illness in the biological and adoptive families of adopted schizophrenics." *American Journal of Psychiatry* 128 (3): 302–6.

Koch, Christof. 2004. *The Quest for Consciousness: A Neurobiological Approach*. Denver, CO: Roberts.

Labrie, V., S. Pai, and A. Petronis. 2012. "Epigenetics of major psychosis: Progress, problems and perspectives." *Trends in Genetics* 28 (9) (September): 427–35. doi:10.1016/j.tig.2012.04.002.

Law, A. J., Y. Wang, Y. Sei, P. O'Donnell, P. Piantadosi, F. Papaleo, R. E. Straub, et al. 2012. "Neuregulin 1-ErbB4-PI3K signaling in schizophrenia and phosphoinositide 3-kinase-p110delta inhibition as a potential therapeutic strategy." *Proceedings of the National Academy of Sciences USA* 109 (30): 12165–70. doi:10.1073/pnas.1206118109.

Lee, S. H., N. R. Wray, M. E. Goddard, and P. M. Visscher. 2011. "Estimating missing heritability for disease from genome-wide association studies." *American Journal of Human Genetics* 88 (3): 294–305. doi:10.1016/j.ajhg.2011.02.002.

Levy, D. L., M. J. Coleman, H. Sung, F. Ji, S. Matthysse, N. R. Mendell, and D. Titone. 2010. "The genetic basis of thought disorder and language and communication disturbances in schizophrenia." *Journal of Neurolinguistics* 23 (3): 176. doi:10.1016/j.jneuroling.2009.08.003.

Lewis, D. A. 2014. "Inhibitory neurons in human cortical circuits: Substrate for cognitive dysfunction in schizophrenia." *Current Opinion in Neurobiology* 26C: 22–26. doi:10.1016/j.conb.2013.11.003.

Lewis, D. A., A. A. Curley, J. R. Glausier, and D. W. Volk. 2012. "Cortical parvalbumin interneurons and cognitive dysfunction in schizophrenia." *Trends Neurosciences* 35 (1): 57–67. doi:10.1016/j.tins.2011.10.004.

Lewis, D. A., and J. A. Lieberman. 2000. "Catching up on schizophrenia: Natural history and neurobiology." *Neuron* 28 (2): 325–34.

Lewis, D. A., and R. A. Sweet. 2009. "Schizophrenia from a neural circuitry perspective: Advancing toward rational pharmacological therapies." *Journal of Clinical Investigation* 119 (4): 706–16. doi:37335 [pii] 10.1172/JCI37335.

Lieberman, J. A., and W. S. Fenton. 2000. "Delayed detection of psychosis: Causes, consequences, and effect on public health." *American Journal of Psychiatry* 157 (11): 1727–30.

Lilienfeld, S. O., and L. Marino. 1995. "Mental disorder as a Roschian concept: A critique of Wakefield's 'harmful dysfunction' analysis." *Journal of Abnormal Psychology* 104 (3): 411–20.

Lilienfeld, S. O., and L. Marino. 1999. "Essentialism revisited: Evolutionary theory and the concept of mental disorder." *Journal of Abnormal Psychology* 108 (3): 400–411.

Machamer, P., L. Darden, and C. Craver. 2000. "Thinking about mechanisms." *Philosophy of Science* 67: 1–25.

Machery, Edouard. 2009. *Doing without Concepts*. New York: Oxford University Press.

Malaspina, D., R. R. Goetz, S. Yale, A. Berman, J. H. Friedman, F. Tremeau, D. Printz, et al. 2000. "Relation of familial schizophrenia to negative symptoms but not to the deficit syndrome." *American Journal of Psychiatry* 157 (6): 994–1003.

McGue, M., and I. I. Gottesman. 1989. "Genetic linkage in schizophrenia: Perspectives from genetic epidemiology. *Schizophrenia Bulletin* 15 (3): 453–64.

McGuffin, P., A. Farmer, and I. I. Gottesman. 1987. "Is there really a split in schizophrenia? The genetic evidence." *British Journal of Psychiatry* 150: 581–92.

Merker, B. 2013. "Cortical gamma oscillations: The functional key is activation, not cognition." *Neuroscience and Biobehavioral Reviews* 37 (3): 401–17. doi:10.1016/j.neubiorev.2013.01.013.

Mill, J., T. Tang, Z. Kaminsky, T. Khare, S. Yazdanpanah, L. Bouchard, P. Jia, et al. 2008. "Epigenomic profiling reveals DNA-methylation changes associated with major psychosis." *American Journal of Human Genetics* 82 (3): 696–711. doi:10.1016/j.ajhg.2008.01.008.

Moises, H. W., T. Zoega, and I. I. Gottesman. 2002. "The glial growth factors deficiency and synaptic destabilization hypothesis of schizophrenia." *BMC Psychiatry* 2 (1): 8.

Morar, B., M. Dragovic, F. A. Waters, D. Chandler, L. Kalaydjieva, and A. Jablensky. 2011. "Neuregulin 3 (NRG3) as a susceptibility gene in a schizophrenia subtype with florid delusions and relatively spared cognition." *Molecular Psychiatry* 16 (8): 860–66. doi:10.1038/mp.2010.70.

Neale, B. M., and P. C. Sham. 2004. "The future of association studies: Gene-based analysis and replication." *American Journal of Human Genetics* 75 (3): 353–62.

Nelson, B., A. Fornito, B. J. Harrison, M. Yucel, L. A. Sass, A. R. Yung, A. Thompson, S. J. Wood, C. Pantelis, and P. D. McGorry. 2009. "A disturbed sense of self in the psychosis prodrome: Linking phenomenology and neurobiology." *Neuroscience and Biobehavioral Reviews* 33 (6): 807–17. doi:10.1016/j.neubiorev.2009.01.002.

Nelson, B., J. Parnas, and L. A. Sass. 2014. "Disturbance of minimal self (ipseity) in schizophrenia: Clarification and current status." *Schizophrenia Bulletin* 40 (3): 479–82. doi:10.1093/schbul/sbu034.

Nicodemus, K. K., A. J. Law, E. Radulescu, A. Luna, B. Kolachana, R. Vakkalanka, D. Rujescu, et al. 2010. "Biological validation of increased schizophrenia risk with *NRG1*, ERBB4, and AKT1 epistasis via functional neuroimaging in healthy controls." *Archives of General Psychiatry* 67 (10): 991–1001. doi:10.1001/archgenpsychiatry.2010.117.

Oh, G., and A. Petronis. 2008. "Environmental studies of schizophrenia through the prism of epigenetics." *Schizophrenia Bulletin* 34 (6): 1122–29. doi:10.1093/schbul/sbn105.

Owen, M. J., M. C. O'Donovan, and I. I. Gottesman. 2002. "Schizophrenia." In *Psychiatric Genetics and Genomics*, edited by P. McGuffin, M. J. Owen, and I. I. Gottesman, 247–66. New York: Oxford University Press.

Owen, M. J., N. M. Williams, and M. C. O'Donovan. 2004. "Dysbindin-1 and schizophrenia: From genetics to neuropathology." *Journal of Clinical Investigation* 113 (9): 1255–57. doi:10.1172/JCI21470.

Parnas, J., P. Moller, T. Kircher, J. Thalbitzer, L. Jansson, P. Handest, and D. Zahavi. 2005. "EASE: Examination of Anomalous Self-Experience." *Psychopathology* 38 (5): 236–58. doi:10.1159/000088441.

Petronis, A. 2001. "Human morbid genetics revisited: Relevance of epigenetics." *Trends in Genetics* 17 (3): 142–46.

Petronis, A., I. I. Gottesman, T. J. Crow, L. E. DeLisi, A. J. Klar, F. Macciardi, M. G. McInnis, et al. 2000. "Psychiatric epigenetics: A new focus for the new century." *Molecular Psychiatry* 5 (4): 342–46.

Petronis, A., I. I. Gottesman, P. Kan, J. L. Kennedy, V. S. Basile, A. D. Paterson, and V. Popendikyte. 2003. "Monozygotic twins exhibit numerous epigenetic differences: Clues to twin discordance?" *Schizophrenia Bulletin* 29 (1): 169–78.

Pidsley, R., and J. Mill. 2011. "Epigenetic studies of psychosis: Current findings, methodological approaches, and implications for postmortem research." *Biological Psychiatry* 69 (2): 146–56. doi:10.1016/j.biopsych.2010.03.029.

Pitcher, G. M., L. V. Kalia, D. Ng, N. M. Goodfellow, K. T. Yee, E. K. Lambe, and M. W. Salter. 2011. "Schizophrenia susceptibility pathway neuregulin 1-ErbB4 suppresses Src upregulation of NMDA receptors." *Nature Medicine* 17 (4): 470–78. doi:10.1038/nm.2315.

Rakyan, V. K., T. A. Down, D. J. Balding, and S. Beck. 2011. "Epigenome-wide association studies for common human diseases." *Nature Reviews Genetics* 12 (8): 529–41. doi:10.1038/nrg3000.

Regier, D. A., C. T. Kaelber, M. T. Roper, D. S. Rae, and N. Sartorius. 1994. "The ICD-10 clinical field trial for mental and behavioral disorders: Results in Canada and the United States." *American Journal of Psychiatry* 151 (9): 1340–50.

Ripke, S., C. O'Dushlaine, K. Chambert, J. L. Moran, A. K. Kahler, S. Akterin, S. E. Bergen, et al. 2013. "Genome-wide association analysis identifies 13 new risk loci for schizophrenia." *Nature Genetics* 45 (10): 1150–59. doi:10.1038/ng.2742.

Ripke, S., A. R. Sanders, K. S. Kendler, D. F. Levinson, P. Sklar, P. A. Holmans, D. Y. Lin, et al. 2011. "Genome-wide association study identifies five new schizophrenia loci." *Nature Genetics* 43 (10): 969–76. doi:10.1038/ng.940.

Risch, N. 1990. "Linkage strategies for genetically complex traits. II. The power of affected relative pairs." *American Journal of Human Genetics* 46 (2): 229–41.

Risch, N., and K. Merikangas. 1996. "The future of genetic studies of complex human diseases." *Science* 273 (5281): 1516–17.

Rosch, E. 1975. "Cognitive representation of semantic categories." *Journal of Experimental Psychology* 104: 192–233.

Rosch, E., and C. B. Mervis. 1975a. "Family resemblances: Studies in the internal structure of categories." *Cognitive Psychology* 7 (4): 573–605.

Rosenthal, David. 1963. *The Genain Quadruplets: A Case Study and Theoretical Analysis of Heredity and Environment in Schizophrenia*. New York: Basic Books.

Rush, A. John, Michael B. First, and Deborah Blacker. 2008. *Handbook of Psychiatric Measures*. 2nd ed. Washington, DC: American Psychiatric Publishing.

Saunders, Alan R., Jubao Duan, and Pablo V. Gejman. 2012. "Schizophrenia genetics." In *The Origins of Schizophrenia*, edited by Alan S. Brown and Paul H. Patterson, 210–26. New York: Columbia University Press.

Schaffner, K. F. 1980. "Theory structure in the biomedical sciences." *Journal of Medicine and Philosophy* 5: 57–97.

Schaffner, K. F. 1986. "Exemplar reasoning about biological models and diseases: A relation between the philosophy of medicine and philosophy of science." *Journal of Medicine and Philosophy* 11 (1): 63–80.

Schaffner, K. F. 1993. *Discovery and Explanation in Biology and Medicine*. Chicago: University of Chicago Press.

Schaffner, K. F. 1994. "Psychiatry and molecular biology: Reductionistic approaches to schizophrenia." In *Philosophical Perspectives on Psychiatric Diagnostic Classification*, edited by J. Sadler, O. Wiggins, and M. Schwartz, 279–94. Baltimore: Johns Hopkins University Press.

Schaffner, K. F. 2006. "Reduction: The Cheshire cat problem and a return to roots." *Synthese* 151 (3): 377–402.

Schaffner, K. F. 2008. "Etiological models in psychiatry: Reductive and nonreductive." In *Philosophical Issues in Psychiatry*, edited by K. S. Kendler and J. Parnas, 48–90. Baltimore: Johns Hopkins University Press.

Schaffner, K. F. 2011. "Reduction in biology and medicine." In *Philosophy of Medicine*, edited by Fred Gifford, 137–57. Amsterdam: Elsevier.

Schaffner, K. F. 2012. "Ernest Nagel and reduction." *Journal of Philosophy* 109 (8–9): 534–65.

Schaffner, K. F. 2014. "Neuroethics." In *Scientism: The New Orthodoxy*, edited by Richard N. Williams and Daniel N. Robinson, 147–76. London: Bloomsbury Publishing.

Schaffner, K. F., and P. D. McGorry. 2001. "Preventing severe mental illnesses: New prospects and ethical challenges." *Schizophrenia Research* 51 (1): 3–15.

Schaffner, K. F, and K. C. Tabb. 2014. "Varieties of social constructionism and the problem of progress in psychiatry." In *Philosophical Issues in Psychiatry III*, edited by K. S. Kendler and J. Parnas. New York: Oxford University Press, 83–115.

Shields, J., I. I. Gottesman, and L. L. Heston. 1975. "Schizophrenia and the schizoid: The problem for genetic analysis." In *Genetic Research in Psychiatry*, edited by R. R. Fieve, D. Rosenthal, and H. Brill, 167–97. Baltimore: Johns Hopkins University Press.

Sideli, L., A. Mule, D. La Barbera, and R. M. Murray. 2012. "Do child abuse and maltreatment increase risk of schizophrenia?" *Psychiatry Investigation* 9 (2): 87–99. doi:10.4306/pi.2012.9.2.87.

St Clair, D., M. Xu, P. Wang, Y. Yu, Y. Fang, F. Zhang, X. Zheng, et al. 2005. "Rates of adult schizophrenia following prenatal exposure to the Chinese famine of 1959–1961." *JAMA* 294 (5): 557–62.

Stefansson, H., J. Sarginson, A. Kong, P. Yates, V. Steinthorsdottir, E. Gudfinnsson, S. Gunnarsdottir, et al. 2003. "Association of neuregulin 1 with schizophrenia confirmed in a Scottish population." *American Journal of Human Genetics* 72 (1): 83–87.

Stefansson, H., E. Sigurdsson, V. Steinthorsdottir, S. Bjornsdottir, T. Sigmundsson, S. Ghosh, J. Brynjolfsson, et al. 2002. "Neuregulin 1 and susceptibility to schizophrenia." *American Journal of Human Genetics* 71 (4): 877–92.

Stone, W. S., X. Hsi, A. J. Giuliano, L. Tan, S. Zhu, L. Li, L. J. Seidman, and M. T. Tsuang. 2012. "Are neurocognitive, clinical and social dysfunctions in schizotaxia reversible pharmacologically? Results from the Changsha study." *Asian Journal of Psychiatry* 5 (1): 73–82. doi:10.1016/j.ajp.2011.12.001.

Straub, R. E., Y. Jiang, C. J. MacLean, Y. Ma, B. T. Webb, M. V. Myakishev, C. Harris-Kerr, et al. 2002. "Genetic variation in the 6p22.3 gene DTNBP1, the human ortholog of the mouse dysbindin gene, is associated with schizophrenia." *American Journal of Human Genetics* 71 (2): 337–48. doi:10.1086/341750.

Tansey, K. E., M. J. Owen, and M. C. O'Donovan. 2015. "Schizophrenia genetics: Building the foundations of the future." *Schizophrenia Bulletin* 41 (1): 15–19. doi:10.1093/schbul/sbu162.

Thiselton, D. L., B. T. Webb, B. M. Neale, R. C. Ribble, F. A. O'Neill, D. Walsh, B. P. Riley, and K. S. Kendler. 2004. "No evidence for linkage or association of neuregulin-1 (*NRG1*) with disease in the Irish study of high-density schizophrenia families (ISHDSF)." *Molecular Psychiatry* 9 (8): 777–83; image 729.

Tyler, Leona E. 1956. *The Psychology of Human Differences.* New York: Appleton-Century-Crofts.

van Os, J., B. P. Rutten, and R. Poulton. 2008. "Gene-environment interactions in schizophrenia: Review of epidemiological findings and future directions." *Schizophrenia Bulletin* 34 (6): 1066–82. doi:10.1093/schbul/sbn117.

Visscher, P. M., M. E. Goddard, E. M. Derks, and N. R. Wray. 2012. "Evidence-based psychiatric genetics, AKA the false dichotomy between common and rare variant hypotheses." *Molecular Psychiatry* 17 (5): 474–85. doi:10.1038/mp.2011.65.

Wakefield, J. C. 1992. "The concept of mental disorder: On the boundary between biological facts and social values." *American Psychologist* 47 (3): 373–88.

Walker, R. M., A. Christoforou, P. A. Thomson, K. A. McGhee, A. Maclean, T. W. Muhleisen, J. Strohmaier, et al. 2011. "Association analysis of Neuregulin 1 candidate regions in schizophrenia and bipolar disorder." *Neuroscience Letters* 478 (1): 9–13. doi:S0304-3940(10)00517-3 [pii] 10.1016/j.neulet.2010.04.056.

Weinberger, D. R. 1986. "The pathogenesis of schizophrenia: A neurodevelopmental theory." In *The Neurology of Schizophrenia,* edited by R. A. Nasrallah and D. R. Weinberger, 387–405. Amsterdam: Elsevier.

Weinberger, D. R. 1987. "Implications of normal brain development for the pathogenesis of schizophrenia." *Archives of General Psychiatry* 44 (7): 660–69.

Weinberger, Daniel R., and P. J. Harrison. 2011. *Schizophrenia.* 3rd ed. Hoboken, NJ: Wiley-Blackwell.

Williams, N. M., M. C. O'Donovan, and M. J. Owen. 2005. "Is the dysbindin gene (DTNBP1) a susceptibility gene for schizophrenia?" *Schizophrenia Bulletin* 31 (4): 800–805.

Wong, A. H., I. I. Gottesman, and A. Petronis. 2005. "Phenotypic differences in genetically identical organisms: The epigenetic perspective." *Human Molecular Genetics* 14 (Spec. No. 1): R11–R18.
Woo, T. U., K. Spencer, and R. W. McCarley. 2010. "Gamma oscillation deficits and the onset and early progression of schizophrenia." *Harvard Review of Psychiatry* 18 (3): 173–89. doi:10.3109/10673221003747609.
World Health Organization. 1992. *The ICD-10 Classification of Mental and Behavioural Disorders: Clinical Descriptions and Diagnostic Guidelines.* Geneva: World Health Organization.
Wyatt, R. J. 1991. "Early intervention with neuroleptics may decrease the long-term morbidity of schizophrenia." *Schizophrenia Research* 5 (3): 201–2.
Yung, A. R., and P. D. McGorry. 1996. "The prodromal phase of first-episode psychosis: Past and current conceptualizations." *Schizophrenia Bulletin* 22 (2): 353–70.
Zachar, Peter. 2000. *Psychological Concepts and Biological Psychiatry: A Philosophical Analysis.* Philadelphia: J. Benjamins.

8

What's Genetic, What's Not, and Why Should We Care?

In the years 2000–2003 the Hastings Center, the major institution examining bioethics in the United States, working jointly with the American Association for the Advancement of Science, conducted a project entitled Tools for a Public Conversation about Behavioral Genetics. This large project was generously funded by the Ethical, Legal, and Social Implications (ELSI) division of the National Genome Research Institute of the NIH. The project's participants included bioethicists, social scientists, lawyers, scientists, and philosophers, including myself. Two books and a special supplement of the *Hastings Center Report* constitute the permanent records of the project's results (Parens 2004; Parens, Chapman, and Press 2006; Baker 2004).

Toward the end of the project, one of the participants, Rick Weiss, who was then the senior science writer for the *Washington Post,* observed that none of the presentations and analyses that the project members had given and heard from guest speakers zeroed in on what disturbed most people about the science of behavioral genetics. That issue, he said, was free will and whether the advances in genetics were soon to show that free will was an illusion, and that we were all, in reality, driven by our genes. Weiss was far from alone in this concern: *The Economist* issue of September 14, 1996, had a figure of a human being as puppet on its cover, with the controlling strings depicted as DNA double helices. And the article in the magazine stated that the Human Genome Project "in the longer run . . . is in danger of creating a philosophical misconception of its own: that men's actions are determined by their genes, not by their own free will."

This current book has re-presented many of the themes that the NIH project was concerned with, and has also gone beyond them in terms of philosophical details and arguments, as well as describing significant scientific changes that have taken place in behavioral and psychiatric genetics since that project and its publications were finalized. Those publications did in point of fact contain some articles that addressed the free will issue, but in part because of scientific advances, including some perspectival shifts from a focus on genetics per se to a combined discipline of *neuro*genetics, a new look is warranted, and it is this chapter's goal to provide such an account.

Table 8.1 Relations of Free Will and Forms of Determinism

Determinism		Free will	
		Free will does not exist	**Free will exists**
	The universe is determined	Hard determinism	Compatibilism
	The universe is indetermined	Hard incompatibilism	Libertarianism

As background to this discussion, it will be useful to distinguish among three general views of the way that physical (including biological and genetic) determinism can be parsed in relation to the issue of free will. Often this is done using a simple table (table 8.1), and the interrelations among these concepts and positions should be evident from that table. In some of the discussion below, terms such as "compatibilist" will be used in attributing positions to philosophers and scientists.

WHAT HAVE PHILOSOPHERS MEANT BY "FREE WILL"?

The question whether our actions are "free" in the sense that they are "our own" and not a product of unknown forces that still permit the "illusion" of freedom is a perennial one. There are nuanced discussions of the issues in Aristotle and in the writings of many subsequent philosophers. The topic has an obvious theological dimension, including various "predestination" doctrines. The issue of responsibility is closely coupled with the notion, and in addition to moral responsibility, the legal aspect has generated a set of diverse approaches to legal guilt as well as exculpatory conditions.

This book is not the place to summarize and parse the huge literature on the various aspects of free will, but it will be useful for the readers to have a sense of some recent positions taken by philosophers and legal scholars that can constitute the beginnings of a framework for the behavioral and psychiatric genetic findings presented in the earlier chapters. In this section I will briefly outline several positions that make sense to me in this complex area.

An accessible introductory article on free will is available to point interested readers to some of the extensive literature on the topic (O'Connor 2011). That article suggests we might think of free will as involving dimensions of agency, ownership, and deliberation as key features. Generally we believe that actions follow on desires coupled with motives and intentions. If the desires are considered compatible with the (suitably reflective) self, such as the desire to have a hearty breakfast, we take that desire and the subsequent feasting as "free." This view assumes the absence of a contrary desire to limit food intake due to health or appearance reasons. The presence of potential conflicting desires, and their joint presence in the "self," is the kind of puzzle that led philosopher Harry Frankfurt to propose his influential account of free will.

Frankfurt's view draws a distinction between human and animal activity involved in desire satisfaction. The human has the capacity to entertain what Frankfurt calls "second-order desires" of the type that might be invoked by the actor deliberating about the hearty breakfast. When an individual acts so that both first- and second-order desires agree, that individual exhibits "free action." Another way of looking at that agreement is that the action reflects the individual's "true self," and that the second-order desire is one with which the individual *identifies* and is happy to *own*.

In Frankfurt's analysis there is a distinction between desires that are effective and noneffective. For first-order desires, a first-order effective desire is one that has motivated, is motivating, or will in the future motivate the desirer to act. A noneffective desire is one that does *not* result in an action. For second-order desires, there are two types: desires to have a first-order desire simpliciter versus a desire to have an *effective* first-order desire. It is the second sense, for an effective first-order desire, that Frankfurt also calls a volition.

To make these distinctions more concrete, consider the case of the psychiatrist and the young anorexic patient that Cohen presents so vividly in his book that develops and applies McHugh and Slavney's perspectives model. (For the case see Cohen 2003, chap. 3; for the perspectives framework see McHugh and Slavney 1998; for a similar, more philosophical example see Fischer 1986, 43–44. Details of the case are not needed here.) The psychiatrist may want to have the desire to be anorexic (he actually has no first-order desire *to be* anorexic) so as to empathetically treat his young patient (a second-order desire), but not wish that such a desire be effective (thus he has no second-order volition). The patient, with a new insight into her disorder, may wish not to starve herself and to eat normally (thus a second-order desire) but may not (yet anyway) be able to put that second-order desire into effect (and thus not have a second-order volition). In Frankfurt's view, this patient would be an "unwilling" anorexic and also lack freedom of the will. "It is in securing the conformity of [her] will to [her] second-order volitions, then, that a person exercises freedom of the will" (Frankfurt 1968, 331). Frankfurt also proposes that a strong "identification" with a first-order desire can be sufficient for a free will.

Frankfurt takes "freedom of action" to be a weaker notion in which one is free to act on a first-order desire, and distinguishes that notion from the stronger "freedom of the will" that requires freedom of action *and* the capacity to have the type of will that one wants to have. Moral responsibility is a bit tricky for Frankfurt, but essentially if the agent has a *volition* to act, even though the agent could not have done otherwise, the agent is responsible for the act. Thus a *willing* drug addict is responsible for taking a drug, even though addicted. (For an analysis of the neuroscience of addiction in agreement with this responsibility view see Hyman 2007.)

Finally, we should note that Frankfurt is what we termed above a compatibilist. He writes: "My conception of the freedom of the will appears to be neutral with regard to the problem of determinism. It seems conceivable that it should be causally determined that a person is free to want what he wants to want. If this is conceivable, then it might be causally determined that a person enjoys a free will" (Frankfurt 1968, 336).

This general view of Frankfurt's is what some have called a "hierarchical" approach to the free will problem, and it has been criticized in the philosophical literature. One critique, by Watson, raises the question of a further regress (to a third level of desires), but also develops an alternative view regarding different sources of desires (rather than levels), such as desires deriving from the passions and desires based on reason (and even acculturation) (Watson 1975). (For another important analysis of free will that develops a "reason" view, see Wolf 1990.)

Though the Frankfurt view has generated a large critical literature (see references in O'Connor 2011; Widerker and McKenna 2003), the view seems to the first approximation to be a plausible one, and one that is consistent with what we know about the role of biology in general and genetics in particular. As such, it seems to be a reasonable one to hold in assessing the issues raised in the present book, as well as additional information related to genetics and neuroscience, to be introduced below.

GENETIC DETERMINISM

The General Idea of Genetic Determinism

A form of determinism of special interest to the readers of this book goes by the term "genetic determinism." That notion can be defined in several ways, but the brief quote from Griffiths captures several of the salient features of this concept:

> Genetic determinism is the idea that significant human characteristics are strongly linked to the presence of certain genes; that it is extremely difficult to attempt to modify criminal behavior or obesity or alcoholism by any means other than genetic manipulation. . . . In contemporary popular discourse, a trait that is supposedly characteristic of one sex, of some ethnic group, or of humanity in general is said to be in "in the genes" just as in previous centuries such traits were said to be "in the blood." Individual differences that might once have been said to "run in the family" are now attributed to genes. The popular concept of a genetic trait is the latest expression of the ancient idea that some traits are inborn and unalterable expressions of an organism's nature, whilst others are acquired, malleable effects of experience. (Griffiths 2006, 177)

The notion has been fine-tuned to reflect a more dimensional notion of genetic determinism, and one that with some further development might be applied in more contemporary contexts.[1] One such more nuanced analysis is provided in this quotation describing roughly three different form of genetic determinism:

Strong genetic determinism: gene G almost always leads to the development of trait T. (G increases the probability of T and the probability of T, given G, is 95% or greater.)

Moderate genetic determinism: more often than not G leads to the development of T. (G increases the probability of T and the probability of T, given G is greater than 50%.)

Weak genetic determinism: G sometimes leads to the development of T. (G increases the probability of T, but the probability of T is still less than 50%.) (Resnik and Vorhaus 2006, 3).

Genetic determinism, particularly of the moderate to strong form, when invoked in connection with genetic differences between ethnic groups, has produced the periodic firestorms noted in chapters 1 and 2, and others may resurface again, for example, following the recent publication of Wade's *A Troublesome Inheritance* (2014). General interest in the notion has also generated a number of deep analytical and "philosophical" attempts to clarify exactly what we might mean when we refer to a trait as genetic, or strongly genetically influenced. The literature is fairly extensive, and in what follows in this section I will need to be selective, but will I think be comprehensive enough so that we can obtain some useful results from this literature.

Philosophical Analyses of Genetic Determinism

GIFFORD

I begin with a proposal made in 1990 by Fred Gifford. Gifford first notes the socially contentious backdrop involved in the genetic-nongenetic distinction, including matters of IQ and race, and proposes a two-part analysis. The first is essentially populational and is termed the DF criterion, since it appeals to a gene as a "differentiating factor." More specifically:

> (DF) A trait is genetic (with respect to population *P*) if it is genetic factors which "make the difference" between those individuals with the trait and the rest of population *P*. (Gifford 1990, 333)

This approach is similar to what quantitative geneticists tried to capture with their notion of "heritability," and though Gifford was sensitive to the complexities of this approach as well as problems with the heritability notion, these do not figure very largely in his analysis. One problem that does require fine-tuning of the population notion contained in the above are fixed traits, such as the opposable thumb, which do *not* vary in a species such as humans, and thus have a heritability of zero. Gifford proposes that expanding the population, say to include primates, might help resolve attributing this trait to being genetic. There are other complications, involving a mixture of social environments, that can raise concerns and may imply normative considerations should be involved, but Gifford prefers to keep these at bay and resolve such conundrums by appealing to more direct genetic causes.

The need to appeal to a notion of directness arises in a different way and motivates Gifford's second criterion for what is genetic, which can apply to individuals, and not only populations. This he calls a proper individuation (PI) criterion, and it is stated thus:

(PI) For a trait to be genetic, the gene (or set of genes) must cause that trait *as described*. The trait must be individuated in such a way that it matches what some genetic factors cause specifically. (Gifford 1990, 343)

The notion of specificity is invoked to handle the intuition that often traits might be defined too broadly, such as high cholesterol, or alternatively too narrowly, such as the ability to speak French. This problem is more generally related to forming the proper contrast class, as in van Fraassen's model of explanation (Van Fraassen 1980). The main problem, as I see it here, is that the terms in this second criterion are notoriously vague, and Gifford does not really provide a set of application instructions for PI. This and related problems with Gifford's analysis are explored in considerable detail in Smith's comment on Gifford (Smith 1992).

Smith has in fact contributed several articles to this debate, but this 1992 essay is one mainly directed to Gifford's account noted above and raises problems for the DF criterion, but especially for the PI analysis. Regarding Gifford's PI definition, Smith says it "fails utterly as a criterion for delineating 'genetic trait' because it is incapable of making the cause/condition (genetic/epigenetic [= environmental]) distinction stick" (Smith 1992, 346). Essentially Smith argues that context determines whether we call a trait genetic or not, and he illustrates his point with an example from a simple operon model of genetic regulation, which I find suggestive but not compelling. However, Smith generalizes from this simple example and characterizes what he terms "the new problem of genetics" as follows: "For any analysis regardless of the criterion) which reveal genes as the true cause of a trait, there will be a complementary analysis showing the trait to be *epigenetically* caused" (346). The spirit of the criticism is akin to those we examined in our analysis of DST proponents in chapter 3, and I will not rehearse them here. Within the framework of an explanatory model that recognizes context including interests, whether a DNA variation is focused on as the important cause, or whether some environmental variation is seen as more significant, is typically clear to the researcher. This is an issue we will return to again in relation to Waters's work. Smith has also written a more recent paper proposing a fourfold approach to characterizing what is genetics, and I shall come back to that briefly in another section below.

Sarkar

In his extensively detailed 1998 monograph on reduction in biology, Sarkar provides a concluding chapter that addresses the question "What's genetic?" (Sarkar 1998). The main theme of Sarkar's book is an examination of various approaches to reduction, including criticism of the Nagel-Schaffner type of theory reduction, but also an in-depth critique of the notion of heritability. Sarkar's preferred approach to reduction, a rather restrictive one, is an important aspect in this debate about the definition of whether a trait is genetic, because his definition strongly depends on the possibility of such a reduction (or explanation) of the trait in terms of genetics.[2] Sarkar comments briefly on Gifford's analysis (as well as briefly on Smith's comment on Gifford), but thinks the attempt to clarify genetic in terms of cause is a nonstarter. Like Smith, Sarkar suggests the entanglement of genetic and environmental will defeat such an approach (see Sarkar 1998, 4–5 and 192). Sarkar provides his own general idea as follows: "to call a trait genetic if and only if its occurrence is best explained on the basis only of the properties of the genes (specific alleles at specific loci)," adding, "Thus this strategy is parasitic on the account of genetic reductionism given in [his] chapter 5" (Sarkar 1998, 181). As background, Sarkar also sharply distinguishes four notions that he says need to be kept separate: biology, genetics, inheritance, and heritability, since they are far from coextensive in meaning and application.

More specifically Sarkar's definition of genetic goes as follows:

> A trait will be called "genetic" if and only if the following three conditions are satisfied:
>
> (i) the trait is under control of a few loci
> (ii) the trait always (that is, in all populations) shows high expressivity
> (iii) the immediate products of the alleles at these loci form part of the biochemical characterization (that is the description at the biochemical level) of the trait. (Sarkar 1998, 182)[3]

This was admitted by Sarkar to be "highly restrictive," and he noted that it contained some vague terms, such as the meaning of "few" modifying loci, as well as how high the expressivity was to be set. In the light of the developments in human behavioral genetics as sketched in chapters 1 and 2 and as detailed in chapters 6 and 7, including the GWAS results, missing heritability, and the era of Big Data (thousands of genes of tiny effect size affecting simple and complex traits), this definition that was promising in 1998 can no longer work. Something different is needed, not only for human genetics, but even for the complications we have noted in simple organisms such as the worm.

I do sympathize with Sarkar's view that *if* we had a sweeping kind of reduction of biology to genetics, we would be able to identify more precisely what types of traits were genetic and which were not. In that world, genetics would be more fundamental, much as electromagnetic theory identified a science that was more fundamental than a reduced optics (see chapter 5 and also Schaffner 2012). However that world is not this world.

Kitcher

In a fascinating paper in a volume dedicated to examining the philosophical contributions of Richard Lewontin, Philip Kitcher criticizes genetic determinism as well as Lewontin's proposal for a dialectical biology as his way to avoid genetic determinism. Perhaps anticipating the recent vampire craze as manifest in the media successes of the *Twilight* and *Southern Vampire/True Blood* series, Kitcher titles his article "Battling the Undead: How (and How Not) to resist Genetic Determinism" (Kitcher 2001).

Kitcher's initial stab at a definition of genetic determinism is remarkably close to Gifford's and Sarkar's definition of a genetic trait. Kitcher writes:

> To suppose that a particular trait in an organism is genetically determined is to maintain that there is some gene, or group of genes, such that any organism of that species developing from a zygote that possessed a certain form (set of forms) of that gene (or a certain set of forms of those genes) would come to have the trait in question, whatever the other properties of the zygote and whatever the sequence of environments through which the developing organism passed. (Kitcher 2001, 397)

The idea is that something like this definition captures the inevitability that affects carriers of the Huntington's disease (HD) allele, but qualifications are quickly added in Kitcher's text. These are needed since the carrier could die before the disease developed (HD does not usually appear until after age 40). Other qualifications are considered, with Kitcher reserving his main exposition for alternative forms of "norms of reaction," of the type we described back in chapter 1 using the classic example of *Achilla* strains at different elevations.

Kitcher, however, is concerned mostly with various kinds of interactionism, and is critical of the hyperinteractionism championed by developmental systems proponents (again see chapter 3 above for examples), but also of Lewontin's concerns that interactionism itself does not go far enough to cure our Cartesian hang-ups. For Lewontin, only a Marxist-like dialectical biology will do so (Levins and Lewontin 1985), a thesis with which Kitcher strongly disagrees.

For my purposes, however, what is most interesting about Kitcher's general view here is his continuing belief in an adequate "gene for" analysis, one that he and Sterelny (Sterelny and Kitcher 1988) had previously published as part of an early critique of a DST paper (Griffiths and Gray 1994). This characterization is another way to capture what is genetic about a trait, and Kitcher phrases it in this way: "We can speak of genes for X if substitutions on a chromosome would lead, in the relevant environments, to a difference in the X-ishness of the phenotype" (Kitcher 2001, 348). Again notice the parallels of this "gene for X" with Gifford's and Sarkar's definitions of a genetic trait.

The notion of the relevant environment needs to be qualified: it becomes part of the usual or standard conditions and ones that do not substantively decrease the fitness of the organisms of interest. These qualifications lead Kitcher into further elaborations to counter possible objections by DSTers that I will not review here. It is worth noting one additional point concerning genetic research. Kitcher writes that investigators "begin with genetic causes not because they are convinced that these are the most important . . . but because they want to unravel the neurochemistry, and they see the investigation of genotypes as a thread that will lead them into the tangle" (409). This is because they can sequence DNA and know enough about the way proteins are synthesized that they can begin to construct a biochemical pathway that leads to brain variation associated with diseases, such as addiction. Thus, Kitcher adds: "There is no question of 'privileging' the genes in this kind of inquiry but rather a pragmatic criterion for using a particular type of model and a readily comprehensible, even admirable, medical motivation" (409).

In this, Kitcher seems to agree with the pragmatic prioritization of DNA and genes developed in my responses to DST challenge in chapter 3 above. Kitcher's criticisms of the DST view were then subsequently replied to by Paul Griffiths in his extensive critique defending some views of Oyama ([1985] 2000), and further developing the notion of "genetic information." The topic of various analyses of genetic information would take us beyond the scope of the present chapter, though there have been a number of accounts that raise topics of general interest related to the role of genetics in biology (see, for example, Godfrey-Smith 2000 on the notion of coding and also teleosemantics). However, these discussions of genetic information have largely been developed prior to the major reorientations of behavioral and psychiatric genetics from focus on single genes during the past five years, influenced by GWAS studies, and will need some overhauling.

Kendler

Returning to the "gene for" topic championed by Kitcher above, in 2005 Kendler published an analysis of the "gene for" phrase that focused on the effects of genes on psychiatric disorders. Kendler introduced his analysis with some history of genetics that singled out the "preformationist" approach as the main backdrop for this kind of genetic determinism of a trait. One can describe this view as postulating what I've previously termed "traitunculi" (or little preexisting traits) in the germ line (Schaffner 1998), which inexorably unfold to produce the trait of interest.

Kendler proposed five criteria for X is a gene for Y:

> 1) strength of association of X with Y, 2) specificity of relationship of X with Y, 3) noncontingency of the effect of X on Y, 4) causal proximity of X to Y, and 5) the degree to which X is the appropriate level of explanation for Y (1243).

Kendler added: "If gene X has a strong, specific association with disease Y in all known environments and the physiological pathway from X to Y is short or well-understood, then it may be appropriate to speak of X as a gene for Y."

In his further analysis of research and hypothetical examples in contemporary genetics for each of these five criteria, Kendler finds them especially deficient in the area of psychiatric genetics. His arguments on these points are similar to many of the examples from simple organisms, discussed in chapters 3–5 above. In general in behavioral and psychiatric genetics, the strength of association is modest or weak (odds ratios < 1.5); the specificity of relationship is not one-to-one but typically many-to many. Genes work in complex contexts, and gene-gene and gene-environment interactions are frequent. Furthermore biochemical chains of influence are long and complex (recall the Lewis example for schizophrenia from chapter 7). Finally the results that seem well supported are not at the level of clinical descriptions of such disorders, but are rather phrased in terms of endophenotypes that may eventually be identified with higher-level disorders.

The bottom line of Kendler's reconstruction of "gene for" locution, then, is that for the area of this book's major interest—behavioral and psychiatric genetics—nothing like Kitcher's "gene for" analysis is likely to be relevant. Furthermore, insofar as Kitcher has strong analogies with Gifford and Sarkar, these analyses also fail to materially illuminate a genetic characterization of behavioral and psychiatric disorders.

Kendler has also recently addressed the issue of what might be called "top down" influence on gene action, namely the roles that individual desires, as well as familial acts and societal interventions, can have on genetic physiological pathways (Kendler 2013). Kendler presents four concrete scenarios in which deliberative human decision-making can interpose itself on putatively autonomous gene-to-phenotype pathways and be a "difference maker." These examples resonate well with the picture developed in the present chapter about free will and genetic determinism.

In another recent article Kendler (2012) asks whether an examination of psychiatric disorders such as alcohol dependence (AD) and major depression (MD) and a medical disease, cystic fibrosis (CF), can disclose a privileged level at which one might look for strong etiologies in psychiatry.[4] This essay extends his five criteria cited above to seven, namely "(i) strength, (ii) causal confidence, (iii) generalizability, (iv) specificity, (v) manipulability, (vi) proximity and (vii) generativity." The latter notion "reflects the probability that the explanatory variables identified would have the potential to lead to further fruitful etiologic understanding of the disease" (Kendler 2012, 14). Thus, this notion is similar to Kitcher's "pragmatic criterion" for focusing on the genetic level for some disorders, but for Kendler the most appropriate level of etiological focus for psychiatric disorders will frequently not be genetic, or even biological, but may be psychological, social, or even legal. In point of fact, Kendler quotes a statement by Dretske as an epigraph to this article: "I have nothing particularly original to say about how one identifies the cause of something from among the many events and conditions on which it depends. It seems fairly clear that this selection is often a response to the purpose and interests of the one doing the describing" (Dretske 1988, 24). Toward the end of his article, Kendler states that his review of AD, MD, and CF underscores the correctness of Dretske's view, adding, "There is no *a priori* way to pick a single level of explanation on which to base an etiological nosology" (17).

We might ask, however, if some more recent philosophically technical ways of parsing the relation between genes and traits could rescue this notion of a fairly simple and prioritized genetic character of interesting traits. For this I look next at some proposals regarding causality made by Waters and followed up by Tabery.

Waters

The philosophy of biology community has tended to accept variants of Hull's original and now classical argument against the reduction of classical genetics to molecular genetics. In addition to Hull's original arguments (1974), similar antireductionist positions were published by Kitcher (1984) and Rosenberg (1985). Those objections were countered by Schaffner (1993, 432–87), but it was Waters, who in addition to also defending a reductionist line, coined the expression "the reductionist anticonsensus" to reflect his as well as my own views (Waters 1990). In addition, Waters has proposed several definitions of a "gene," in attempts to clarify that notion at a general level, and counter various arguments by DST-oriented proposals that the gene concept has become so fragmented that it cannot serve as a fundamental concept or foundation for genetic prioritization (Waters 1994, 2004).

It was consistent with this type of proreductionist research program that Waters, in a lengthy article in 2007, developed an analysis of genetic causation that argued against the parity thesis for genes, or typically DNA. This view is similar in spirit to my conclusions against parity developed in chapter 3 above, though Waters wishes to make his argument even stronger than the heuristic thesis urged in this book. In fact, Waters wants to argue that genes have a privileged ontological status as causes of phenotypic properties, in distinction to other factors, such as enzymes and a hospitable environment. The argument depends on attributing to genes the status of "actual difference makers," in contrast to accessory factors that would be mere conditions. Woodward's theory of causation provides the framework for articulating the actual difference maker concept. More specifically Waters characterizes this notion as follows:

> X is the actual difference maker with respect to Y in a population p if and only if:
>
> (i) *X* causes *Y* (in the sense of Woodward's manipulability theory).
> (ii) The value of *Y* actually varies among individuals in *p*.
> (iii) The relationship expressed by "*X* causes *Y*" is invariant with respect to the variables that actually vary in *p* (over the spaces of values those variables actually take in *p*).
> (iv) Actual variation in the value of *X* fully accounts for the actual variation of *Y* values in population *p* (via the relationship *X* causes *Y*). (Waters 2007, 567)

This definition is applied to Waters's main example of fly color as investigated by Morgan in the first quarter of the last century, namely that homozygous red-eyed females crossed with purple-eyed males will produce all red-eyed offspring (the red-eyed gene being dominant). This application shows, Waters claims, that the real cause (the genetic alleles)—ontologically distinct from other potential causes—is the actual cause of the trait, and thus has a privileged position in contrast to other potential causes that co-act with genes to produce traits. Further application of this actual difference maker to traditional examples, such as identifying the lit match rather than the ambient oxygen or kindling material as "*the* actual difference making cause" (569, my emphasis) of a fire, is also outlined. In this latter case the struck match is considered as drawn from a population (of unstruck matches) and thus the striking is the difference, and *the* cause, rather than the presence of oxygen, which is merely A cause being present in all cases.

Though genetic variation is the important contrast conditions in the fly example, as the matches are in the traditional example, it is difficult to see that the argument does not beg the question in an important way. Consider the example presented in chapters 3 and 5 above of the worm feeding behavior. In that example, initially (in 1998) the main contrast was with an *npr-1* allelic difference. The wild type of worm was a solitary eater, and a single mutation resulting in an amino acid change at position 215 in the associated NPR-1 protein from valine to phenylalanine position resulted in a worm that dined gregariously. However, further investigations indicated that the *npr-1* genes actually functioned in a more complex network, and that there were *several pathways* by which feeding behavior could be altered. In these later studies, the main trigger affecting feeding behavior was shown to be the oxygen percentage in the environment (see figure 5.3 in chapter 5). Thus the major contrast case resulting in solitary or gregarious feeding (in the wild types) was the presence of a sufficient amount of ambient oxygen, though disruptions in any one of several genes in the network could override a standard response to the oxygen signal. Even more recently, attention has shifted to CO_2 as playing a primary role, though oxygen still figures in the behavior of the worm's response network, which again utilizes several pathways. Whether the focus shifts to O_2 or more recently to CO_2, what was initially thought to be a fundamental *genetic* difference for clumping turned out to be a fundamental *environmental* difference, with genetics playing a contributory background condition role.[5]

But the point that I think needs to be made is *more general* than suggested by the worm example. That point is that one can *always* so choose a set of known jointly acting factors with a difference in one of those factors so as to identify the *selected difference factor* as the fundamental cause of the result in that setup reflecting the main interests of the investigators.[6] This point seems to resonate strongly with the Kendler (2012) analysis of levels summarized above, and with Kendler's endorsement of the Dretske position.

Water's analysis of the primacy of DNA as an actual specific difference maker is the subject of an extended critique by Griffiths and Stotz (2013, primarily pages 78–84 but also recurring in other parts of their book). I view their analysis as in the same spirit as my suggestions above, though I do not, as indicated in chapter 3, accept the stronger parity position of the developmental systems theorists. Again, as proposed in chapter 3, DNA is heuristically primus inter pares in many contexts. I suspect that this might partly be because of Woodwardian "stability" of the use of DNA as described in my chapter 3, rather than its "specificity," which Waters wishes to emphasize. (For an account of Woodwardian specificity and stability see Woodward 2010.)

It might be too quick to infer, as Mill seems to have done along with Mackie, and possibly Hart and Honore (1985), that the selection of the factor chosen for contrast is open to *any interest whatsoever.* In worm research, DNA variations are important to study for many reasons, though ultimately those reasons may not differ all that much from the Kitcher or Kendler opinions noted earlier, or the various reasons regarding DNA priority discussed in chapter 3 in this book. For suggestions along these lines see the fourfold typology outlined by Smith (2007) and also Northcott's (2012) suggestions for establishing cause-condition contrasts in the area of genetics.

This criticism of Waters analysis should not be interpreted as suggesting it is not a helpful advance in our discussions of genetics in general, and its application to behavioral genetics. As Tabery has written, the actual difference maker and its related potential difference maker concepts can illuminate the way that quantitative behavioral genetics focuses on individual differences in its analysis (Tabery 2014), and Woodward has noted that the Waters analysis is "a real advance over previous discussions of the parity thesis" (Woodward, personal communication). That said, the notions do not seem to resolve the traditional conundrum of distinguishing causes versus conditions in multifactorial causal situations, nor does the analysis so prioritize genetics as against environment that it privileges genetics in biology. A powerful reduction, perhaps along the lines that Sarkar speculated about, would provide such a privileged role for genetics, but as shown in chapter 5, that kind of sweeping reduction has been elusive, and later discussions about reduction summarized in this book have not changed that likelihood.

Some Additional Moral and Legal Aspects of Genetic Determinism

As the Human Genome Project (HGP) ramped up in the 1990s, various government agencies involved in the project, either as funders or evaluators, raised the question whether the genetic results of the project might not threaten our concepts of free will and human responsibility (US Department of Health and Human Services and Department of Energy 1990). Philosopher P. S. Greenspan was an early commentator on this issue, but she argued that the HGP actually did not raise any special new issues for human freedom and responsibility (Greenspan 1993). She indicated that "what we are worried about . . . is an alternative, motivational rather than causal version of the free will question: free will versus internal constraint" (39). But we always encounter various constraints, often external, with the genetic possibilities acting as internal limitations indicating difficulties for doing otherwise, but not necessarily for an excuse for moral irresponsibility. Greenspan has elaborated her views of how other forms of constraint, including behavior control and emotions, can be better understood within her account of free will. One additional article that develops some of her later ideas on genetics and free will appeared in 2001.

More has probably been written on the legal implications of genomics than the moral implications. These issues are surveyed in an excellent multiauthored book edited by legal scholar and ethicist Nita Farahany (2009). Denno reviews the details of a number of legal cases involving evidence from behavioral genetics (Denno 2009). One case is based on a Dutch family with a null mutation (actually a stop codon) in the gene for MAOA (Brunner et al. 1993) (and thus related to the Caspi and Moffitt first G × E paper on MAOA alleles). The legal case is known generically as the Mobley case, and it has generated much comment in legal and ethical circles. David Goldman, a geneticist at the National Institutes of Health who was asked to be a court consultant and to test Mobley for the null MAOA mutation (he declined), discusses the case in his very recent book (see Goldman 2012 as well as Denno 2009 for details).

The gist of the case involves Stephen Mobley, who in 1991 murdered a Domino's pizza deliverer. After his 1994 conviction for the murder and in the sentencing phase of his case his lawyers argued that Mobley, who came from a family with a history of criminal behavior, might have the null MAOA mutation and that this should be taken account of to mitigate his sentence. Actually the Mobley family pedigree did not match that of the Dutch family, the judge refused to order a state-paid genotyping, and Mobley was not tested. He was subsequently executed in 2005. There have been similar types of appeals to that in the Mobley case, and in an Italian court one has succeeded in obtaining a sentence reduction (Feresin 2009). Goldman reports Farahany's comment that there have been over 200 attempts to introduce DNA evidence into a trial and that several have been successful (Goldman 2012, 31).

In the past 10 years, appeals solely to genetics to account for behavior (and for exculpation) have waned and have been superseded by either neurogenetics and even mainly neurological and neuroscience drivers of behavior. This period has also seen the emergence of a new academic discipline termed "neuroethics" (Farah 2005) as well as a "neurological defense" in legal matters (Restak 1993). This is not surprising, and could have been predicted given the view that geneticists have been maintaining that genes do not determine behavior in isolation, but do so via the construction, jointly with the actions of the environment, of a neural network that is responsible for behavior (see chapter 3 for details).

Though neuroethical issues are a topic for other books, I will close this chapter noting that free will has been investigated as an emerging set of research programs in the brain and neuroscience disciplines, and it is worth providing a brief summary of some of those projects.

BRAIN AND NEUROSCIENCE STUDIES OF FREE WILL

Discussions about the role of neuroscience and brain research almost always begin with references to the Libet experiments on volition. These experiments, which Libet began nearly 30 years ago (Libet et al. 1983), have been thought to show that the brain unconsciously generates plans for movements that only later (albeit only about a second later) become conscious and are attributed to being initiated by a conscious decision. Many authors have inferred from such experiments that "free will" is an illusion, a confabulation that the conscious mind creates for reasons that remain obscure. Wegner (2002) provides a recent book-length development of this idea that free will is an illusion.

The Libet experiment asks subjects to initiate a simple voluntary action, such as pressing a button, when they feel like doing so. The subject's brain activity is measured, initially using an EEG, but more recently fMRI has been used. The subjects simultaneously are observing a clock and are asked to record the time when they first "feel the urge" or their conscious intention to press the button. This "urge" moment is referred to as W, standing for the subject's "will."

The experiment revealed that brain activity in the frontal motor areas began several hundred milliseconds prior to the W event, suggesting that a decision to press the button was really an unconscious one and only subsequently became conscious after the "decision" had been made in the brain. Like the *Economist* figure described in the introductory section to this chapter, the inference is that the unconscious brain is the true puppet master, and our conscious selves are but puppets of the unconscious brain.

An extensive literature both defending and critiquing the Libet experiment has developed that cannot be surveyed in this current book. The field of neuroscience has also moved on. And currently there exist much more sophisticated methods, including the ability to record from single neurons in the brain, to test the timing and more specific localization of the origination of "voluntary movements" and conscious decision-making. Suffice it to say that though the technology has advanced considerably, the implications of even recent studies are equivocal whether the "conscious mind" is the initiator of such bodily movements. A good summary of some of these recent experiments can be found in Haggard's survey (2011). One very interesting finding comes from the considerably higher resolution spatial and temporal this approach permits, which enables more specific localization of the W process in a brain region known as the midline supplementary motor area, or SMA, which in monkeys contains a map of the body. Relatedly, it is noted:

> This finding suggests a revision of how we interpret the [Libet] W judgment. It is clearly wrong to think of W as a prior intention, located at the very earliest moment of decision in an extended action chain. Rather, W seems to mark an intention-in-action, quite closely linked to action execution. The experience of conscious intention may correspond to the point at which the brain transforms a prior plan into a motor act through changes in activity of SMA proper. (Haggard 2011, 405)

The relation between brain phenomena, consciousness, and the experiencing self that makes choices is only beginning to become clarified in spite of an enormous growth in brain and neuroscience research activities and literature. Some 20 years ago Dan Dennett in his influential book on consciousness (Dennett 1991) offered an insightful perspective on the way that our prima facie serial mind might relate to the enormously complex parallel hardware we first commented on back in chapter 2. Dennett wrote:

> Since any computing machine at all can be imitated by a virtual machine on a von Neumann architecture, it follows that if the brain is a massively parallel processing machine, it too can be perfectly imitated by a von Neumann machine....
>
> Now we are ready to turn this standard idea upside-down. Just as you can simulate a parallel brain on a serial von Neumann machine, you can, also, in principle, simulate (something like) a von Neumann machine on parallel

> hardware and that is just what I am suggesting. Conscious human minds are more or less serial virtual machines implemented—inefficiently—on the parallel hardware that evolution has provided for us. (Dennett 1991, 217–18)

As we struggle with the issues discussed in this chapter, it might be useful to keep in mind that *we* are the *constructs* that are built by our genes (and inherited protoplasm) working through neural nets honed by a complex environment including a set of diverse cultures.

FREE WILL: THE BOTTOM LINE

In this chapter I have considered why genetics seems to generate such strong concerns about its implications. A major theme has been the context in which genetics seems to abrogate the essential human feature of "free will." But another concern among scientists and philosophers is the apparent priority and deterministic aspects that genetic explanations seem to have in contemporary biology. Those two themes interact synergistically to each strengthen the concerns produced by the other.

Earlier in this chapter I mentioned that recent scientific advances in biology and behavioral and psychiatric genetics warranted another look at these deterministic implications that raise the concerns discussed in this chapter. From the earlier chapters of this book, including those on personality genetics and schizophrenia, it should be evident from GWAS results that most genetic influences related to behaviors are based in literally thousands of genes, each with small effects. It will be a long time before we can even identify those genes with any security, given all the presumably missing heritability involved. We also examined that "heritability" notion, pointing out that it is mainly a heuristic notion, and conditional on the environment, a concept that itself also needs further development and a decent theory of its own.

Further, even though the candidate gene approach that initially uncovered the interesting work published by Caspi and Moffitt's group has been called into question by GWAS, the probability of complex interactions of these small-effect genes with the environment is not seriously questioned. One of the recurring messages from even simple systems discussed in chapters 3–5 indicates that the relations of genes and phenotypes is many-many, and that genes work through partially probabilistic neural networks. The recent psychiatric and neuroscience investigations of the self, some discussed in chapter 7, and some covered in the article by Haggard noted above, suggest that a very complex brain can be the seat of a deliberative self, one characterized by agency, ownership, and a hierarchical freedom of the will that is worth having. The implications of that view, and a set of practical tests, are noted in the recommendation I provide below from Dennett.

In this chapter, I also sketched a compatibilist account of free will, largely following Frankfurt. And though I noted that his account has been critiqued by other philosophers, I think it survives as one of the best that seems to accord with our considered sense of self as well as with the broadly deterministic features of our world. I also analyzed various attempts to characterize the deterministic aspects of genetics and reasons for their causal priority, and found that it is mainly human interests that lie at the root of such accounts. This permits both environmental and fundamental characteristics of the developed self to play roles in the causal trajectories that affect our lives. Expansion of genetics to include the more recent neurogenetics focus of recent work on free will, as well as the genetic and neurological implications for legal contexts, does not seem to alter this conclusion.

Libertarians such as Kane will take issue with these conclusions (Kane 1996), but it seems to me that they wish for a view of agency that is incompatible with what we currently know on the basis of contemporary science. I take a more practical view of the kind of free will that seems to be available to us. As additional findings are developed in human behavioral and psychiatric genetics, one might do well to test them against the wise counsel proposed back in 1984 by Dennett, and subject any existing and any new arguments against the kind of compatibilist view urged in this chapter (and by many other philosophers)[7] by following this process. Dennett wrote:

> First, inquire closely about just what variety of free will is supposedly jeopardized by the argument. Is it, in fact, a variety worth caring about? Ask yourself whether you have any clearly statable reason to hope you have that variety, any reason to fear that you might not. Would lacking this freedom really be like being in prison, or like being a puppet? For perhaps the conclusion of the new argument is only that no one could have some metaphysical property that is of academic interest at best. Or worse, the yearned-for freedom in question may turn out on inspection to be an incoherently conceived blessing. Ask yourself: can I even conceive of beings whose wills are freer than our own? What regrettable feature of our lot as physical organisms is not a feature of their lot? If the ideal of freedom we hold out for is simply self-contradictory, we should hardly feel bereft when we learn we cannot have it. There's no sense wringing our hands because we can't undo the past, and can't prevent an event that actually happens, and can't create ourselves *ex nihilo,* and can't choose both alternatives at a decision point, and can't be perfect. If the proposed argument passes all these tests, so that a dire conclusion does seem to be in the offing, then and only then must we take it very seriously and see exactly what the argument is. (Dennett 1984, 172)

References

Baker, Catherine. 2004. *Behavioral Genetics.* Washington, DC: American Association for the Advancement of Science.

Bretscher, A. J., K. E. Busch, and M. de Bono. 2008. "A carbon dioxide avoidance behavior is integrated with responses to ambient oxygen and food in *Caenorhabditis elegans.*" *Proceedings of the National Academy of Sciences USA* 105 (23): 8044–49. doi:10.1073/pnas.0707607105.

Bretscher, A. J., E. Kodama-Namba, K. E. Busch, R. J. Murphy, Z. Soltesz, P. Laurent, and M. de Bono. 2011. "Temperature, oxygen, and salt-sensing neurons in *C. elegans* are carbon dioxide sensors that control avoidance behavior." *Neuron* 69 (6): 1099–113. doi:10.1016/j.neuron.2011.02.023.

Brunner, H. G., M. Nelen, X. O. Breakefield, H. H. Ropers, and B. A. van Oost. 1993. "Abnormal behavior associated with a point mutation in the structural gene for monoamine oxidase A." *Science* 262 (5133): 578–80.

Cohen, Bruce J. 2003. *Theory and Practice of Psychiatry*. New York: Oxford University Press.

Dennett, Daniel C. 1984. *Elbow Room: The Varieties of Free Will Worth Wanting*. Cambridge, MA: MIT Press.

Dennett, Daniel C. 1991. *Consciousness Explained*. Boston: Little, Brown.

Denno, D. 2009. "Behavioral genetics evidence in criminal cases: 1994–2007." In *The Impact of Behavioral Sciences on Criminal Law*, edited by Nita A. Farahany, 317–54. New York: Oxford University Press.

Dretske, Fred I. 1988. *Explaining Behavior: Reasons in a World of Causes*. Cambridge: MIT Press.

Dupré, John. 2012. *Processes of Life*. Oxford: Oxford University Press.

Farah, M. J. 2005. "Neuroethics: The practical and the philosophical." *Trends in Cognitive Sciences* 9 (1): 34–40. doi:10.1016/j.tics.2004.12.001.

Farahany, Nita A. 2009. *The Impact of Behavioral Sciences on Criminal Law*. New York: Oxford University Press.

Feresin, E. 2009. "Lighter sentence for murderer with 'bad genes.'" *Nature*. doi:10.1038/news.2009.1050.

Fischer, John Martin. 1986. *Moral Responsibility*. Ithaca, NY: Cornell University Press.

Frankfurt, Harry. 1968. "Freedom of the Will and the Concept of a Person." *Journal of Philosophy* 68: 5–20.

Gifford, Fred. 1990. "Genetic traits." *Biology and Philosophy* 5: 327–47.

Godfrey-Smith, Peter. 2000. "On the theoretical role of 'genetic coding." *Philosophy of Science* 67: 26–44.

Goldman, David. 2012. *Our Genes, Our Choices: How Genotype and Gene Interactions Affect Behavior*. Boston: Elsevier Academic Press.

Greenspan, P. S. 1993. "Free will and the Genome Project." *Philosophy and Public Affairs* 22 (1): 31–43.

Greenspan, P. S. 2001. "Genes, electrotransmitters, and free will." In *Genetics and Criminal Behavior*, edited by David Wasserman and Robert Wachbroit, 243–58. New York: Cambridge University Press.

Griffiths, P. E. 2006. "The fearless vampire conservator: Philip Kitcher, genetic determinism and the informational gene." In *Genes and Development: Rethinking the Molecular Paradigm*, edited by Eva M. Neumann-Held and Christoph Rehmann-Sutter, 175–98. Durham, NC: Duke University Press.

Griffiths, P. E. 2009. "The distinction between innate and acquired characteristics." In *The Stanford Encyclopedia of Philosophy*, edited by Edward N. Zalta. Fall 2009 ed.

Griffiths, P. E., and R. D. Gray. 1994. "Developmental systems and evolutionary explanation." *Journal of Philosophy* 91: 277–304.

Griffiths, P. E., and K. Stotz. 2013. *Genetics and Philosophy*. New York: Cambridge University Press.

Haggard, P. 2011. "Decision time for free will." *Neuron* 69 (3): 404–6. doi:10.1016/j.neuron.2011.01.028.

Hart, H. L. A., and Tony Honore. 1985. *Causation in the Law*. 2nd ed. Oxford: Clarendon Press; New York: Oxford University Press.

Hull, David L. 1974. *Philosophy of Biological Science*. Englewood Cliffs, NJ: Prentice-Hall.

Hyman, S. E. 2007. "The neurobiology of addiction: Implications for voluntary control of behavior." *American Journal of Bioethics* 7 (1): 8–11. doi:10.1080/15265160601063969.

Kane, Robert. 1996. *The Significance of Free Will*. New York: Oxford University Press.

Kendler, K. S. 2005. "Psychiatric genetics: A methodologic critique." *American Journal of Psychiatry* 162 (1): 3–11.

Kendler, K. S. 2012. "Levels of explanation in psychiatric and substance use disorders: Implications for the development of an etiologically based nosology." *Molecular Psychiatry* 17 (1): 11–21. doi:10.1038/mp.2011.70.

Kendler, K. S. 2013. "Decision making in the pathway from genes to psychiatric and substance use disorders." *Molecular Psychiatry* 18 (6): 640–45. doi:10.1038/mp.2012.151.

Kitcher, Philip. 1984. "1953 and all that: A tale of two sciences." *Philosophical Review* 93: 335–73.

Kitcher, Philip. 2001. "Battling the undead: How (and how not) to resist genetic determinism." In *Thinking about Evolution: Historical, Philosophical, and Political Perspectives*, edited by Rama S. Singh, Costas B. Krimbas, Diane B. Paul, and John Beatty, 396–414. New York: Cambridge University Press.

Levins, Richard, and Richard C. Lewontin. 1985. *The Dialectical Biologist*. Cambridge, MA: Harvard University Press.

Libet, B., C. A. Gleason, E. W. Wright, and D. K. Pearl. 1983. "Time of conscious intention to act in relation to onset of cerebral activity (readiness-potential): The unconscious initiation of a freely voluntary act." *Brain* 106 (Pt. 3): 623–42.

Linquist, S., E. Machery, P. E. Griffiths, and K. Stotz. 2011. "Exploring the folk biological conception of human nature." *Philosophical Transactions of the Royal Society of London B: Biological Sciences* 366 (1563): 444–53. doi:10.1098/rstb.2010.0224.

McHugh, Paul R., and Phillip R. Slavney. 1998. *The Perspectives of Psychiatry*. 2nd ed. Baltimore: Johns Hopkins University Press.

Northcott, Robert. 2012. "Genetic traits and causal explanation." In *Philosophy of Behavioral Biology*, edited by Kathryn S. Plaisance and Thomas A. C. Reydon, 65–82. New York: Springer.

O'Connor, Timothy. 2011. "Free will." In *The Stanford Encyclopedia of Philosophy*, edited by Edward N. Zalta. Fall 2008 ed.

Oyama, Susan. [1985] 2000. *The Ontogeny of Information: Developmental Systems and Evolution*. 2nd ed. Durham, NC: Duke University Press.

Parens, Erik. 2004. "Genetic differences and human identities: On why talking about behavioral genetics is important and difficult." *Hastings Center Report* 34 (1): S4–S35.

Parens, Erik, Audrey R. Chapman, and Nancy Press, eds. 2006. *Wrestling with Behavioral Genetics: Science, Ethics, and Public Conversation*. Baltimore: Johns Hopkins University Press.

Resnik, D. B., and D. B. Vorhaus. 2006. "Genetic modification and genetic determinism." *Philosophy, Ethics, and Humanities in Medicine* 1 (1): E9. doi:10.1186/1747-5341-1-9.

Restak, R. 1993. "The neurological defense of violent crime: 'Insanity defense' retooled." *Archives of Neurology* 50 (8): 869–71.

Rosenberg, Alexander. 1985. *The Structure of Biological Science*. New York: Cambridge University Press.

Sarkar, Sahotra. 1998. *Genetics and Reductionism*. New York: Cambridge University Press.

Schaffner, K. F. 1993. *Discovery and Explanation in Biology and Medicine*. Chicago: University of Chicago Press.

Schaffner, K. F. 1998. "Model organisms and behavioral genetics: A rejoinder." *Philosophy of Science* 65: 276–88.

Schaffner, K. F. 2012. "Ernest Nagel and reduction." *Journal of Philosophy* 109 (8–9): 534–65.

Smith, Kelly. 1992. "The new problem of genetics: A response to Gifford." *Biology and Philosophy* 7: 331–48.

Smith, Kelly. 2007. "Towards an adequate account of genetic disease." In *Establishing Medical Reality: Essays in the Metaphysics and Epistemology of Biomedical Science*, edited by Harold Kincaid and Jennifer McKitrick, 83–110. Dordrecht: Springer.

Sterelny, Kim, and Philip Kitcher. 1988. "Return of the gene." *Journal of Philosophy* 85: 339–61.

Tabery, J. 2014. *Beyond versus: The Struggle to Understand the Interaction of Nature and Nurture*. Cambridge, MA: MIT Press.

US Department of Health and Human Services and Department of Energy. 1990. *Understanding Our Genetic Inheritance*. Springfield, VA: National Technical Information Service.

van Fraassen, Bas C. 1980. *The Scientific Image*. Oxford: Clarendon Press; New York: Oxford University Press.

Wade, N. 2014. *A Troublesome Inheritance: Genes, Race and Human History*. New York: Penguin.

Waters, C. K. 1990. "Why the antireductionist consensus won't survive the case of classical genetics." *Proceedings of the Philosophy of Science Association* 1: 125–39.

Waters, C. K. 2004. "What concept analysis in philosophy of science should be." *History and Philosophy of the Life Sciences* 26 (1): 29–58.

Waters, C. K. 2007. "Causes that make a difference." *Journal of Philosophy* 104 (11): 551–79.

Waters, C. K. 1994. "Genes made molecular." *Philosophy of Science* 61: 163–85.

Watson, Gary. 1975. "Free Agency." *Journal of Philosophy* 72: 205–20.

Wegner, Daniel M. 2002. *The Illusion of Conscious Will*. Cambridge, MA: MIT Press.

Widerker, David, and Michael McKenna. 2003. *Moral Responsibility and Alternative Possibilities: Essays on the Importance of Alternative Possibilities*. Burlington, VT: Ashgate.

Wolf, Susan R. 1990. *Freedom within Reason*. New York: Oxford University Press.

Woodward, James. 2010. "Causation in biology: Stability, specificity, and the choice of levels of explanation." *Biology and Philosophy* 25: 287–318.

9

Summary and Conclusion

The general purpose of this book is to provide an overview of aspects of the recent history and methodology of behavioral genetics, as well as of psychiatric genetics, with which it shares both its history and its methodology. The perspective is mainly philosophical and addresses a wide range of issues, including genetic reductionism and determinism, as well as the behavioral genetic implications for "free will." Some chapters can be easily read by individuals without a preexisting philosophical background, such as chapters 1, 2, 6, and 8. Other chapters dig deeper into both the genetics and the philosophy. Chapters 3, 4, and 5 examined criticisms of standard views of behavioral genetics, and did so by making use of what we have learned about the simplest biological system with a neural network, *C. elegans*, or "the worm," as it is often affectionately called. As the reader should have discovered, though the worm is a "simple system," the molecular and neurogenetics of it are far from simple. Many of the arguments of these three chapters are related to issues of reduction and the nature of explanations in genetics. Chapters 6 and 7 looked closely at human behavioral genetics, both at normal behaviors and mental dispositions via personality theories, and at psychopathology, especially at depression, psychosis, and schizophrenia. Chapter 8 considered more general philosophical issues as to why there is so much concern about, and general interest in, behavioral and psychiatric genetics. These more general issues included introductions to issues of free will, the fundamentality of genetic explanations, and implications for ethical and legal dimensions of behavioral genetics.

The philosophy material in this text is mainly analytical and pragmatic in orientation but is open to alternative philosophical perspectives, including the phenomenological approaches represented by Parnas's writings (in chapter 7). In chapters 1, 2, 6, and 7 the philosophy was largely implicit, akin to the approach that Ronald Dworkin took, which he called doing "philosophy from the inside out" (1994, 28–29). Chapters 3, 4, 5, and 8 are written more in the explicit analytical tradition of philosophy of science. Readers interested in the role that classical pragmatism, of the type associated with James, Peirce, and Dewey, might play explicitly in psychiatry can consult my article on problems of validity(Schaffner 2012).

What I think might be a useful and fairly detailed summary of this book follows, organized in general by chapter numbers.

CHAPTERS 1 AND 2

In order to provide a common basis for reasoning about as well as critically evaluating behavioral genetics, the first two chapters contain three dialogues between a behavioral geneticist and an appeals court judge who wishes to find out more about behavioral genetics. As I was writing up a long introductory essay for a large interdisciplinary project on behavioral genetics and its implications,[1] it slowly became evident to me that the best pedagogical way to introduce the field was via the give and take that can only be presented in dialogue form.

The first dialogue introduced the basic the notions of the more traditional "quantitative" (often called "epidemiological" or "classical") approaches to understanding the influence of genes on behavior. The second dialogue (in chapter 2) introduced the newer, "molecular" approaches to understanding the genetics of behavior, including results from more recent genome-wide association studies (GWAS) and analyses of the effects of copy number variants (CNVs). In the third dialogue, I introduced two hypothetical legal cases involving testing for genes implicated in IQ and ADHD, in order to focus much of the discussion from dialogues 1 and 2. Finally, I concluded with a preliminary projection of where the current debates and new methodologies may lead regarding genetics and the understanding of human behavior in a combined nature-nurture perspective. Chapters 1 and 2 also outlined some additional tools used in behavioral genetics, such as the ACE diagram, and examined briefly some additional key concepts that did not easily fit into the dialogues, such as various subtle distinctions in the analysis of the concepts of the shared and nonshared environments.

These two chapters and the dialogues, especially those in chapter 2, had to be extensively rewritten from their initial versions (which appeared in Parens, Chapman, and Press 2006). These changes were required by what has amounted to a major revolution occasioned by the development of GWAS methods. The application of GWAS methods to a trait as noncontroversial and prima facie "simple" as human height, which has a heritability of about 80%, has resulted in an extensive paradigm shift in the genetics of both "simple" and more complex traits. In 2008, three groups of researchers examined large numbers of humans (the largest study involved 30,000 people) for genetic variants related to height differences. They found about 40 genes. But as Maher reported in the journal *Nature* that year, "There was a problem: the variants had tiny effects. Altogether, they accounted for little more than 5% of height's heritability—just 6 centimetres [out of an expected 27] by the calculations above" (Maher 2008, 1235). This generated the problem of "missing heritability," but more importantly the finding led to the clearer realization that genes typically will have tiny effects, and there will be a huge number of them. In a way, this view of genes had been anticipated by Fisher nearly a century ago in his classical paper (Fisher 1918), but Fisher never thought there would be thousands of genes affecting relatively simple traits.

Two commentators have characterized this realization as a major change in our view of the relation of genetics and traits, saying, we now live in an era of "big data and small effects" (Duncan and Keller 2011, 1041).

Thus the purpose of these first two chapters was to introduce the subject of behavioral genetics in a gradual manner, building on clear, albeit somewhat oversimplified, definitions, and then introducing the methods and some results. My intent was to demystify some of the basic ideas for people who are not behavioral geneticists, but who would like to know what the results of behavioral genetics do and do not mean. But too often it looks like behavioral genetics results are pulled like a rabbit from a magician's hat. One way of looking at these two chapters is that they allow a peek inside the hat. That peek indicates that there are many assumptions that behavioral geneticists make, and also that their findings are often small ones, as well as hard to replicate. Characterizing the limitations in these disciplines and limits in what we can expect to infer from the disciplines' findings is the first step in a wise application of them.

CHAPTERS 3, 4, AND 5

As already noted, chapters 3, 4, and 5 are more philosophical and also dig much deeper into the specifics of behavioral genetics, as well as molecular biology and the neurosciences.

Among the lessons of chapter 2, and leading toward the next three chapters, was the proposal that what behavioral geneticists needed to do was to seek useful and robust simplifications in the vast neural network that constitutes our nervous systems and brain. Such simplifications will be needed in order to connect the variations in the heritable elements, the genes (more accurately the different forms of the gene—the alleles), with specific behavior types. That "simplifying" analysis typically will have to be done with the help of new scientific tools, drawn in no small part from the neurosciences, which will assist in at least partially decomposing the huge network and localizing functional parts. This, it turns out, is very difficult to do, even with a very simple organism like the common roundworm known as *C. elegans*.

In chapter 3 we discussed how "the worm," as this organism is affectionately known, has been approached to look at the relation of its genes and its behaviors. We needed to start with one of the simplest of organisms that is just complex enough to have both a genome and a nervous system. The backdrop to the analysis of the worm was a set of issues that arises in the context of debates between orthodox developmental studies and those I termed the "developmentalists"—a set of debates that posed a series of quite new philosophical problems.

In point of fact, the "developmentalist challenge" affects a far broader area than behavioral genetics—it has relevance for any claims about the separable effects of genes and environment on any traits—but it has its greatest force and has been applied most vigorously to behavioral traits (Lewontin 1992, 1995). There, it attacks the traditional "nature-nurture" distinction initially discussed in chapter 1, and also directs some powerful criticisms against the "innate-learned behavior" dichotomy.

In my analysis I discerned 11 theses. I divided these up into what developmentalists seem to think are seven deadly sins about trait causation, and another four major mistakes that classical approaches to the study of nature and nurture have made. Several of these theses also implied a view of how organisms' behavior can be appropriately studied.

The approach in this chapter was initially bottom-up, in the sense that it proceeded from an account of how a number of contemporary scientists were developing explanations of behavior in simple living systems, frequently called "model organisms."

I drew from the analysis a number of general principles, which I called "rules," governing the relation between genes and behavior that are discernible in the investigations of this extraordinarily well-worked-out simple organism. (I use the term "rules" for these principles, and not laws, because in some cases they admit of exceptions, but hold generally, and I think are default assumptions for all organisms.) There were eight rules identified that instantiated a "many-many" set of relations between genes and behaviors, and which anticipated the difficulties subsequently presented for human genetics in chapters 6 and 7.

These rules suggested a "network" type of genetic explanation probably would hold for most behaviors, including even more complex organisms than worms and fruit flies, such as mice and humans. But I leavened the pessimism with some optimism, via an analysis of a set of five core concepts, which I extracted from the developmentalists' response to the seven deadly sins and four major mistakes of traditional approaches to behavioral genetics. Earlier it had been important to cite the specific sources of these criticisms.

The concepts were those of parity, nonpreformationism, contextualism, indivisibility, and unpredictability. I asked what successful research programs in the *C. elegans*' area tell us about the soundness and applicability of these concepts, and concluded that several of the claims of the developmentalist challenge were doubtful, among them the indivisibility and unpredictability theses. In addition, a thesis of emergence—at least in any strong and "mysterious" sense—was also viewed as not supported. The parity thesis was given a somewhat complex heuristic and methodological reading: genes are special, but are at best "necessary condition explainers," and genes, through the analysis of mutations, offer powerful tools for investigating behavior.

Chapter 4 both asked and answered the set of issues raised in the chapter title: "What's a Worm Got to Do with It? Model Organisms and Deep Homology." The themes expressed in this chapter were provoked by three sets of comments (Gilbert 1998; Griffiths 1998; Wimsatt 1998) generated by my original article against the developmentalist challenge, which appeared in 1998. But chapter 4 also introduced some newer philosophical themes that were helpful as we looked at more complex genetic systems, including humans. One theme was the way in which biological knowledge in general is structured. The contrast is with physics, where the main "explainers" are theories viewed as collections of a small number of interrelated universal statements (e.g., Newton's three or four laws, Maxwell's four or six equations, or the three-axiom version of quantum mechanics), a notion I have called the "Euclidean Ideal" (Schaffner 1986). In biology, with a few (important) exceptions, biological knowledge and biological explanations seem to be framed around a few exemplar subsystems in specific organisms, and perhaps even in specific strains. Examples include the Jacob-Monod *lac* operon in *E. coli* K12, Mendel's pea "factors," Morgan's white-eyed male mutant in *Drosophila*, and Kandel's *Aplysia* model for learning in neurobiology. These exemplar subsystems are used as (interlevel) *prototypes* to organize information about other similar (overlapping) models to which they are related more by analogical reasoning rather than by deductive elaboration. *C. elegans* as presented in chapter 3 is a source of a number of these prototypes, and the social versus solitary feeding exemplar from 1998 was shown to have become more complex and relevant for a discussion of partial reduction in the following chapter 5. In chapter 7, I extended this notion of prototypes to include psychiatric disorders such as schizophrenia.

Another recurrent theme of this chapter 4 was how we can generalize from some specific and simpler prototypes, but the next obvious question was, "On what grounds?" The answer was to appeal to homologies, especially the idea of "deep homology," and conserved genetic features.

I also briefly considered the related notion of a "high connectivity model." This is a term initially introduced by Morowitz's 1985 report on models in biomedical research. In such models, knowledge gained in one area of research ultimately "connects" with research in other areas. This connectivity both expands and reinforces understanding and speeds research progress, as has been noted by worm researchers. For example, Riddle and others have cited "parallels between the development of the body plan in nematodes, flies, and mice" and also the similarity of proteins used for programmed cell death in both nematodes and humans (Riddle 1997, 6).

In chapter 5 I began by proposing two theses, and then examined what the consequences of those theses were for reduction and emergence. The first thesis was that what have traditionally been seen as robust reductions of one theory or one branch of science by another more fundamental one are a largely a myth in biology, though some rare, but extraordinarily important, instances of them can be found in physics. On closer inspection, and particularly in biology, these reductions seem to fade away, like the body of the famous Cheshire cat, leaving only a grin, or a smile. The second thesis was that these "smiles" are fragmentary, patchy explanations, and often partial reductions, and though patchy and fragmentary, they are very important, potentially Nobel Prize–winning advances.

To get the best grasp of these patchy reductions, I argued that we needed to return to the roots of discussions and analyses of scientific explanation more generally, and not focus mainly on reduction models, though three conditions based on earlier reduction models were retained in the analysis. This led us through a brief history of explanation and its relation to reduction models, such as Nagel's, and through a brief account of my own evolving views in this area. In this chapter I characterized my favored approach to explanation as involving a field-and-focus approach: more specifically field elements and a preferred causal model system abbreviated as FE-PCMS. This FE-PCMS account was then applied to a more recent set of results than presented in chapter 3 of neurogenetic papers on the worm's foraging behaviors: solitary and social feeding. One of the preferred model systems from a 2002 *Nature* paper was used to illustrate the FE-PCMS analysis in detail and was characterized as a partial reduction. Other later papers amplifying on further developments in this area were also briefly described, including a post-2004 shift to the role of environmental oxygen, and then environmental carbon dioxide, as key determiners of the worm's solitary or social foraging.

One important philosophical lesson based on the worm not made explicit in the chapters 3 and 4 was that what I call in this chapter the PCMS typically will utilize a number of subtly different types of causal sequences. One type of causal sequence that is frequently cited in the biomedical sciences is a "pathway." A review of a comprehensive summary of research in *C. elegans* suggests that a pathway is a coordinated causal sequence that may contain entities at different levels of aggregation (and not necessarily molecular) with a defined endpoint, which may be a behavior or facet/component of behavior. Often these are called signaling or transduction pathways. They also may be termed regulatory or adaptive, depending on their function. Pathways can have genes as "entry points." Pathways are typically not fully detailed and (especially initially) may contain place markers and gaps. Interestingly, some proponents of the "new mechanism" tradition prefer to see pathways as "mechanism schemas" (Machamer, personal communication; Craver and Darden 2013), but that seems to me to expand the biologists' notion of "mechanism" beyond its typical usage.

The core meaning of the PCMS is that of a "model," which we can think of as representing or capturing one or more pathways and which usually provides additional unification beyond a pathway leading to a defined endpoint. A model can be at any level and is typically interlevel. The "model" notion is more dynamic than a "circuit" (see below), which seems to be more structural. Models are abstract—often in different degrees—and are typically idealized (see Schaffner 1993, 98; Bogen, personal communication). The sense of "model" here is reasonably distinct from the sense of "model" in "model organisms," though there are some overlapping features. Finally, the PCMS may also be termed a network, which can be characterized as an integrated set of pathways often involving adaptive or regulatory functions and oriented toward some general goal/behavior. Accordingly, one investigator may term a structure a network, while another may call the same structure a model. In some instances, a network may be identified with a circuit, also seen as a model to account for a key behavior (for example, figure 7 in Rogers et al. 2006 on *C. elegans* aggregation and O_2 response, and also Chang et al. 2006 for a "distributed network of oxygen-sensing neurons"). A circuit, however, is usually thought of as an anatomically existing structure built up from neurons and interneurons and excitatory/inhibitory connections (synapses) containing (implicitly) various pathways and mechanisms. The neurons may have "acting genes" as part of them, or even acting ion channels, such as K+ types or cGMP gated types.

I should add here that though "mechanism" language has become widely utilized in recent philosophy of science, it seems to me that the distinctions captured by the pathway, model, and network expressions are more faithful to the variety of ways that biologists think of their work.[2] To me, and I think most biologists, a mechanism is a highly specific set of interactions (one could perhaps use the term "connecting activities" in place of interactions; Machamer, personal communication) — interactions that are typically "molecular" but which may be at a higher level of aggregation (e.g., cellular), and which may be found operating in various and quite diverse different types of pathways or models.

This partial reduction for a very simple model organism does have some lessons for reductions related to far more complex human behavior, including psychiatric genetics. These additional lessons for reduction are best presented on the basis of some recent schizophrenia studies, and were deferred until we had a chance to rediscuss psychiatric genetics, including the genetics of schizophrenia, in more depth (in chapter 7).

CHAPTERS 6, 7, AND 8

Chapters 6, 7, and 8 represent a transition back to human behavioral and psychiatric genetics last analyzed in depth in chapters 1 and 2. In chapter 6 I revisited the novelty-seeking gene work mentioned earlier, reconsidered the serotonin transporter gene, and discussed some further subtleties regarding replication problems in these two examples and more generally. In that context, I also considered difficulties with a simple linear main effect approach to behavioral genetics, as well as some of the gene-environment interactions that initially appeared to be reestablishing a more optimistic view in current behavioral and psychiatric genetics, but have become more controversial in the past few years. These interaction approaches point behavioral and psychiatric genetics toward both environmental studies, in which genetics interacts significantly with the environment over humans' lifetimes, and toward more complex strategies that can still be characterized as reductionistic.

Some of these reductionistic strategies have recently begun to deal with epistatic (gene-gene) interactions that do not fit simpler linear models, and in one case, which I discussed in more detail in the following chapter, epigenetic mechanisms may be involved. Often these reductionistic approaches are using tools from the neurosciences, endophenotypes including brain imaging, conjointly with genetic research designs. I also briefly discussed some of the newer GWAS results in the area of personality genetics, including the claim that most, perhaps 95%, of the "candidate gene" (cG) results are false positives. In the following chapter, I turn my attention to schizophrenia, a field of investigation in which there have been some very recent promising, though still perplexing, results on both the genetics and the neuroscience fronts.

The complex reductionistic strategies described in this chapter are, I think, at the present time rather mixed in terms of their results and their methodological approach. Their strengths and shortcomings point the way toward deeper issues that are certainly in the short run and maybe ultimately nonreductionistic, multilevel, and multiperspectival. The importance of these latter approaches was noted in several preceding chapters on *C. elegans*, and was also revisited in chapters 7 and 8.

Chapter 6 noted there are probably more than two dozen different theories of personality that attempt to capture and systematize in a general scheme those traits that we think of as associated with the lay meaning of the term "personality." In the area of personality genetics, however, the focus has been on what are called "trait" theories. It is the "Big Five," or perhaps more specifically the "five-factor model" (FFM), that some influential contributors to the personality literature have recently characterized as representing a "paradigm shift" to the personality taxonomy (John, Naumann, and Soto 2008). However it was Cloninger's well-known biologically based personality theory that initially measured three domains of temperament (and later seven), and was developed in the middle to late 1980s, that was of particular interest to us in the first few sections of this chapter. This interest arose in part because of the theory's close historical association with the *DRD4* gene, which constitutes a (mainly cautionary) "teaching moment" in personality genetics. In addition, Cloninger has recently interpreted his theory within a more interactionist genetic framework, as well as speculated on some intriguing philosophical implications of his developed account. In addition, my informal *PubMed* review (in August of 2012, updated in June 2014) of recent studies in personality genetics suggested that in spite of the "Paradigm Shift to the Integrative Big Five Trait Taxonomy" (John, Naumann, and Soto 2008) noted above, about half of the genetic studies utilize the Cloninger approach, and the other half varieties of the FFM (approximately 150 studies each).

An important set of conclusions regarding personality genetics is not only cautionary, but at present rather pessimistic. With the advent of GWAS in 2005, as discussed in chapters 2, 6, and 7, an alternative method to linkage and candidate gene association studies became available, supplemented with the even more recent GCTA methods. Though some previously discovered genes related to traits and diseases have been confirmed by GWAS results, for example in Crohn's disease, in the behavioral and psychiatric disorders this has not been the case. I discussed this problem in the area of schizophrenia research in the following chapter, but here want to resummarize some of the developments that affected areas more related to personality genetics and the issue of replication. In a recent review of candidate gene studies, which appeared in the *American Journal of Psychiatry* in late 2011, the authors cast serious suspicion on the vast majority of results in the personality genetics area (Duncan and Keller 2011, 1047; also in Duncan, Pollastri, and Smoller 2014). That said, in a long article published in 2010, also in the *American Journal of Psychiatry* (Caspi et al. 2010), Caspi, Moffitt, and their colleagues anticipated and responded to many of the problems that Duncan and her coauthors summarized. Caspi et al. argued that excess reliance on "a purely statistical (theory free) approach that relies wholly on meta-analysis," that is, GWAS studies, does not take into account an alternative "cultural" approach involving "construct validation," which is "theory guided" (521). This debate represented by the Duncan and her coauthors, as well as several others articles, on one side and the Caspi et al. 2010 article on the other, will almost certainly continue to generate extensive dialogue in the behavioral genetics area for several years to come.

In a quite recent article, two researchers who have contributed extensively to personality studies and personality genetics write, "Even though molecular genetic approaches hold tremendous promise for unraveling the distal etiology of conscientiousness (one of the FFM traits), it is important to remember that molecular personality genetics is in its infancy as a field" (South and Krueger 2014, 1367). These investigators add that GWAS findings suggest that there are likely thousands of genes of small effect size that influence personality and that at present these results "tell us little about the biological pathways involved in personality and psychopathology."

It is also worth considering the (theoretical) possibility that the trait categories studied in personality theories such as the FFM, as plausible and well developed as they are, are still too close to folk psychology origins, and do not parse the emotional and cognitive worlds of humans in quite the right way to disclose real effects at the genetic level. However, because there have been so many replicating studies at still-sound classical (heritability and multivariate) levels, another possibility suggests itself. This is that the personality studies represent an area of "emergent simplifications," due to the robust nature of personality dimensions, though multiple pathways at the specific genetic level yield those robust simplifications (see chapter 5 for a discussion of emergent simplifications, or for more details Schaffner 2008). Relatedly, a gloomy implication of this thesis for personality genetics research would be the possibility, raised to an extent by South and Krueger, that there are so many genes of small effect working together in complex ways that despite the moderate heritability of the traits, gene-finding methods will fail to find individual gene effects on personality. Munafo and Flint suggest that there may well be "thousands" of variants with small effects" involved in influencing personality (Munafo and Flint 2011). Neuroendophenotypes may possibly provide stronger gene signals, as suggested in various areas of psychology and psychiatry (also see chapter 7), though integration of those signals with the mental life of the subjects will still be a sine qua non for useful explanations and interventions in psychology and psychiatry.

The chapter on schizophrenia genetics looked in depth at the experimental basis of searches for schizophrenia susceptibility genes, at some new ways to think about different types of schizophrenia-related genes, and also, again, at some preliminary results of the post-2005 genome-wide association studies (GWAS) results. In addition I discussed some epigenetic approaches to schizophrenia genetics and considered some general theories of schizophrenia's etiology. In connection not only with the latter, but also because of relations with experimental strategies, it seemed to be useful to consider what kind of a thing schizophrenia is, that is, as a mental disorder, an analysis that not only could be extended to other mental disorders, but also resonated with the general notion of theory structure in biology sketched in chapters 4 and 5.

The position developed there has been written on extensively in my previous articles (Schaffner 1994, 2008, 1980, 1986) and my book (Schaffner 1993, chap. 3), but in this chapter it is instantiated in a dimensional account of a quite specific disorder. I also situate that discussion in the context of classical behavioral and psychiatric genetics, some probable developments of DSM-5, and recent molecular genetics results.

I examined the core features of schizophrenia, but then expanded the analysis to cover the spectrum approaches to the disorder. This was a backdrop to the classical studies and a lead-in to the molecular approaches to schizophrenia. The complications in these studies introduced by GWAS serve again to indicate what we already encountered in personality genetics. However, the situation in schizophrenia seems more promising. I present the Lewis model of schizophrenia, which focuses on more recently developed cognitive aspects of the disorder, but which also employs some genetic results, as well as multilevel components, such as cells and circuits. Other models that resonate with recent neuroscience, such as the Harrison-Weinberger approaches, are also briefly profiled. The Lewis model was also used to illustrate more specifically how the partial reduction account from chapter 5 developed initially for the worm could be extended to cover human psychiatry. Also, I discussed how genes versus SNPs and so on should be thought of, and addressed the endophenotypes concept so prevalent in current schizophrenia investigations. Some speculation and studies involving epigenetics were considered, as were some ways to approach different types of schizophrenia-related genes.

This chapter was concluded as additional papers regarding specific genes that may be contributing to schizophrenia, and also bipolar disorder, were being announced. I have not tried to summarize very recent studies since they will require further replications. At present, these recent announcements involve genes that are still not appropriately embedded in reasonable pathways, so much more work will need to be done in this area.

My chapter 8 was the last substantive chapter in this book, and it is in several senses the most general philosophical one. The issues raised there are direct offshoots of the genetic analyses of the first seven chapters, but have each been the subject of many recent books and articles. The general question of this chapter was why genetics seems to generate such strong concerns about its implications. A major subtheme was the context in which genetics seems to abrogate the essential human feature of "free will." But another concern among scientists and philosophers is the apparent priority and deterministic aspects that genetic explanations seem to have in contemporary biology. Those two themes appear to interact synergistically to each strengthen the concerns produced by the other.

Early in this chapter I noted that recent scientific advances in biology and behavioral and psychiatric genetics warranted another look at the deterministic implications that raise the concerns discussed here. From the earlier chapters of this book, including those on personality genetics and schizophrenia, it should be evident from GWAS results that most genetic influences related to behaviors are based in literally thousands of genes, each with very small effects. This seems to be the case even with disorders such as schizophrenia, though it is likely that in some *rare cases* we will find genetic elements of reasonably large effect. For normal and near-normal personality variants, however, this seems quite unlikely based on current knowledge. And it will be a long time before we can even identify those genes with any security, given all the presumably missing heritability involved. We had earlier also examined that "heritability" notion, pointing out that it is mainly a heuristic notion, has very few if any implications for personal and policy considerations, and is conditional on the environment, a concept that itself also needs further development and a decent theory of its own.

Further, even though the candidate gene approach that initially uncovered the interesting work published by Caspi and Moffitt's group has been called into question by GWAS, the probability of complex interactions of these small-effect genes with the environment is not seriously questioned. One of the recurring messages from even simple systems discussed in chapters 3–5 indicates that the relations of genes and phenotypes is many-many, and that genes work through partially probabilistic neural networks. Coupled with the recent psychiatric and neuroscience investigations of the self, some discussed in chapter 7 and some covered in the article by Haggard discussed in chapter 8 suggest that a very complex brain can be the seat of a deliberative self, one characterized by agency, ownership, and a hierarchical freedom of the will that is worth having. The implications of that view, and a set of practical tests, are noted in the recommendation I provided from Dennett.

In this chapter, I also sketched a compatibilist account of free will, largely following Frankfurt, and though I noted that his account has been critiqued by other philosophers, I think it survives as one of the best that seems to accord with our considered sense of self as well as with the broadly deterministic features of our world. I also analyzed various attempts to characterize the deterministic aspects of genetics and reasons for their causal priority, and found that it is mainly human interests that lie at the root of such accounts. This permitted both environmental factors and fundamental characteristics of the developed self to play roles in the causal trajectories that affect our lives. Expansion of genetics to include the neurogenetics focus of recent work on free will, as well as the genetic and neurological implications for legal contexts, does not seem to alter this conclusion.

As noted earlier, the general purpose of this book is to provide an overview of aspects of the recent history and methodology of behavioral genetics, as well as of psychiatric genetics, with which it shares both its history and its methodology. In a number of places, the overview did drill down into details, though this is not a textbook, and other sources are easily available for broader accounts of behavioral and psychiatric genetics (see, for example, Plomin et al. 2013, especially the appendix in that book, for details on statistical methods in the area). The present book is much more philosophical, and I believe appropriately sensitive to the major shifts in thinking about the subject that have recently occurred due to GWAS and related advances. The current state of the field is struggling with the realization that genetics is far from the "simple" tool that it was once hoped to be in the early and mid-1990s, but is optimistic about its future, as it now seems to be in possession of the biomedical methods and technology that will in the long run provide ways to attend to human illnesses and thus better human existence.

References

Baker, Catherine. 2004. *Behavioral Genetics*. Washington, DC: American Association for the Advancement of Science.

Caspi, A., A. R. Hariri, A. Holmes, R. Uher, and T. E. Moffitt. 2010. "Genetic sensitivity to the environment: The case of the serotonin transporter gene and its implications for studying complex diseases and traits." *American Journal of Psychiatry* 167 (5): 509–27. doi:10.1176/appi.ajp.2010.09101452.

Chang, A. J., N. Chronis, D. S. Karow, M. A. Marletta, and C. I. Bargmann. 2006. "A distributed chemosensory circuit for oxygen preference in *C. elegans*." *PLoS Biology* 4 (9): e274. doi:10.1371/journal.pbio.0040274.

Craver, Carl, and Lindley Darden. 2013. *In Search of Mechanisms: Discoveries across the Life Sciences*. Chicago: University of Chicago Press.

Duncan, L. E., and M. C. Keller. 2011. "A critical review of the first 10 years of candidate gene-by-environment interaction research in psychiatry." *American Journal of Psychiatry* 168 (10): 1041–49. doi:10.1176/appi.ajp.2011.11020191.

Duncan, L. E., A. R. Pollastri, and J. W. Smoller. 2014. "Mind the gap: Why many geneticists and psychological scientists have discrepant views about gene-environment interaction (G × E) research." *American Psychologist* 69 (3): 249–68. doi:10.1037/a0036320.

Dworkin, Ronald. 1994. *Life's Dominion: An Argument about Abortion, Euthanasia, and Individual Freedom*. New York: Vintage Books.

Fisher, R. A. 1918. "The Correlation between Relatives on the Supposition of Mendelian Inheritance." *Transactions of the Royal Society of Edinburgh* 52: 399–433.

Gilbert, S. F. and E. M. Jorgensen. 1998. "Wormholes: A commentary on K. F. Schaffner's 'Genes, Behavior, and Developmental Emergentism.'" *Philosophy of Science* 65: 259–66.

Griffiths, P. E., and R. D. Knight. 1998. "What is the developmentalist challenge?" *Philosophy of Science* 65: 253–58.

John, Oliver P., Laura P. Naumann, and Christopher J. Soto. 2008. "Paradigm shift to the integrative Big Five trait taxonomy: History, measurement, and conceptual issues." In *Handbook of Personality: Theory and Research*, edited by Oliver P. John, Richard W. Robins, and Lawrence Pervin, 114–58. New York: Guilford Press.

Lewontin, Richard C. 1992. *Biology as Ideology: The Doctrine of DNA*. New York: HarperPerennial.

Lewontin, Richard C. 1995. *Human Diversity*. New York: Scientific American Library, distributed by W.H. Freeman.

Maher, B. 2008. "The search for genome 'dark matter' moves closer." *Nature*, November 17. doi:10.1038/news.2008.1235.

Morowitz, H., and the National Research Council Committee on Models for Biomedical Research. 1985. *Models for Biomedical Research: A New Perspective*. Washington, DC: National Academy of Sciences Press.

Munafo, M. R., and J. Flint. 2011. "Dissecting the genetic architecture of human personality." *Trends in Cognitive Sciences* 15 (9): 395–400. doi:10.1016/j.tics.2011.07.007.

Parens, Erik. 2004. "Genetic differences and human identities: On why talking about behavioral genetics is important and difficult." *Hastings Center Report* 34 (1): S4–S35.

Parens, Erik, Audrey R. Chapman, and Nancy Press, eds. 2006. *Wrestling with Behavioral Genetics: Science, Ethics, and Public Conversation*. Baltimore: Johns Hopkins University Press.

Plomin, Robert, John C. DeFries, Valerie S. Knopik, and Jenae M. Neiderhiser. 2013. *Behavioral Genetics*. 6th ed. New York: Worth Publishers.

Riddle, Donald L. 1997. C. elegans *II*. Plainview, NY: Cold Spring Harbor Laboratory Press.

Rogers, C., A. Persson, B. Cheung, and M. de Bono. 2006. "Behavioral motifs and neural pathways coordinating O_2 responses and aggregation in *C. elegans*." *Current Biology* 16 (7): 649–59. doi:10.1016/j.cub.2006.03.023.

Schaffner, K. F. 1980. "Theory structure in the biomedical sciences." *Journal of Medicine and Philosophy* 5: 57–97.

Schaffner, K. F. 1986. "Exemplar reasoning about biological models and diseases: A relation between the philosophy of medicine and philosophy of science." *Journal of Medicine and Philosophy* 11 (1): 63–80.

Schaffner, K. F. 1993. *Discovery and Explanation in Biology and Medicine*. Chicago: University of Chicago Press.

Schaffner, K. F. 1994. "Psychiatry and molecular biology: Reductionistic approaches to schizophrenia." In *Philosophical Perspectives on Psychiatric Diagnostic Classification*, edited by J. Sadler, O. Wiggins, and M. Schwartz, 279–94. Baltimore: Johns Hopkins University Press.

Schaffner, K. F. 2008. "Etiological models in psychiatry: Reductive and nonreductive." In *Philosophical Issues in Psychiatry*, edited by K. S. Kendler and J. Parnas, 48–90. Baltimore: Johns Hopkins University Press.

Schaffner, K. F. 2012. "A philosophical overview of the problems of validity for psychiatric disorders." In *Philosophical Issues in Psychiatry II: Nosology*, edited by K. S. Kendler and J. Parnas, 167–89. New York: Oxford University Press.

South, S. C., and R. F. Krueger. 2014. "Genetic strategies for probing conscientiousness and its relationship to aging." *Developmental Psychology* 50 (5): 1362–76. doi:10.1037/a0030725.

Wimsatt, W. 1998. "Simple systems and phylogenetic diversity." *Philosophy of Science* 65: 267–75.

Notes

CHAPTER 1

1. This emphasis on *individual differences* in quantitative genetics is an important point. Though mainly capturing *variation* in populations, the concept has analogies with philosophers' emphasis on the roles of Mendelian genes as makers of phenotypic *difference*, rather than phenotype makers. (On this see Waters 1994.) In order to explain how to make a phenotype, a developmental explanation is needed—a huge undertaking (see Kitcher 1984). Some critics of behavioral genetics demand a developmental explanation in all cases, whereas behavioral geneticists argue they can provide important results that bypass such demands, and explore differences in a trait in a population at the more abstract level, similar to what Mendelian geneticists do, although quantitative genetics does not seek single-locus prototypical Mendelian traits. A developmental explanation separates the effects of genes and environments only with great difficulty (see (Schaffner 1998). Behavioral genetics can make a good case that such gene-environmental effect separation can be done analytically for differences (see Plomin et al. 2001, 87–88; Plomin et al. 2013, 89–95).
2. Technically this is true only of the "probandwise concordance rate," a detail that need not concern us. Also, concordance rates can be extended to other family members' risk as well, such as parents, children, and sibs.
3. Relatedly, less than one-third of the patients recently seen in early detection (prodromal and first-episode psychosis) programs (see Schaffner and McGorry 2001) have first- or second-degree relatives diagnosed with a psychotic disorder (L. J. Phillips, T. H. McGlashan, personal communications).
4. It was Mark Frankel's suggestion that I imagine the audience for this chapter as including a judge. (Relatedly, Franklin Zweig educated judges for a number of years in genetics as part of his Einstein Institute programs, largely supported by DOE grants.) I decided to put much of the material in these two chapters into dialogue form after multiple attempts to write more traditional simple text explanations. Traditional exposition did not work in a natural way since the concepts had to be revisited a number of times to make them clear. A dialogue format seems to do that more easily and more readably.

5. The dialogue starts with what is termed "broad" heritability (symbolized by H^2) and later distinguishes it from narrow heritability (h^2). The use of the squared term seems to come from the fact that statistics defines a correlation measure symbolized by r (dealt with in the dialogue below), and also introduces a statistic r^2 that can be interpreted as the proportion of variance in the dependent variable that is contained in the independent variable; r^2 is misleadingly, perhaps, called the "coefficient of determination"—misleadingly, since it's only hypothetically a causal, rather than a correlational notion, and at values less than 1 it is probabilistic, and not deterministic. Later in the dialogue after the concept of the correlation coefficient is introduced, another way to understand heritability in terms of r^2 will be discussed.
6. For those who wish to be reminded of their elementary statistics course, the variance of a population is the average of the squared differences between the mean of the population and the individual values of the measured variable.
7. This comment, though helpful, cannot be strictly true, since if the genes were entirely irrelevant to the trait of interest, they would not affect the variance of the trait either. Perhaps it's more accurate to say such genes can have vanishingly small, or empirically undetectable, effects on the mean.
8. Acid phosphatase may regulate the intracellular concentrations of flavin coenzymes that are electron carriers in the oxidative phosphorylation pathway.
9. In the next chapter I will refer to recent genome-wide association studies (GWAS) that indicate empirically that genetic contributions to human height involve hundreds if not thousands of genes.
10. Except males have one X and one Y chromosome.
11. The variance of the total distribution is 607.8 from all of the phenotypic data. But the average of the environmental variances within each genotype is only 310.7. Thus v_g is equal to $607.8 - 310.7 = 297.1$, and $h^2 = 297.1/607.8 = 49\%$.
12. The arguments and how additive effects work in a linear way are shown explicitly in Plomin et al. 2013, 33–34 and appendix, 379–80.
13. There are more complex models that can check for dominance, but more complex forms of gene interactions such as epistasis are harder to model.
14. Critics of genetic influence often question this EEA assumption, but it has been defended using empirical data. On the critical side see Lewontin, Rose, and Kamin 1984 and Joseph 2001, who argue we need a better set of controls. See also Jonathan Beckwith's essay (2006). Compare this view with the defense of EEA by Hettema, Neale, and Kendler (1995) and Kendler et al. (1994, 1993). Feldman has pointed to specific violations of the EEA in IQ studies (personal communication). Perhaps the most persuasive data supporting the EEA comes from twin studies in families that have misidentified the zygosity of their twins and treated monozygotic twins as dizygotic, and vice versa (see Kendler et al. 1993). Though a systematic and in-depth unbiased general review of the EEA would be extremely useful, this author is unaware of any such review authored by a disinterested party—that is, one who has neither ideological nor genetically based research-program interests involved. (A point-by-point debate of the EEA would be a good alternative to an unbiased review, but I do not know of any parties that have done this either.) Prima facie, the EEA issue *seems* to be of critical importance, since if overlooked environmental correlations are truly responsible for observed similarities in identical twins, then this will result in higher reported concordances and an overestimation of the true genetic effects. But the difference if EEA does not hold is estimated not to be a huge change in heritabilities, probably a reduction from 50% to 35% (for an h^2 of 0.5)

(Kendler, personal communication). This estimate needs to be quantitatively verified, however, using some empirical data.

15. More needs to be said about the prima facie tautological character of the characterizations of shared and nonshared environment, and about some ways that tautology can be clarified to have empirical content. This is covered toward the end of this chapter.
16. The nonshared environmental term also includes an implicit error term—an important point that will be noted again toward the end of this chapter and also in the discussion of Devlin, Daniels, and Roeder's (1997) article on IQ in dialogue 3.
17. See LeDoux's discussion of the similar fearful rat phenotype in relation to neuroscience 1996, (135).
18. This comes from Falconer 1996, 186.
19. This condition seems not to distinguish "shared" from "nonshared" effects but rather to stipulate a general plausibility condition for any environmental trait.
20. A reference to why the model is called biometric—it is within the "biometric" tradition—follows further below.
21. There are some exceptions to this remarkable conclusion, including such variant ones as interest in music and delinquent behavior where c^2 is high (see Rowe and Jacobson 1999, 44; Caspi et al. 2000). Also contrast the results found in the important Nonshared Environment in Adolescent Development (NEAD) study where shared environment often had large effects (Reiss et al. 2000).
22. Twelve years ago, in preparation for a AAAS-sponsored presentation to two US congressional members (Reps. Slaughter and Morella) and their staffs on genes and environments, related to Rep. Slaughter's bill on genetic discrimination, I conducted a Medline/PubMed search of citations with the breast cancer gene (BRCA1) in its title or abstract, and another search with BRCA1 *and* environment in the title or abstract. The hits were about 2,500 for the first search and 44 for the second search (repeated June 24, 2002). This suggested the major focus is on the genetics, not on the gene-environment interaction in breast cancer studies, even though there is strong evidence for environmental effects in breast cancer. A further look for programs that were prioritizing environmental studies in cancer showed that the National Cancer Institute subsequently identified this as an "extraordinary opportunity" area for special attention back in 2003: see http://plan.cancer.gov/scipri/genes.htm (accessed June 24, 2002). A repeat test of these searches on October 29, 2009, yielded about 6,900 for the first search and 120 for the second search; a search done on May 13, 2014, produced 10,500 for search 1 versus 185 for search 2. These results were obtained despite the fact that since the original search, considerably more interest has been expressed in the roles of environment for many diseases and disorders. Readers should know that the Slaughter bill eventually became law in 2008 as the Genetic Information Nondiscrimination Act (GINA).

Chapter 2

1. The field is constantly changing, so new methods and refinements of old methods are continually appearing. Both Lander's group and Roses have argued that explorations using single nucleotide polymorphisms (SNPs) and haplotypes—chunks of chromosomes with small amounts of genetic variation—may afford what is known as linkage disequilibrium studies more power. For recent work see the discussion of GWAS above, and for earlier SNP comments see Daly et al. 2001; Roses

2000. A discussion of the differences between linkage and linkage disequilibrium, however, is beyond this chapter.

2. An influential paper by Risch and Merikangas (1996) urged this strategy, and Risch (2000) presented additional refinements for this approach, which he views as being more congruent with the old Galtonian-Pearsonian biometrical approach than with a Mendelian tradition. As noted in the second dialogue, and in more detail in chapter 6, the candidate gene approach has come in for extensive criticism since the advent of GWA analyses.
3. Essential completion of the Human Genome Project was announced on April 14, 2003, apparently timed to coincide with the 50th anniversary of Watson and Crick's discovery of the double helix—see http://web.ornl.gov/sci/techresources/Human_Genome/project/press4_2003.shtml (accessed March 10, 2015).
4. The same allele for this dopamine receptor of the fourth type is also sometimes referred to in the literature as D4DR; the long form is also referred to as the 7-repeat allele of the *DRD4* gene.
5. There were several more meta-analyses of DRD4 covering some 30 studies that have appeared in the literature including those by Kluger, Siegfried, and Ebstein (2002) and by Schinka, Letsch, and Crawford (2002, 647). The first finds essentially no effect, though the second analysis states, "Although the associations between DRD4 and NS are small, our results suggest that they are real, at least for the long repeat and -521 C/T genotypes." A somewhat skeptical meta-analysis in 2008 concluded that "the DRD4 gene may be associated with measures of novelty seeking and impulsivity but not extraversion. The association of the C-521T variant with these measures, if genuine, may account for up to 3% of phenotypic variance" (Munafo et al. 2008, 197). Also see chapter 6 for additional updates on DRD4.
6. This second study of Caspi et al. has been questioned in a more recent meta-analysis (Risch et al. 2009), but when related neuroscientific data are taken into account (Munafo, Brown, and Hariri 2008), the result appears to be reasonably well supported. More details about this study and later articles are presented in chapter 6 of this book along with "dueling meta-analyses" that were done in 2012.
7. http://www.complextraitgenomics.com/software/gcta/ (accessed March 10, 15).
8. Some of these other AD genes are discussed further below. Earlier good reviews of gene effects in late-onset AD can be found in Bertram and Tanzi 2001 and in Nussbaum and Ellis 2003; a more recent review is by Tanzi (2012). It should also be noted that there are well-confirmed genes (alleles) that have a protective effect in alcoholism (alcohol and aldehyde dehydrogenase polymorphisms, ADH2(2) and ALDH2(2), respectively), though as mentioned in the text, this is largely confined to some subgroups with Asian ancestry.
9. Again, for a current review of the various genetic contributions to AD, see Tanzi 2012 and also Kim et al. 2014. For a GWAS analysis that supported APOE but added another locus on chromosome 12 see Beecham et al. 2009.
10. There is a brief history of these attempts in Thomas et al. 1998.
11. For a brief discussion on this point and references see box 2 in Munafo and Flint 2011, 398.
12. Roses and his colleagues wrote: "Several independent lines of evidence led us to examine apolipoprotein E in later onset familial Alzheimer's Disease [FAD]. We observed several proteins in CSF that bound to immobilized amyloid beta-peptide

[found in AD-affected brains] with high affinity. Microsequencing and Western blotting techniques identified ApoE as one of these proteins. *APOE* was known to be localized to the region of chromosome 19, which in previous studies had shown possible linkage to late-onset FAD. Furthermore, antisera to ApoE stained senile plaques, neurofibrillary tangles, and cerebral vessel amyloid deposits in AD brains" (Saunders et al. 1993, 1468).

13. The most frequently used control uses parental controls and is known as the TDT test (for transmission disequilibrium test). It is too technical a subject to develop in this chapter, but see Risch and Merikangas 1996 and also McGue's meta-analysis of *DRD4* that indicates TDT was not used in all novelty-seeking studies (McGue 2001).
14. The cost of full genome sequencing continues to drop, and a number of commentators have anticipated "the $1,000 dollar genome (Sharp 2011).
15. The topic of missing heritability has exploded in the last four years, and chapters 6 and 7 cover those discussions in more depth.
16. Endophenotypes are looked for extensively in the schizophrenia area and are beginning to be examined using GWAS approaches in the Alzheimer disease area literature; for example see Melville et al. 2012.
17. See chapter 1 and the (two-dimensional) diagram in figure 1.2 of the reaction range, which showed that the same seeds of *Achillea* would grow to different heights in different environments.
18. The US Supreme Court *Daubert* decision—*Daubert v. Merrell Dow Pharmaceuticals, Inc.*, 113 S.Ct. 2786 (1993)—requires the trial judge to admit only reliable and accepted scientific evidence. The Frye rule, which is more limited, requires generally accepted evidence. The terms are not well defined, and the legal discussion remains contentious.
19. The heritability argument was made by the trial court Judge (Parslow), who referred to the twin studies that attributed about 70% of the observed variation in IQ to genetic factors. Two subsequent appeal court decisions, however, utilized different nonhereditarian grounds in their reasoning. See Krim 1996 and also Annas's (1991) comments on this case.
20. Only the more complex reaction norm or reaction surface we considered in an earlier meeting can represent the broader range of differing environmental and genetic influences on a trait, such as IQ. And in virtually all cases we just do not have the data on humans to be able to produce the graphs for the reaction norms or surfaces.
21. A good introduction to the different studies can be found in a book chapter by McGue and his colleagues (1993). Also see chapter 6 on recent GWAS approaches to IQ.
22. The Devlin et al. analysis also provides corrections for assortative or nonrandom mating and uses an advanced form of model fitting employing Bayesian inference. Details are not appropriate for discussion in the dialogue. An overview of the methodology can be found in the appendix to Daniels et al. 1997.
23. For the unpublished critique by Bouchard see http://genetsim.org/bgnews/1997/msg00167.html. For the backstory on why this was ultimately not published in *Nature* see http://genetsim.org/bgnews/1997/msg00165.html.
24. This can be tested using identical twin data; see Devlin, Daniels, and Roeder 1997 for specifics.

25. The relative risk is about 50% in various meta-analyses, but the general population probability is only about 14%; see Faraone et al. 2001 and Waldman and Rhee 2002.
26. For some details of this case see http://www.biojuris.com/buddha/criminal.html.
27. Readers should not be swayed to believe that the free will issue has been settled because the environment is involved, and not just genetics, or because "thoughts can turn genes on and off." See S. Begley, *Wall Street Journal*, June 24, 2002, B1. Thoughts could be determined, too, but this gets into more difficult philosophical areas. For my views on this topic, see my "Neuroethics" essay in Marcus 2002 as well as the discussion of free will in chapter 8 below.
28. It is startling to discover that in yeast, with its 6,000 genes, about 66% of them are turned on and off by the environment—factors involving temperature, salinity, and food supply. See Causton et al. 2001.
29. In 2002 Hamer wrote: "The results [in human behavioral and psychiatric genetics] have been disappointing and inconsistent. Large and well-funded linkage studies of the major psychiatric disorders including schizophrenia, alcoholism, Tourette syndrome, and bipolar disorder have come up empty-handed; not a single new gene has been conclusively identified. Most candidate gene findings have failed consistent replication, and even those that have been verified account for only a small fraction of total variation. Meanwhile, the statisticians who are supposed to be guiding and evaluating the research are unable to agree on how to design experiments or to interpret the results; their advice has proven as faddish (and useful) as the Hula-Hoop" (Hamer 2002, 71).
30. From the WHO World Health Report on Mental Health, 2001, at http://www.who.int/whr/2001/en/whr01_ch1_en.pdf (accessed June 4, 2014).
31. Hamer's 2002 diagnosis and treatment of the problems cited in note 29 above are instructive. He wrote: "What's the problem? It's not the basic premise of linkage and candidate gene analysis; these approaches have identified dozens of genes involved in inherited diseases. Nor is it the lack of DNA sequence information; virtually the entire code of the human genome is now known. The real culprit is the assumption that the rich complexity of human thought and emotion can be reduced to a simple, linear relation between individual genes and behaviors (see the figure in Hamer [2002]). This oversimplified model, which underlies most current research in behavior genetics, ignores the critical importance of the brain, the environment, and gene expression networks" (71).
32. See Bechtel and Richardson 1993 or their second edition (Bechtel and Richardson 2010) for some philosophical discussion of how this is done, and how difficult it might be to do in neural networks.
33. For an application of microarrays to the worm, discussed more extensively in the next three chapters, see Zhang et al. 2002 and also De et al. 2013.

Chapter 3

1. See Bechtel and Richardson 1993 and 2010 for some philosophical discussion of how this is done, and how difficult it might be to do in neural networks; also see chapter 5 for additional related comments on Bechtel and Richardson 1993.
2. See Nelkin and Lindee 2004 for an in-depth account of this oversimplified use of genetics in the media.

3. Authors such as Gottlieb (1992), Gray (1994), and Griffiths and Gray (1994) stress the *evolutionary* implications of their developmentalist/interactionist position. In this chapter these implications are not explicitly considered.
4. The term "social constructionism" covers a number of related approaches. For an accessible introduction to some see Hacking 1999; Schaffner and Tabb 2014.
5. That these terms are heavily freighted with complex metaphors is a thesis that Van der Weele (1995) explores in depth.
6. Debates revolving around the term of "innateness" continue to appear in the biological and philosophical literature. See, for example, the recent book by Goldhaber (2012).
7. I emphasize *molecular* genetics here because of the limitations of classical behavioral genetics to populations, and because the classical techniques do not tell us anything about which specific genes influence behavior or how they do so. Compare the quotation from Greenspan (1995) on this point below.
8. *C. elegans* sequencing was conducted as a joint project at Washington University, St. Louis, and the Sanger Centre (UK). "An essentially complete C. elegans sequence was published in Science in December 1998 and the last remaining gap in the sequence was finished in October 2002" (from http://www.sanger.ac.uk/research/projects/caenorhabditisgenomics/, where the complete sequence is available; accessed June 4, 2014).
9. See http://www.genome.gov/10001837 (accessed June 4, 2014).
10. Brenner published his letter to Max Perutz of June 5, 1963, in which he develops this belief in his foreword (Brenner 1988) to the Wood (1988) reference volume on the nematode. In that same foreword Brenner also includes portions of his October 1963 proposal to the Medical Research Council laying out the reasons why the nematode (though at this point it was *C. elegans*'s cousin, *C. briggsiae*, that was mentioned) would be a model organism for these studies. At the time I first contacted Brenner, he indicated that though several people had told him they planned to write a history of *C. elegans* and its community of researchers, none to his knowledge had yet been done (S. Brenner, personal communication, March 1995). This oversight was first remedied when Rachel Ankeny completed her PhD dissertation in 1997 at the University of Pittsburgh on the worm (Ankeny 1997). In 2002 Brenner shared the Nobel Prize for Medicine (see below) and has published additional reflections in his Nobel address available at http://nobelprize.org/nobel_prizes/medicine/laureates/2002/brenner-lecture.pdf; accessed June 4, 2014. The Nobel award for worm research then stimulated the analysis of Brown (2003), who provided a general historical account of the three Nobel Prize winners' work.
11. The term "brute force" as used here is Horace Judson's (personal communication).
12. A current snapshot of its extensive resources is available on the Internet at http://www.wormbase.org/#012-3-6 (accessed June 4, 2014). Wormbase is a successor to Leon Avery's homepage at http://elegans.swmed.edu/, itself formerly designated (but now closed) by the more colorful name http://eatworms.swmed.edu.
13. Bargmann quotes figures from Durbin 1987: "For any synapse between two neurons in any one animal, there was a 75% chance that a similar synapse would be found in the second animal . . . [and] if two neurons were connected by more than two synapses, the chances they would be interconnected in the other animal increased greatly (92% identity)" (Bargmann 1993, 49).

14. Ideally one would like to study individual neural cells (neurons) in terms of their biochemistry and their electric physiology—what currents there are and how they are controlled. But as Chalfie and White had noted back in 1988, "because of the small size of the animal, it is at present impossible to study the electrophysiological or biochemical properties of individual neurons" (Chalfie 1988, 338). There was a way around this problem of the small size of the cells (and the problem that the cells are sealed in a cuticle, so they can easily burst if probed by tiny wires). One uses a another closely related nematode, *Ascaris suum,* with much larger neurons, that permits some analogical inferences about *C. elegans*'s neurons. More about this later.
15. The claim that these should be default assumptions—those things we should believe unless we learn something specific to the contrary—partly follows from the simplicity of the organism investigated. The rules delineated below suggest that even in *C. elegans* the relations between genes and behavior are quite complex. A preliminary analysis of *Drosophila* supports these rules as well. It is therefore very unlikely that still more complex organisms such as mice and humans will *generally* conform to simple relations such as one gene type, one behavior type. I discuss some exceptions to the complex rules further below.
16. The dauer stage of development refers to an alternative developmental pathway brought on by a limited food supply available to larvae. In such a state, *C. elegans* can survive up to three months without food (Wood 1988, 14–15).
17. These one-many, many-one, and ultimately, many-many relations are akin to a thesis advanced independently by Hull (1974) and Fodor (1975), developed by Rosenberg (1985 and again in Rosenberg 1994). These authors, whose views I have extensively critiqued (1993, esp. chap. 9), infer biological unpredictability and antireductionist themes from such relations, whereas I infer a manageable complexity (see below, and also chapter 5).
18. Necessary condition forms of explanation once received considerable attention from philosophers of science, and especially from philosophers of biology, since they seemed biologically distinctive. As I argue (1993, chap. 7), however, in part following Beckner 1959, these types of "explanation" are not distinctive and are essentially just empirically weak types of explanation.
19. Wicks and Rankin point out (1995) that "hypotheses about what polarity configurations might best account for behavioral observations . . . are difficult given the complexity of the [tap withdrawal] circuit. However these hypotheses can be aided by the formulation of an appropriate computational model of the circuitry" (2443). Such a model has been published in Wicks et al. 1996.
20. See this program's website, http://www.neuron.yale.edu/neuron/ (accessed March 10, 2015).
21. I thank David Touretzky for referring me to these programs.
22. See http://www.genesis-sim.org/GENESIS and http://www.genesis-sim.org/GENESIS/G3/index.html for details (accessed March 10, 2015).
23. The early embryology of *C. elegans* is becoming better understood, but is still not as well characterized as is *Drosophila*'s; compare Roush 1996 and Gonczy and Rose 2005.
24. Also see http://www.nobel.se/chemistry/laureates/2003/adv.html for further background on ion channels at the atomic level.

25. Additional information about environmental effects can be found in a discussion of the dauer state of the worm—see Riddle 1997; Schaffner 1998a.
26. It is of interest to note that for many years maternal effects were disregarded and interpreted as "random noise that tended to obscure the genetic variation we were interested in" (Pennisi quoting T. Mosseau in Pennsini 1996, 1334). This suggests that one should be *very* cautious in accepting any hypothesis appealing to an irreducible "developmental noise."
27. See dictionary.com.
28. Why these are default assumptions is defended in note 15 above. An anonymous referee for the original publication of these rules posed a point that takes a rather different tack, suggesting that these eight rules at the level of the generalizations as presented in table 8.1 are simply "common sense," generally accepted by "sophisticated geneticists and philosophers of biology long before any particular examination of the work on *C. elegans*." The same referee urges that "the rich detail behind" these rules should find its way into a "sharper instrument" to use in addressing the philosophical problems of this chapter. I agree that it would be useful to formulate what I would see not necessarily as a "sharper," but as a more robust, "model," intermediate between the rich details discussed in the text and the eight generalizations, but I believe this must wait on further comparative examination of other organisms, such as *Drosophila* and *Mus*. It is possible that a more developed instrument could be found if rules for highly conserved molecular machines and conserved circuits were articulated, but the diversity that evolution has introduced into biology may argue against broad substantive generalizations at these levels. This, however, will be an empirical issue, and it is conceivable that circuits might be strongly conserved that would license a more developed set of rules. The eight rules and their implications are, however and in point of fact, used in addressing the developmentalist challenge in the following section, where they appear sufficiently "sharp" to perform this function. Moreover, many sophisticated geneticists will probably find them too strong.
29. Soccer (or "football") fans may remember that in 2010, the German zoo octopus Paul correctly predicted the outcome of eight straight World Cup games, including the final victory by Spain.
30. I first heard this point about the significance of DNA's "linearity" as a basis for the importance of genetic (DNA) considerations in biology from Allan Tobin (personal communication), though here the citation is to Albert et al.'s influential text on the molecular biology of the cell. (The Alberts et al. quote is from the 1994 third edition, and is not evident—in so many words anyway—in the later editions, including Alberts 2008. Tobin also suggested another, more sociological reason, for the importance of DNA in contemporary biology, namely that many biologists believed that "really smart people" work(ed) on DNA (personal communication, October 1996).
31. Greg Morgan has suggested to me (personal communication) that it is also the fact that the genome is sequestered from somatic cells that permits this genomic focus.
32. Paul Griffiths has suggested to me that genes (and DNA) have a special heuristic value is "*because* they have been so thoroughly investigated" (personal communication, December 1996). This is a nice (partly sociological?) parry to Alberts et al.'s view, quoted above, that we do "understand more about genetic mechanisms

than about most other biological processes," perhaps because the molecule is one-dimensional.

One of the anonymous referees of an earlier version of this chapter suggested that the account developed here is compatible with (some) genes being "master molecules," citing Gehring's group's work on the *eyeless* (*ey*) gene in *Drosophila* (Halder, Callaerts, and Gehring 1995). This group does use language referring to *ey* as a "master control gene," but, in my view, this is not entirely on point regarding the causal parity issue. Gehring's' master control gene, like the properties of the *fru* gene in Drosophila, is a gene that acts first in a linear branch of development and seems to code for a transcription factor that regulates genes that are distal in the pathway. Sometimes there are simple linear cascades with a "master control gene" at "the top," but more frequently there are not. In *C. elegans* there are very few such "master control genes," at least recognized at present (Chalfie, personal communication), though the worm's sex development cascade seems a promising source of gene effects that have significant distal effects. The typical genetic programs are apparently largely regulated by a more complex accumulative and combinatoric logic, as in Chalfie's touch circuit described earlier, though there is a "sex determination hierarchy" that has been identified in *C. elegans* (see Ryner and Swain 1995, for an overview). Even the *ey* gene acts *in concert with* protein synthesis machinery, which reaffirms the point about causal parity. For additional discussion of the "master molecule" concept, and some of the colorful metaphors that are associated with that notion, see Van der Weele 1995, esp. 15–16.

33. Also compare Waters's (1994) analysis of gene effects as primarily involving "differences" of phenotypes.
34. Rosenberg would surely disagree, at least if he still follows his earlier (1994) views, and see Stent's point as supporting philosophical instrumentalism, but I do not believe that Stent would have drawn such an inference.

Chapter 4

1. Hedges (2002) suggests that genomics, and also economics, has influenced what we identify as model organisms: "In past decades, the term 'model organism' has been narrowly applied to those species—such as mouse or Drosophila—that, because of their small size and short generation times, facilitate experimental laboratory research. However, in the past decade, with the increase in the number of genome-sequencing projects, this definition has broadened. For example, researchers have focused attention on some organisms, such as the tiger pufferfish, because of unique aspects of their genome rather than their feasibility for experimental studies, and referred to them as 'genomic' models. . . . In most cases, economics has had a large part in the choice of organism to study, such as the agriculturally important species (for example, rice) and those related to human health (for example, the malarial parasite *Plasmodium*). All these species are receiving an unusually large amount of attention from the research community and fall under the broad definition of 'model organism'" (Hedges 2002, 838).
2. The question could be framed more broadly than I have the possibility of doing here, to include historical and sociological factors—perhaps a variant of Gilbert and Jorgensen's question, "Why do genes sell?"—and to ask, "Why do model organisms sell?" These broader types of questions have and will continue to be investigated in science studies (see Clarke and Fujimura 1992 and most recently

Ankeny's [2011] detailed discussion of model organisms). In a somewhat curious June 2010 talk, Harvard Medical School's Gary Ruvkun suggested the term "model organisms" be replaced by the term "cardinal organisms." As reported in genomeweb's *The Daily Scan*:

> "The reason I don't like 'model organisms,'" he told the audience of geneticists, is because it "sounds like model airplanes—it just sounds tiny. . . . If we renamed it 'cardinal organisms,' that sounds much more important," Ruvkun told attendees. "In navigation, you have these cardinal sites that tell you where to go"—which is appropriate because geneticists who study model systems are "really the beacons for genetic analysis in humans," he said. (https://www.genomeweb.com/blog/gsa-cardinal-organisms-human-biology, accessed December 13, 2015)

In my view, this suggestion relies on a misleading and weak analogy and disregards history.

3. This is a view I have held for over 30 years, since I first wrote on the special character of theory structure in biology (Schaffner 1980). An updated version, and an extension of the view, appeared in chapters 3 and 5 of my book (Schaffner 1993).
4. The sense of "mechanism" used here is generic. In chapter 5 I address the more developed concept of "mechanism" that has been the focus of a large number of recent articles about the "new mechanism."
5. Wimsatt suggests (1998, 272) that one of my heuristics for finding common pathways where only a few genes will explain behavior will not work unless these pathways are "widely distributed phylogenetically." But this misunderstands my sense of "common," which at that point is meant to describe a coming together of diverse inputs into a common pathway that may account for those rare (?) circumstances in which one or a few genes have a strong effect on a trait of interest. Moreover, these pauci-genetic explanations could be extremely important even if not "widely distributed phylogenetically," if they accounted for a human disease, such as schizophrenia.
6. See Gunther Wagner's (1994) discussion of this.
7. More accurately, this is a *partly* empirical investigation, given the extensive number of methodological and philosophical assumptions that underlie phylogeny. This has been written on extensively in philosophy of biology by David Hull, Elliott Sober, and many others, but this is not the place to discuss this.
8. I thank Manfred Leibichler for bringing this article to my attention.
9. There are two factual points where Gilbert and Jorgensen (1998) provide appropriate corrections to statements I made in my paper. (1) I used the *odr-7* gene as an example of gene-neuron behavior specificity, whereas they cite the *odr-10* locus. It has turned out that *odr-7* is a control gene for *odr-10* that codes for a receptor, so though the former has the specific effects on the AWA neuron I mentioned, the site of action is earlier in the pathway (see Bargmann and Mori 1997). (2) Gilbert and Jorgensen say (261) it is sometimes unclear whether I am talking about a worm's attractant *detection* ability or a worm's ability to *move* toward an attractant. I do not believe that I conflate these two notions, and allow the context to make the distinction. But that distinction could be masked in the experimental population if the appropriate controls were not provided. See Sengupta 1994 for an account

of those comparative controls and possible alternative explanations, esp. 971–73, 975, and 977.

10. The issues of information and important "differences" relates in part to the general issue of "parity," discussed in chapter 3 and returned to again in chapter 8 below.
11. See the argument for more such research in Freimer and Sabatti 2003. There are several projects that may obtain some relevant phenomic data, including Biobank in the UK (at http://www.ukbiobank.ac.uk/) and the curious Personal Genome Project (at http://www.personalgenomes.org/). There is also a recent NIH-funded Consortium for Neuropsychiatric Phenomics at UCLA (see (Bilder et al. 2009). The Phenome project might turn out to become embedded in a still broader "Human Variome Project"; for this possibility see Oetting et al. 2013.
12. A similar criticism has been directed at DST experimental inquiries by Helen Longino in chapter 6 of her recent book (Longino 2013).
13. See chapters 1 and 2 above as well as the earlier versions of these chapters in Schaffner 2006, 2008, as well as other chapters in those volumes for examples.

Chapter 5

1. These difficulties were systematically developed in the writings of Feyerabend and Kuhn about this time, in the 1960s and 1970s; for a discussion of their work and early influences see Schaffner 1967, 1993a.
2. In Schaffner 2012 I develop this example in considerable detail, as well as the conditions for such a reduction, and the ways in which it is still partial.
3. Compare Ernst Mayr on the distinction between explanatory and theory reduction (1982, esp. 60–63); and also Sarkar 1998.
4. Hempel and Oppenheim cite Mill as well as Popper and a number of other authors as sources of their model. For additional discussion of the D-N model in this context, see Schaffner 2006.
5. In actual reductions, this takes place at a considerably more specified level of detail. See, for example, the discussion in Watson 1987 of the *lac* operon (476–80), which specifies operator and promoter DNA sequences.
6. By substance pluralism, I mean the existence of two independent substances, such as mind and matter were for Descartes, or matter and field seemed to be for Einstein—see his comments on Maxwell in his "Autobiographical Notes" (Einstein 1949).
7. I found Salmon's discussion of marks, forks, and interactions not fully satisfactory. For a recent summary of Salmon's approach, and criticism, see Woodward 2003, chap. 8.
8. Partial decomposability has been discussed by Simon (1981) and by Wimsatt (1976a).
9. One possibility that retains what Nagel called a correspondence rule interpretation of these connectability assumptions is to use a causal sequence interpretation of the logical empiricists' correspondence rules. For how this might be further analyzed see my paper (1969) on correspondence rules and also Suppe's discussion of this view (1977, 104–6).
10. This second type of (global) explanation involves what I call in chapter 5 of my 1993 book "temporally extended theories" that allow for replacement in some circumstances. Using such temporally extend theories is too complex for a first cut

at getting back to the explanatory roots of reductions. This global type of explanation also involves issues of "global evaluation" (transtheoretical criteria) that need to bracketed for another discussion, though a list of those criteria and a Bayesian analysis of how they work can be found in chapter 5 of Schaffner 1993a.

11. I have debated whether this aspect should be best characterized as logical or epistemological. It seems to involve a logic of weighing and comparing, but the aspect also indicates varying strengths of warranted belief. Further below, I will describe subscribing to a type of causality as the "epistemological" aspect of the second substantive component.
12. A still further illustration of a partial reduction involving schizophrenia will be taken up in chapter 7.
13. How to best define a field and a discipline is likely to require considerably more analysis than I provide in this chapter, and there may be historical and sociological dimensions that need to be taken into account to provide an adequate characterization of these terms and their relations. That neuroscience is extraordinarily interdisciplinary is a point that has been stressed by several commentators, including Craver and Bickle—see Bickle 2005; Craver 2005, 2007.
14. In medicine, an analogy to a list of alternative hypotheses is what is known as a "differential diagnosis" for the cause of an illness that afflicts a particular patient or population of patients; it is a list of possible diseases.
15. The list of alternatives and a comparison of their strengths and weaknesses can include alternative possible states in which a mechanism might be can also be evaluated in this approach. The original suggestion for this evaluative dimension is due to van Fraassen (1980), who asked, e.g., why is this circuit off rather than on? why is this patient sick rather than healthy?
16. To examine what types of metascientific vocabulary were used in presenting *C. elegans* results in some major journals, I systematically searched for terms such as "mechanism," "model," "pathway," "circuit," and "network" in several publications from Bargmann's and de Bono's labs. A list of those publications and a summary of terms is available from the author on request.
17. For an excellent example of a model that uses multiple equations see Bogen's (2010) account of the classic 1952 Hodgkin-Huxley paper on action potentials in Bogen; also see Schaffner 2008b.
18. Again I term this aspect an "epistemological" aspect, though causal claims involve, in addition, logical dimensions (at least in the sense of types of conditionals) and also metaphysical dimensions (in the sense of ontic claims and process metaphysics). For details of my view on these dimensions see Schaffner 1993a, 298–307, and for a slightly later discussion of a manipulation interpretation of causation see Schaffner 1993b.
19. Some explanation may claim to be noncausal, e.g., unificatory, but I find this claim (by Kitcher) questionable (see Schaffner 1993a, chap. 6), but for a possibly more positive view see Woodward 2003, chap. 8.
20. Probabilistic explanation using infinite classes is deductively elaboratable. The logic in some cases might even be abductive in some instances, maybe in "inference to the best explanation," though abductive inference (and logic) is even less well understood than inductive.
21. A quantitative derivation of a path of *C. elegans* motion that agrees with the experimentally observed path can be computed based on neural theory, though

the explanation quickly becomes extraordinarily complicated—see my summary (2000) of Lockery's results using this type of approach.

22. The following philosophically general account parallels the discussion in my 1993 book. It assumes an analysis of biological explainers as involving models representable as a collection of generalizations of variable scope instantiated in a series of overlapping mechanisms as developed in chapter 3 of the 1993 book. We can, as described in that chapter, employ a generalization of Suppe's set-theoretic approach and also follow Giere (1984) in introducing the notion of a "theoretical model" as "a kind of system whose characteristics are specified by an explicit definition" (1984, 80). Here entities $\eta_1 \ldots \eta_n$ will designate neurobiological objects such as neuropeptide receptors, the Φs such *causal* properties as "ligand binding" and "neurotransmitter secretion," and the scientific generalizations $\Sigma_1 \ldots \Sigma_n$ will be of the type "This odorant activates a G-protein cascade opening an ion channel." Then $\Sigma_i(\Phi(\eta_1 \ldots \eta_n))$ will represent the ith generalization, and

$$\prod_{i=1}^{n}\left[\Sigma_i\left(\Phi\left(\eta_1 \ldots \eta_n\right)\right)\right]$$

will be the conjunction of the assumptions (which we will call Π) constituting the preferred model system or PCMS. Any given system that is being investigated or appealed to as explanatory of some explanandum is such a PCMS if and only if it satisfies Π. We understand Π, then, as implicitly defining a kind of natural system, though there may not be any actual system that is a realization of the complex expression Π. To claim that some particular system satisfies Π is a theoretical hypothesis, which may or may not be true. If it is true, then the PCMS can serve as an explanandum of phenomena such as "social feeding." (If the PCMS is potentially true and well confirmed, then it is a "potential and well-confirmed explanation" of the phenomena it entails or supports.)

23. Subsequently, de Bono's lab showed that the internal circuit involves soluble guanylate cyclases in that pathway. See Cheung et al. 2004. These appear to be activated by oxygen—see Gray et al. 2004.

24. Another paradigmatic example of how generalizations, and even simplified "laws," are involved in the articulation of a model or mechanism can be found in Hodgkin and Huxley's (1952) classic article on the action potential in the giant squid axon. Bogen (2010) analyzes Hodgkin and Huxley's model construction as not supporting a typical generalization account, but I read their paper differently.

25. A full quotation of the statement of the operon model from Jacob and Monod 1961 can be found in Schaffner 1993a, 158–59.

Chapter 6

1. The instruments employed in the studies of normal subjects, and especially those used in clinical studies, for assessing personality and also some key symptoms that are facets of personality categories are another complexity, and both areas constitute a veritable alphabet soup of abbreviations. As one author

notes, (O'Connor 2002) these include the following: ASI = Anxiety Sensitivity Index; BAI = Beck Anxiety Inventory; BHS = Beck Hopelessness Scale; CDI = Children's Depression Inventory; CES-D = Center for Epidemiological Studies Depression Scale; MMPI-Adolescent = Minnesota Multiphasic Personality Inventory Adolescent Content Scales; MMPIContent = Minnesota Multiphasic Personality Inventory Content Scales; NEO-PI-R = Revised NEO Personality Inventory; STAI-J = State-Trait Anxiety Inventory, Japanese version; TPQ = Tridimensional Personality. In this chapter *some* of the instruments will be introduced and described, but for a fuller account see such references as O'Connor 2002 and Krueger and Tackett 2003, who list many more instruments, including the DAPP-BQ and SNAP.

2. Interestingly when Dean Hamer defended his "gay gene" work, he found he needed to become acquainted with the NEO instrument to rule out his results as not being due to personality characteristics, such as "being against the mainstream" (Hamer, interview, February 9, 2005).
3. Missing was the "openness" to experience, but there are probably good reasons why this dimension was not identified in this study—see Livesley, Jang, and Vernon 1998.
4. A pdf of Carey's personality chapter is also available at http://psych.colorado.edu/~carey/hgss/hgsschapters/HGSS_Chapter22.pdf (accessed June 8, 2014).
5. Cloninger writes that "persistence had originally [in the TPQ] been a subscale of Reward Dependence, but it proved empirically to be independently heritable in a large-scale twin studies in the United States and Australia" (2004, 40). Cloninger has added that the TPQ also did not predict very well, and needed even more scales to capture "what makes us human" (Cloninger, interview, May 18, 2004).
6. In addition to the exon III polymorphism, an effect of the *DRD4* promoter on novelty seeking has been reported as well, this involving an upstream regulatory region usually referred to as the -521C/T promoter polymorphism. For a discussion and references see Ebstein and Auerbach 2002.
7. "Schizotypy" is not a DSM diagnosis, though it shares some features with schizophrenia personality disorder. Rather it is a theoretical proposal due to Paul Meehl that has had considerable influence in discussions about schizophrenia and attenuated forms of that disorder. For a detailed account of Meehl's changing views on schizotypy, see Lenzenweger 2006.
8. This section draws in part on the account of increasing *DRD4* complexity described in chapter 7 of Cloninger 2004.
9. Though classical quantitative studies cannot investigate the interactions of specific genes, classical twin studies can estimate the total proportion of genetic variance that is additive by comparing correlations in identical twins with those of fraternal twins or siblings (see chapter 1). These classical studies imply that much of the genetic variance is in fact additive, thus arguing against interactions. (I owe this point to Ken Kendler.)
10. For a recent meta-analysis that supports the Caspi MAOA results see Byrd and Manuck 2014.
11. For a short history of GWAS as well as several criticisms and a defense of the method see Visscher et al. 2012.

12. The term "cultural" here should be understood in the sense of "methodologically foundational."
13. http://www.complextraitgenomics.com/software/gcta/ (accessed March 9, 2015).

Chapter 7

1. Specific changes in the American Psychiatric Association's DSM-5 related to schizophrenia can be found in at http://www.dsm5.org/Documents/Schizophrenia%20Fact%20Sheet.pdf.
2. http://www.schizophrenia.com/news/costs1.html (accessed March 11, 2015).
3. Weinberger dates the neurodevelopmental hypothesis of schizophrenia to about 1982–1883, though the explicit form of the hypothesis was first published in 1986. Weinberger states that the hypothesis "was based on a lot of early imaging studies." More specifically, Weinberger recalls,

   > There were three main findings that led to a hypothesis. The first was that evidence of some of the brain abnormalities that we were observing in our early CAT scan studies were present at the first episode of illness in young adolescents, and that over a period of 10 years, in most patients, these findings did not change, which suggested that we were looking at a more static process that was dissociated, in its temporal course, from the emergence of the clinical phenomenology. . . . The second was I did a study of looking at the correlation between early childhood social and educational adjustment, and the presence of these CAT scan findings. It was back in 1982 or something. And we found that there was a very strong prediction of the early childhood adjustment, and adult evidence of pathology, which suggested that the pathology might have existed from early in life, but expressed itself clinically in a different form than the adult form. And then the third evidence was that we were looking at a lot of schizophrenic brains, as had many people during the first half of the twentieth century—couldn't find any evidence that the brain was undergoing any sort of destructive change. . . . It may be the pathology that was implicated in schizophrenia affected the development and function of highly evolved, highly complex cortical neural systems, that mediate cognition and motion, but that these are late-maturing neural systems. And the fact that they are destined from early in development to malfunction does not manifest itself in the signature behavioral disorders that we make our diagnoses based on, until those neural systems have reached a certain level of maturational integration."

   Weinberger added: "That was the basis of the neurodevelopmental hypothesis that I published first in 1986, that schizophrenia involves early developmental abnormalities that predispose to the malfunction of late-maturing neural systems, and that the age of onset had nothing to do with when the pathology occurred; it had to do with when the systems matured, such that they now expressed that they were dysfunctional and dismaturational in this new behavioral way" (Weinberger interview, December 2005).
4. In addition there are more general assessment instruments such as the Schedules for Clinical Assessment in Psychiatry (SCAN) containing the Present State Examination (PSE) and two versions of the Structured Clinical Interview for

DSM-IV Axis I Disorders (SCID-I). These tend to use a scale of absent, mildly (below threshold) present, and present.

5. Though twin studies have been the main source of the initial genetic findings in schizophrenia, Gottesman recalls that it was a set of quadruplets that was initially quite influential in his thought. He says: "There's another important book that I try to talk about when I have the chance. It's this one with the name *Quadruplets*. And this book inspired a lot of my and Shields's thinking. It was edited by David Rosenthal. But it was a team approach to understanding these four young women who were identical quadruplets. And they all became schizophrenic. And they were then brought to NIMH to live in for over a year" (Gottesman, interview, October 29, 2003). The book mentioned is Rosenthal 1963. Another book that Gottesman notes influenced him was Anastasi 1958 on individual differences but less so than Tyler (1956). Gottesman's work on the genetics of schizophrenia with Shields was began in 1962, and after a number of articles then culminated in their extraordinarily influential 1972 book (Gottesman and Shields 1972).
6. Even in the mid to late 1980s Farmer and McGuffin (Farmer, McGuffin, and Gottesman 1987), using data gathered by Gottesman and Shields, realized that the strength of the genetic component of schizophrenia (roughly "its heritability") would depend on the definition used. McGuffin recalls: "We were estimating heritabilities, but using it [heritability] in a slightly different way, because we were saying: which definition of schizophrenia looks 'most genetic'? It turned out that it wasn't necessarily the most reliable one, so we found, for example, that Schneider's first rank symptoms, which British psychiatrists are terribly fond of, defined a form of schizophrenia with virtually zero heritability, whereas these newfangled American definitions, like the RDoC, the Research Diagnostic Criteria, [and then DSM-III] actually did rather well in defining a form of schizophrenia with 80% heritability" (interview, June 2004).
7. A number of researchers have suggested that the high heritability of schizophrenia makes it worth searching for genes that influence it. For example, Pogue-Geile commented: "I think it's somewhat useful, for example, in schizophrenia to know that overall schizophrenia is pretty highly heritable. I wouldn't do what I do if it were zero heritability" (interview, February 2004).
8. And in "clearing," the sense was that the number of genes related to schizophrenia would significantly increase from the six or so believed to be known in 2004, but be manageable. For example Karoly Mirnics speculated in 2004 as follows: "I think that in the next 10 years, that we will have at least 20 schizophrenia susceptibility genes. And I think that once you have the 20, and once you look at all those 20 on a panel of thousands of schizophrenics, and ask the question, "Is it—what proportion of schizophrenia is explained in these 2- genes?," I think that you will get a completely interesting 60%, 70%, maybe. So that's at least my bet. I don't think that we have endless number of schizophrenia genes. I do think that their effects will be additive, and I think that we will have very good likelihood to predict with a reasonable accuracy. . . . Probably never in the 99% range, but we can definitely, will be able to, say that somebody has a predisposition to develop schizophrenia or not. The other thing, which is very rarely even mentioned and rarely talked about—are there any protective genes?" (Mirnics interview, February 2004).
9. The dysbindin gene finding came slowly, after extensive linkage analysis of the Irish Study of High-Density Schizophrenia Families by Straub, Kendler, and their colleagues. Kendler recalls that initial analyses showed there was no Mendelian-like

gene, though the investigators did identify a region on the 6P chromosome. Then, Kendler added: "We then said, of course, well, where's the gene? There was no gene. We were finding a linkage and an association, which was a little disturbing. We then went to the mouse genome and found that there was a mouse homolog to the mouse DTMBP1 in that region, and low and behold, there was in fact the human gene there. . . . So we eventually published—I think about now two years ago—the dysbindin gene [finding]" (Kendler, interview, April 2004). For the dysbindin publication reference, see Straub et al. 2002.

10. Originally stated on-line at http://www.dsm5.org/ProposedRevision/Pages/proposedrevision.aspx?rid=411#. But currently this is only available via the Web Archive Service, WayBack, at https://web.archive.org/web/20121031103621/http://www.dsm5.org/ProposedRevision/Pages/proposedrevision.aspx?rid=411 under the "Rationale" tab.
11. In 2004 Ming Tsuang commented on an endophenotypes strategy, saying "We would really like to do genetic studies [of schizophrenia], so we are proposing the endophenotypes: the neurocognitive, and also neurophysiological, and possibly neurochemical markers, to identify those characteristics, even before the onset." Tsuang added, "We wrote the proposal having several universities combined together to look into endophenotypes of schizophrenia. And that is just funded by NIMH. It's a consortium for the Endophenotype Study of Schizophrenia" (Tsuang, interview, May 2004).
12. See, for example, http://www.schizophreniaresearch.net/about-us/ (accessed March 12, 2015).

Chapter 8

1. Genetic determinism is also related to the notion of "innateness" or inborn inherited traits that markedly affect behavioral patterns. That notion has also been analyzed in depth by Griffiths (2009) and, relatedly, has been the subject of an empirical philosophical project by Linquist et al. (2011).
2. In essence, Sarkar's view of reduction is a critical and skeptical one, noting problems with wholesale reductions of the type that Hull and Schaffner, as well as Kitcher, disputed over the course of two decades. Newer methods including allelic association were just developing when Sarkar's book appeared in 1998, and the complications that GWAS has recently introduced were not even contemplated in 1998.
3. Sarkar's requirement (1) involving only "a few loci" seems to have been introduced because of Sarkar's general remarks regarding distinctions between Fisher's biometrical approach involving many genes of small effect and a Mendelian approach. See Sarkar 1998), chap. 5, for details.
4. I thank Maria Kronfeldner for suggesting that I add a summary of Kendler's more recent (2012) article, which amplifies his earlier causal criteria, as well as to note the analogy between Kitcher's pragmatic view and Kendler's position.
5. See Bretscher, Busch, and de Bono 2008 and more recently Bretscher et al. 2011.
6. As I read him, Dupré in his recent analysis of the cause-condition argument in genetics also holds this view. See Dupré 2012, 286–87.

7. Also see another recent account of a compatibilist account developed by Dupré (2012) in the context of analyses of recent genetics, especially his self-revisionist comments on pages 292–94 of that book.

Chapter 9

1. This project, entitled Tools for a Public Conversation about Behavioral Genetics, ran from 2000 to 2003 and was directed by the Hastings Center, the major institution examining bioethics in the United States, working jointly with the American Association for the Advancement of Science (AAAS). Two books and a special supplement of the *Hastings Center Report* constitute the permanent records of the project's results (Baker 2004; Parens 2004; Parens, Chapman, and Press 2006).
2. Informal discussion with scientists about terms such as "mechanism," "model," "pathway," and "network" suggests that the use of the terms can vary among biologists and physicians, and that what fixes the meanings are the specific applications. That said, the terms may, via even simple rational reconstruction, fall into discernible philosophical clusters worth further analysis.

INDEX

Note: Figures and tables are indicated by "f" and "t" respectively.